*SCHAUM'S OUTLINE OF*

# THEORY AND PROBLEMS

**OF**

# SET THEORY
# and Related Topics
## Second Edition

•

**SEYMOUR LIPSCHUTZ, Ph.D.**
*Professor of Mathematics*
*Temple University*

•

**SCHAUM'S OUTLINE SERIES**
McGRAW-HILL
*New York San Francisco Washington, D.C. Auckland Bogotá Caracas Lisbon London Madrid Mexico City Milan Montreal New Delhi San Juan Singapore Sydney Tokyo Toronto*

**SEYMOUR LIPSCHUTZ**, who is presently on the mathematics faculty of Temple University, formerly taught at the Polytechnic Institute of Brooklyn and was visiting professor in the Computer Science Department of Brooklyn College. He received his Ph.D. in 1960 at the Courant Institute of Mathematical Sciences of New York University. Some of his other books in the Schaum's Outline Series are *Beginning Linear Algebra*; *Discrete Mathematics, 2nd ed.*; *Probability*; and *Linear Algebra, 2nd ed.*

Schaum's Outline of Theory and Problems of
SET THEORY



1 2 3 4 5 6 7 8 9 10 11 12 13 14 15 16 17 18 19 20 PRS PRS 9 0 2 1 0 9 8

ISBN 0-07-038159-3

Sponsoring Editor: Barbara Gilson
Production Supervisor: Clara Stanley
Editing Supervisor: Maureen B. Walker
Project Supervision: Keyword Publishing Services Ltd

**Library of Congress Cataloging-in-Publication Data**

*McGraw-Hill*

*A Division of The McGraw-Hill Companies*

# Preface

The theory of sets lies at the foundations of mathematics. Concepts in set theory, such as functions and relations, appear explicitly or implicitly in every branch of mathematics. These concepts also appear in many related fields such as computer science, the physical sciences, and engineering. This text is an informal, nonaxiomatic treatment of the theory of sets.

The material is divided into three Parts, since the logical development is thereby not disturbed while the usefulness as a text and reference book on any of several levels is increased. Part I contains an introduction to the elementary operations of sets and a detailed discussion of the concepts of relation and function. Part II develops the theory of cardinal and ordinal numbers in the classical approach of Cantor. It also considers partially ordered sets, and the Axiom of Choice and its equivalents including Zorn's lemma. Part III treats those topics which are usually associated with the elementary theory of sets, that is, logic and Boolean algebra.

This second edition of *Set Theory* covers more material than the first edition. In particular, it includes a deeper discussion of the real numbers **R** and a more complete discussion of the integers **Z**. Furthermore, it includes a discussion of algorithms and their complexity in the chapter on functions, and it includes new material, including Karnaugh maps, in the chapter on Boolean algebra.

Each chapter begins with clear statements of pertinent definitions, principles, and theorems together with illustrative and other descriptive material. This is followed by graded sets of solved and supplementary problems. The solved problems serve to illustrate and amplify the theory, bring into sharp focus those fine points without which the student continually feels himself on unsafe ground, and provide the repetition of basic principles so vital to effective learning. Numerous proofs of theorems and derivations of basic results are included among the solved problems. The supplementary problems serve as a complete review of each chapter.

Finally, the author wishes to thank the staff of the McGraw-Hill Schaum's Outline Series, especially Barbara Gilson, Mary Loebig Giles, and Maureen Walker, for their excellent cooperation.

SEYMOUR LIPSCHUTZ

*Temple University*

# Contents

## PART I *Elementary Theory of Sets*



## PART II *Cardinals, Ordinals, Transfinite Induction*

# PART I: **Elementary Theory of Sets**

# Chapter 1

## Sets and Basic Operations on Sets

### 1.1 INTRODUCTION

The concept of a *set* appears in all branches of mathematics. This concept formalizes the idea of grouping objects together and viewing them as a single entity. This chapter introduces this notion of a set and its members. We also investigate three basic operations on sets, that is, the operations union, intersection, and complement.

Although logic is formally treated in Chapter 10, we indicate here the close relationship between set theory and logic by showing how Venn diagrams, pictures of sets, can be used to determine the validity of certain arguments. The relation between set theory and logic will be further explored when we discuss Boolean algebra in Chapter 11.

### 1.2 SETS AND ELEMENTS

A *set* may be viewed as any well-defined collection of objects; the objects are called the *elements* or *members* of the set.

Although we shall study sets as abstract entities, we now list ten examples of sets:

(1) The numbers 1, 3, 7, and 10.
(2) The solutions of the equation $x^2 - 3x - 2 = 0$.
(3) The vowels of the English alphabet: a, e, i, o, u.
(4) The people living on the earth.
(5) The students Tom, Dick, and Harry.
(6) The students absent from school.
(7) The countries England, France, and Denmark.
(8) The capital cities of Europe.
(9) The even integers: 2, 4, 6, ....
(10) The rivers in the United States.

Observe that the sets in the odd-numbered examples are *defined*, that is, specified or presented, by actually listing its members; and the sets in the even-numbered examples are defined by stating properties or rules which decide whether or not a particular object is a member of the set.

**Notation**

A set will usually be denoted by a capital letter, such as,

$$A, B, X, Y, \ldots,$$

whereas lower-case letters, $a, b, c, x, y, z, \ldots$ will usually be used to denote elements of sets.

There are essentially two ways to specify a particular set, as indicated above. One way, if possible, is to list its elements. For example,

$$A = \{\text{a}, \text{e}, \text{i}, \text{o}, \text{u}\}$$

means that $A$ is the set whose elements are the letters a, e, i, o, u. Note that the elements are separated by commas and enclosed in braces { }. This is sometimes called the *tabular form* of a set.

The second way is to state those properties which characterize the elements in the set, that is, properties held by the members of the set but not by nonmembers. Consider, for example, the expression

$$B = \{x : x \text{ is an even integer, } x > 0\}$$

which reads:

"$B$ is the set of $x$ such that $x$ is an even integer and $x > 0$"

It denotes the set $B$ whose elements are the positive even integers. A letter, usually $x$, is used to denote a typical member of the set; the colon is read as "such that" and the comma as "and". This is sometimes called the *set-builder form* or *property method* of specifying a set.

Two sets $A$ and $B$ are *equal*, written $A = B$, if they both have the same elements, that is, if every element which belongs to $A$ also belongs to $B$, and vice versa. The negation of $A = B$ is written $A \neq B$.

The statement "$p$ is an element of $A$" or, equivalently, the statement "$p$ belongs to $A$" is written

$$p \in A$$

We also write

$$a, b \in A$$

to state that both $a$ and $b$ belong to $A$. The statement that $p$ is not an element of $A$, that is, the negation of $p \subset A$, is written

$$p \notin A$$

**Remark:** It is common practice in mathematics to put a vertical line "|" or slanted line "/" through a symbol to indicate the opposite or negative meaning of the symbol.

**EXAMPLE 1.1**

(*a*) The set $A$ above can also be written as

$$A = \{x : x \text{ is a letter in the English alphabet, } x \text{ is a vowel}\}$$

Observe that $b \notin A$, $e \in A$, and $p \not\subset$ A.

(*b*) We cannot list all the elements of the above set $B$, although we frequently specify the set by writing

$$B = \{2, 4, 6, \ldots\}$$

where we assume everyone knows what we mean. Observe that $8 \in B$, but $9 \notin B$.

(*c*) Let $E = \{x : x^2 - 3x + 2 = 0\}$. In other words, $E$ consists of those numbers which are solutions of the equation $x^2 - 3x + 2 = 0$, sometimes called the *solution set* of the given equation. Since the solutions are 1 and 2, we could also write $E = \{1, 2\}$.

(*d*) Let $E = \{x : x^2 - 3x + 2 = 0\}$, $F = \{2, 1\}$, and $G = \{1, 2, 2, 1, 6/3\}$. Then $E = F = G$ since each consists precisely of the elements 1 and 2. Observe that a set does not depend on the way in which its elements are displayed. A set remains the same even if its elements are repeated or rearranged.

Some sets of numbers will occur very often in the text, and so we use special symbols for them. Unless otherwise specified, we will let:

$\mathbf{N}$ = the set of nonnegative integers: $0, 1, 2, \ldots$
$\mathbf{P}$ = the set of positive integers: $1, 2, 3, \ldots$
$\mathbf{Z}$ = the set of integers: $\ldots, -2, -1, 0, 1, 2, \ldots$
$\mathbf{Q}$ = the set of rational numbers
$\mathbf{R}$ = the set of real numbers
$\mathbf{C}$ = the set of complex numbers

Even if we can list the elements of a set, it may not be practical to do so. For example, we would not list the members of the set of people born in the world during the year 1976 although theoretically it is possible to compile such a list. That is, we describe a set by listing its elements only if the set contains a few elements; otherwise we describe a set by the property which characterizes its elements.

## 1.3 UNIVERSAL SET, EMPTY SET

All sets under investigation in any application of set theory are assumed to be contained in some large fixed set called the *universal set* or *universe*. For example, in plane geometry, the universal set consists of all the points in the plane, and in human population studies the universal set consists of all the people in the world. We will denote the universal set by

$$\mathbf{U}$$

unless otherwise specified.

Given a universal set $\mathbf{U}$ and a property $P$, there may be no element in $\mathbf{U}$ which has the property $P$. For example, the set

$$S = \{x : x \text{ is a positive integer, } x^2 = 3\}$$

has no elements since no positive integer has the required property. This set with no elements is called the *empty set* or *null set*, and is denoted by

$$\varnothing$$

(based on the Greek letter phi). There is only one empty set: If $S$ and $T$ are both empty, then $S = T$ since they have exactly the same elements, namely, none.

## 1.4 SUBSETS

Suppose every element in a set $A$ is also an element of a set $B$; then $A$ is called a *subset* of $B$. We also say that $A$ is *contained* in $B$ or $B$ *contains* $A$. This relationship is written

$$A \subseteq B \qquad \text{or} \qquad B \supseteq A$$

If $A$ is not a subset of $B$, that is, if at least one element of $A$ does not belong to $B$, we write $A \not\subseteq B$ or $B \not\supseteq A$.

**EXAMPLE 1.2**

(*a*) Consider the sets

$$A = \{1, 3, 5, 8, 9\}, \qquad B = \{1, 2, 3, 5, 7\}, \qquad C = \{1, 5\}$$

Then $C \subseteq A$ and $C \subseteq B$ since 1 and 5, the elements of $C$, are also elements of $A$ and $B$. But $B \not\subseteq A$ since some of its elements, e.g., 2 and 7, do not belong to $A$. Furthermore, since the elements in the sets $A, B, C$ must also belong to the universal set $\mathbf{U}$, it is clear that $\mathbf{U}$ must at least contain the set $\{1, 2, 3, 4, 5, 6, 7, 8, 9\}$.

(*b*) Let $\mathbf{P}, \mathbf{N}, \mathbf{Z}, \mathbf{Q}, \mathbf{R}$ be defined as in Section 1.2. Then:

$$\mathbf{P} \subseteq \mathbf{N} \subseteq \mathbf{Z} \subseteq \mathbf{Q} \subseteq \mathbf{R}$$

(*c*) The set $E = \{2, 4, 6\}$ is a subset of the set $F = \{6, 2, 4\}$, since each number 2, 4, and 6 belonging to $E$ also belongs to $F$. In fact, $E = F$. Similarly, it can be shown that every set is a subset of itself.

The following properties of sets should be noted:

(i) Every set $A$ is a subset of the universal set $\mathbf{U}$ since, by definition, all the elements of $A$ belong to $\mathbf{U}$. Also the empty set $\varnothing$ is a subset of $A$.

(ii) Every set $A$ is a subset of itself since, trivially, the elements of $A$ belong to $A$.

(iii) If every element of $A$ belongs to a set $B$, and every element of $B$ belongs to a set $C$, then clearly every element of $A$ belongs to $C$. In other words, if $A \subseteq B$ and $B \subseteq C$, then $A \subseteq C$.

(iv) If $A \subseteq B$ and $B \subseteq A$, then $A$ and $B$ have the same elements, i.e., $A = B$. Conversely, if $A = B$ then $A \subseteq B$ and $B \subseteq A$ since every set is a subset of itself.

We state these results formally.

**Theorem 1.1:** (i) For any set $A$, we have $\varnothing \subseteq A \subseteq \mathbf{U}$.
(ii) For any set $A$, we have $A \subseteq A$.
(iii) If $A \subseteq B$ and $B \subseteq C$, then $A \subseteq C$.
(iv) $A = B$ if and only if $A \subseteq B$ and $B \subseteq A$.

**Proper Subset**

If $A \subseteq B$, then it is still possible that $A = B$. When $A \subseteq B$ but $A \neq B$, we say that $A$ is a *proper subset* of $B$. We will write $A \subset B$ when $A$ is a proper subset of $B$. For example, suppose

$$A = \{1, 3\}, \qquad B = \{1, 2, 3\}, \qquad C = \{1, 3, 2\}$$

Then $A$ and $B$ are both subsets of $C$; but $A$ is a proper subset of $C$, whereas $B$ is not a proper subset of $C$.

**Disjoint Sets**

Two sets $A$ and $B$ are disjoint if they have no elements in common. For example, suppose

$$A = \{1, 2\}, \qquad B = \{2, 4, 6\}, \qquad C = \{4, 5, 6, 7\}$$

Note that $A$ and $B$ are not disjoint since they both contain the element 2. Similarly, $B$ and $C$ are not disjoint since they both contain the element 4, among others. On the other hand, $A$ and $C$ are disjoint since they have no element in common. We note that if two sets $A$ and $B$ are disjoint sets then neither is a subset of the other (unless one is the empty set).

## 1.5 VENN DIAGRAMS

A Venn diagram is a pictorial representation of sets where sets are represented by enclosed areas in the plane. The universal set $\mathbf{U}$ is represented by the points in a rectangle, and the other sets are represented by disks lying within the rectangle. If $A \subseteq B$, then the disk representing $A$ will be entirely within the disk representing $B$, as in Fig. 1-1($a$). If $A$ and $B$ are disjoint, i.e., have no elements in common, then the disk representing $A$ will be separated from the disk representing $B$, as in Fig. 1-1($b$).

On the other hand, if $A$ and $B$ are two arbitrary sets, it is possible that some elements are in $A$ but not $B$, some elements are in $B$ but not $A$, some are in both $A$ and $B$, and some are in neither $A$ nor $B$; hence, in general, we represent $A$ and $B$ as in Fig. 1-1($c$).

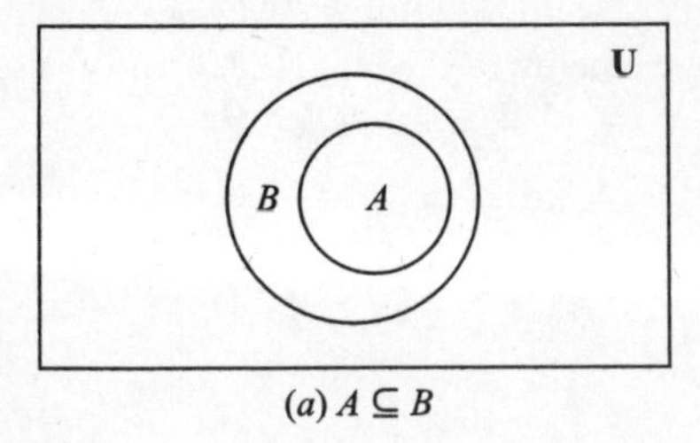

($a$) $A \subseteq B$

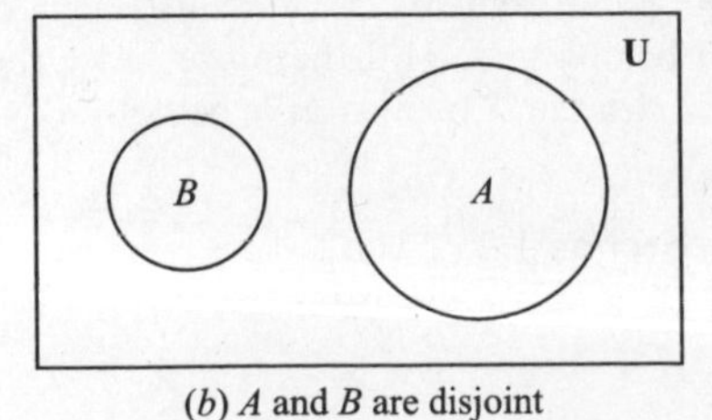

($b$) $A$ and $B$ are disjoint

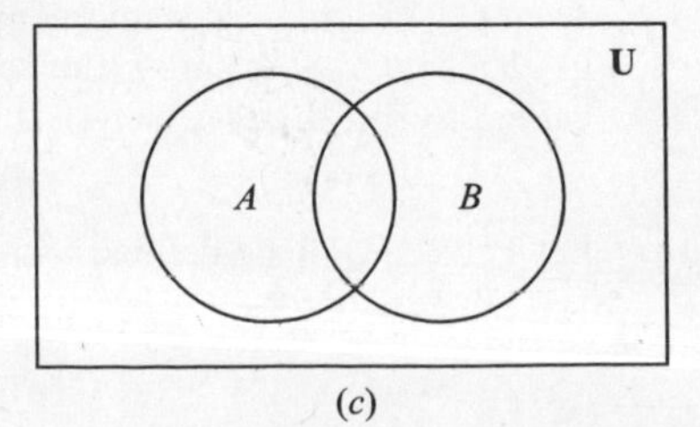

($c$)

**Fig. 1-1**

## 1.6 SET OPERATIONS

The reader has learned to add, subtract, and multiply in the ordinary arithmetic of numbers; that is, to each pair of numbers $a$ and $b$, we assign a number $a+b$ called the *sum* of $a$ and $b$, a number $a-b$ called the *difference* of $a$ and $b$, and a number $ab$ called the *product* of $a$ and $b$. These assignments are called the operations of addition, subtraction, and multiplication of numbers. This section defines a number of set operations, including the basic operations of union, intersection, and difference of sets, where new sets will be assigned to pairs of sets $A$ and $B$. We will see that set operations have many properties similar to the above operations on numbers.

### Union and Intersection

The *union* of two sets $A$ and $B$, denoted by $A \cup B$, is the set of all elements which belong to $A$ or $B$; that is,

$$A \cup B = \{x : x \in A \text{ or } x \in B\}$$

Here "or" is used in the sense of and/or. Figure 1-2($a$) is a Venn diagram in which $A \cup B$ is shaded.

The *intersection* of two sets $A$ and $B$, denoted by $A \cap B$, is the set of all elements which belong to both $A$ and $B$; that is,

$$A \cap B = \{x : x \in A \text{ and } x \in B\}$$

Figure 1-2($b$) is a Venn diagram in which $A \cap B$ is shaded.

Recall that sets $A$ and $B$ are said to be disjoint if they have no elements in common. Accordingly, using the above notation, $A$ and $B$ are disjoint if $A \cap B = \varnothing$, the empty set.

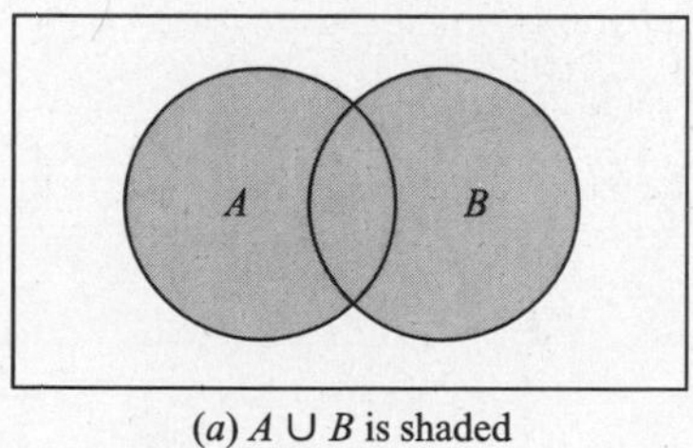

($a$) $A \cup B$ is shaded

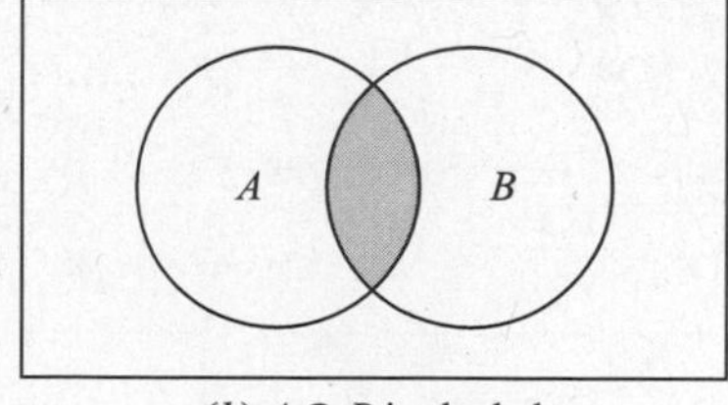

($b$) $A \cap B$ is shaded

**Fig. 1-2**

**EXAMPLE 1.3**

($a$) Let $A = \{1, 2, 3, 4\}$, $B = \{3, 4, 5, 6, 7\}$, $C = \{2, 3, 8, 9\}$. Then

$$A \cup B = \{1, 2, 3, 4, 5, 6, 7\}, \qquad A \cap B = \{3, 4\}$$
$$A \cup C = \{1, 2, 3, 4, 8, 9\}, \qquad A \cap C = \{2, 3\}$$
$$B \cup C = \{2, 3, 4, 5, 6, 7, 8, 9\}, \qquad B \cap C = \{3\}$$

($b$) Let $\mathbf{U}$ denote the set of students at a university, and let $M$ and $F$ denote, respectively, the set of male and female students at the university. Then

$$M \cup F = \mathbf{U}$$

since each student in $\mathbf{U}$ is either in $M$ or in $F$. On the other hand,

$$M \cap F = \varnothing$$

since no student belongs to both $M$ and $F$.

The following properties of the union and intersection of sets should be noted:

(i) Every element $x$ in $A \cap B$ belongs to both $A$ and $B$; hence $x$ belongs to $A$ and $x$ belongs to $B$. Thus $A \cap B$ is a subset of $A$ and of $B$, that is,

$$A \cap B \subseteq A \quad \text{and} \quad A \cap B \subseteq B$$

(ii) An element $x$ belongs to the union $A \cup B$ if $x$ belongs to $A$ or $x$ belongs to $B$; hence every element in $A$ belongs to $A \cup B$, and also every element in $B$ belongs to $A \cup B$. That is,

$$A \subseteq A \cup B \quad \text{and} \quad B \subseteq A \cup B$$

We state the above results formally.

**Theorem 1.2:** For any sets $A$ and $B$, we have

$$A \cap B \subseteq A \subseteq A \cup B \quad \text{and} \quad A \cap B \subseteq B \subseteq A \cup B$$

The operation of set inclusion is also closely related to the operations of union and intersection, as shown by the following theorem, proved in Problem 1.13.

**Theorem 1.3:** The following are equivalent:

$$A \subseteq B, \qquad A \cap B = A, \qquad A \cup B = B$$

Other conditions equivalent to $A \subseteq B$ are given in Problem 1.51.

### Complement

Recall that all sets under consideration at a particular time are subsets of a fixed universal set $\mathbf{U}$. The *absolute complement*, or, simply, *complement* of a set $A$, denoted by $A^c$, is the set of elements which belong to $\mathbf{U}$ but which do not belong to $A$; that is,

$$A^c = \{x : x \in \mathbf{U}, x \notin A\}$$

Some texts denote the complement of $A$ by $A'$ or $\bar{A}$. Figure 1-3($a$) is a Venn diagram in which $A^c$ is shaded.

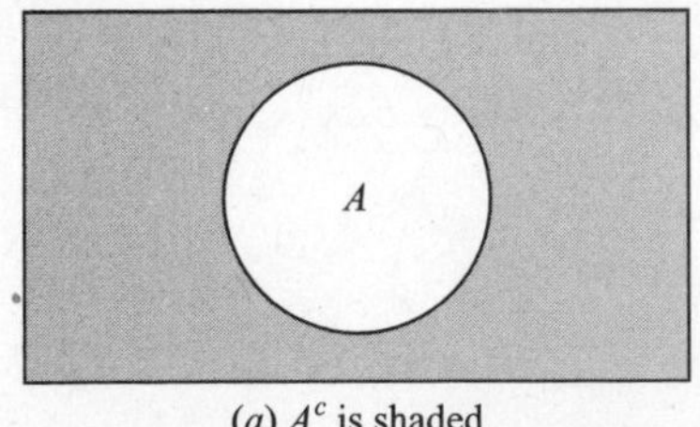


($a$) $A^c$ is shaded

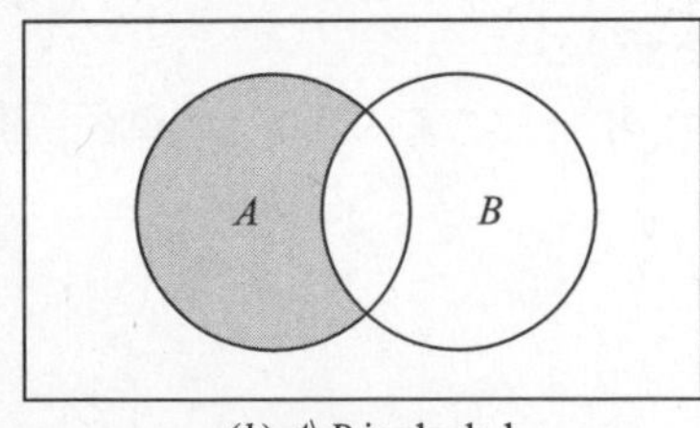


($b$) $A \backslash B$ is shaded

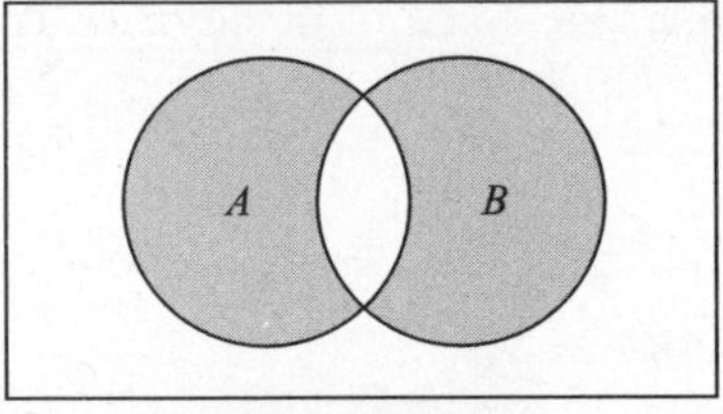


($b$) $A \oplus B$ is shaded

**Fig. 1-3**

**EXAMPLE 1.4**

($a$) Let $\mathbf{U} = \{a, b, c, \ldots, y, z\}$, the English alphabet, be the universal set, and let

$$A = \{\mathrm{a, b, c, d, e}\}, \qquad B = \{\mathrm{e, f, g}\}, \qquad V = \{\mathrm{a, e, i, o, u}\}$$

Then

$$A^c = \{\mathrm{f, g, h, \ldots, y, z}\} \quad \text{and} \quad B^c = \{\mathrm{a, b, c, d, h, i, \ldots, y, z}\}$$

Since $V$ consists of the vowels in $\mathbf{U}$, $V^c$ consists of the nonvowels, called consonants.

($b$) Suppose the set $\mathbf{R}$ of real numbers is the universal set. Recall that $\mathbf{Q}$ denotes the set of rational numbers. Hence $\mathbf{Q}^c$ will denote the set of irrational numbers.

(*c*) Let **U** be the set of students at a university, and suppose $M$ and $F$ denote, respectively, the male and female students in **U**. Then

$$M^c = F \quad \text{and} \quad F^c = M$$

### Difference and Symmetric Difference

Let $A$ and $B$ be sets. The *relative complement* of $B$ with respect to $A$ or, simply, the *difference* of $A$ and $B$, denoted by $A \backslash B$, is the set of elements which belong to $A$ but which do not belong to $B$; that is,

$$A \backslash B = \{x : x \in A, x \notin B\}$$

The set $A \backslash B$ is read "$A$ minus $B$". Many texts denote $A \backslash B$ by $A - B$ or $A \sim B$. Figure 1-3(*b*) is a Venn diagram in which $A \backslash B$ is shaded.

The *symmetric difference* of the sets $A$ and $B$, denoted by $A \oplus B$, consists of those elements which belong to $A$ or $B$ but not to both $A$ and $B$. That is,

$$A \oplus B = (A \cup B) \backslash (A \cap B) \quad \text{or} \quad A \oplus B = (A \backslash B) \cup (B \backslash A)$$

Figure 1-3(*c*) is a Venn diagram in which $A \oplus B$ is shaded. The fact that

$$(A \cup B) \backslash (A \cap B) = (A \backslash B) \cup (B \backslash A)$$

is proved in Problem 1.18.

**EXAMPLE 1.5** Consider the sets

$$A = \{1, 2, 3, 4\}, \qquad B = \{3, 4, 5, 6, 7\}, \qquad C = \{6, 7, 8, 9\}$$

Then

$$A \backslash B = \{1, 2\}, \qquad B \backslash C = \{3, 4, 5\}, \qquad B \backslash A = \{5, 6, 7\}, \qquad C \backslash B = \{8, 9\}$$

Also,

$$A \oplus B = \{1, 2, 5, 6, 7\} \quad \text{and} \quad B \oplus C = \{3, 4, 5, 8, 9\}$$

Note that $A$ and $C$ are disjoint. This means

$$A \backslash C = A, \qquad C \backslash A = C, \qquad A \oplus C = A \cup C$$

## 1.7 ALGEBRA OF SETS, DUALITY

Sets under the above operations of union, intersection, and complement satisfy various laws (identities) which are listed in Table 1-1. In fact, we formally state:

**Theorem 1.4:** Sets satisfy the laws in Table 1-1.

Each of the laws in Table 1-1 follows from an equivalent logical law. Consider, for example, the proof of DeMorgan's law:

$$(A \cup B)^c = \{x : x \notin (A \text{ or } B)\} = \{x : x \notin A \text{ and } x \notin B\} = A^c \cap B^c$$

Here we use the equivalent (DeMorgan's) logical law:

$$\neg (p \vee q) \equiv \neg p \wedge \neg q$$

Here $\neg$ means "not", $\vee$ means "or", and $\wedge$ means "and". Sometimes Venn diagrams are used to illustrate the laws in Table 1-1 (cf. Problem 1.16).

**Table 1-1 Laws of the Algebra of Sets**

| | |
|---|---|
| **Idempotent laws** | |
| (1*a*) $A \cup A = A$ | (1*b*) $A \cap A = A$ |
| **Associative laws** | |
| (2*a*) $(A \cup B) \cup C = A \cup (B \cup C)$ | (2*b*) $(A \cap B) \cap C = A \cap (B \cap C)$ |
| **Commutative laws** | |
| (3*a*) $A \cup B = B \cup A$ | (3*b*) $A \cap B = B \cap A$ |
| **Distributive laws** | |
| (4*a*) $A \cup (B \cap C) = (A \cup B) \cap (A \cup C)$ | (4*b*) $A \cap (B \cup C) = (A \cap B) \cup (A \cap C)$ |
| **Identity laws** | |
| (5*a*) $A \cup \varnothing = A$ | (5*b*) $A \cap \mathbf{U} = A$ |
| (6*a*) $A \cup \mathbf{U} = \mathbf{U}$ | (6*b*) $A \cap \varnothing = \varnothing$ |
| **Involution law** | |
| (7) $(A^c)^c = A$ | |
| **Complement laws** | |
| (8*a*) $A \cup A^c = \mathbf{U}$ | (8*b*) $A \cap A^c = \varnothing$ |
| (9*a*) $\mathbf{U}^c = \varnothing$ | (9*b*) $\varnothing^c = \mathbf{U}$ |
| **DeMorgan's laws** | |
| (10*a*) $(A \cup B)^c = A^c \cap B^c$ | (10*b*) $(A \cap B)^c = A^c \cup B^c$ |

**Duality**

The identities in Table 1-1 are arranged in pairs, as, for example, (2*a*) and (2*b*). We now consider the principle behind this arrangement. Let $E$ be an equation of set algebra. The *dual* $E^*$ of $E$ is the equation obtained by replacing each occurrence of $\cup, \cap, \mathbf{U}, \varnothing$ in $E$ by $\cap, \cup, \varnothing, \mathbf{U}$, respectively. For example, the dual of

$$(\mathbf{U} \cap A) \cup (B \cap A) = A \qquad \text{is} \qquad (\varnothing \cup A) \cap (B \cup A) = A$$

Observe that the pairs of laws in Table 1-1 are duals of each other. It is a fact of set algebra, called the *principle of duality*, that, if any equation $E$ is an identity, then its dual $E^*$ is also an identity.

## 1.8 FINITE SETS, COUNTING PRINCIPLES

A set is said to be *finite* if it contains exactly $m$ distinct elements where $m$ denotes some nonnegative integer. Otherwise a set is said to be infinite. For example, the empty set $\varnothing$ and the set of letters of the English alphabet are finite sets, whereas the set of even positive integers $\{2, 4, 6, \ldots\}$ is infinite. [Infinite sets will be studied in detail in Chapter 6.]

The notation $n(A)$ or $|A|$ will denote the number of elements in a finite set $A$.

First we begin with a special case.

**Lemma 1.5**: Suppose $A$ and $B$ are finite disjoint sets. Then $A \cup B$ is finite and

$$n(A \cup B) = n(A) + n(B)$$

*Proof.* In counting the elements of $A \cup B$, first count those that are in $A$. There are $n(A)$ of these. The only other elements of $A \cup B$ are those that are in $B$ but not in $A$. Since $A$ and $B$ are disjoint, no element of $B$ is in $A$, so there are $n(B)$ elements that are in $B$ but not in $A$. Therefore, $n(A \cup B) = n(A) + n(B)$, as claimed.

**Remark:** A set $C$ is called the *disjoint union* of $A$ and $B$ if

$$C = A \cup B \quad \text{and} \quad A \cap B = \varnothing$$

Lemma 1.5 tells us that, in such a case, $n(C) = n(A) + n(B)$.

### Special Cases of Disjoint Unions

There are two special cases of disjoint unions which occur frequently.

(1) Given any set $A$, then the universal set $U$ is the disjoint union of $A$ and its complement $A^c$. Thus, by Lemma 1.5,

$$n(U) = n(A) + n(A^c)$$

Accordingly, bringing $n(A)$ to the other side, we obtain the following useful result.

**Theorem 1.6:** Let $A$ be any set in a finite universal set $U$. Then

$$\boxed{n(A^c) = n(U) - n(A)}$$

For example, if there are 20 male students in a class of 35 students, then there are $35 - 20 = 15$ female students.

(2) Given any sets $A$ and $B$, we show (Problem 1.37) that $A$ is the disjoint union of $A \backslash B$ and $A \cap B$. This is pictured in Fig. 1-4. Thus Lemma 1.5 gives us the following useful result.

**Theorem 1.7:** Suppose $A$ and $B$ are finite sets. Then

$$\boxed{n(A \backslash B) = n(A) - n(A \cap B)}$$

For example, suppose an archery class $A$ contains 35 students, and 15 of them are also in a bowling class $B$. Then

$$n(A \backslash B) = n(A) - n(A \cap B) = 35 - 15 = 20$$

That is, there are 20 students in the class $A$ who are not in class $B$.

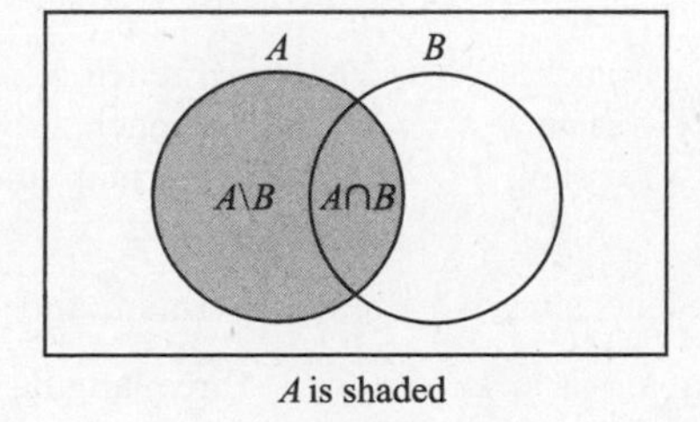


*A* is shaded

**Fig. 1-4**

### Inclusion-Exclusion Principle

There is also a formula for $n(A \cup B)$ even when they are not disjoint, called the *inclusion-exclusion principle*. Namely:

**Theorem 1.8**: Suppose $A$ and $B$ are finite sets. Then $A \cap B$ and $A \cup B$ are finite, and

$$\boxed{n(A \cup B) = n(A) + n(B) - n(A \cap B)}$$

That is, we find the number of elements in $A$ or $B$ (or both) by first adding $n(A)$ and $n(B)$ (inclusion) and then subtracting $n(A \cap B)$ (exclusion) since the elements in $A \cap B$ were counted twice.

We can apply this result to get a similar result for three sets.

**Corollary 1.9**: Suppose $A, B, C$ are finite sets. Then $A \cup B \cup C$ is finite and

$$n(A \cup B \cup C) = n(A) + n(B) + n(C) - n(A \cap B) - n(A \cap C) - n(B \cap C) + n(A \cap B \cap C)$$

Mathematical induction (Section 1.11) may be used to further generalize this result to any finite number of finite sets.

**EXAMPLE 1.6** Consider the following data among 110 students in a college dormitory:

30 students are on a list A (taking Accounting),

35 students are on a list B (taking Biology),

20 students are on both lists.

Find the number of students: (*a*) on list or $B$, (*b*) on exactly one of the two lists, (*c*) on neither list.

(*a*) We seek $n(A \cup B)$. By Theorem 1.8,

$$n(A \cup B) = n(A) + n(B) - n(A \cap B) = 30 + 35 - 20 = 45$$

In other words, we combine the two lists and then cross out the 20 student names which appear twice.

(*b*) List $A$ contains 30 names and 20 of them are on list $B$; hence $30 - 20 = 10$ names are only on list $A$. That is,

$$n(A \backslash B) = n(A) - n(A \cup B) = 30 - 20 = 10$$

Similarly, list $B$ contains 35 names and 20 of them are on list A; hence $35 - 20 = 15$ names are only on list B. That is,

$$n(B \backslash A) = n(B) - n(A \cup B) = 35 - 20 = 15$$

Thus there are $10 + 15 = 25$ students on exactly one of the two lists.

(*c*) The students on neither the $A$ list nor the $B$ list form the set $A^c \cap B^c$. By DeMorgan's law, $A^c \cap B^c = (A \cup B)^c$. Hence

$$n(A^c \cap B^c) = n((A \cup B)^c) = n(U) - n(A \cup B) = 110 - 45 = 65$$

**EXAMPLE 1.7** Consider the following data for 120 mathematics students:

65 study French, 20 study French and German,
45 study German, 25 study French and Russian,
42 study Russian, 15 study German and Russian,
8 study all three languages

Let $F$, $G$, and $R$ denote the sets of students studying French, German, and Russian, respectively.

(*a*) Find the number of students studying at least one of the three languages, i.e. find $n(F \cup G \cup R)$.

(*b*) Fill in the correct number of students in each of the eight regions of the Venn diagram of Fig. 1-5(*a*).

(*c*) Find the number $k$ of students studying: (1) exactly one language, (2) exactly two languages.

(*a*) By Corollary 1.9,

$$\begin{aligned} n(F \cup G \cup R) &= n(F) + n(G) + n(R) - n(F \cap G) - n(F \cap R) - n(G \cap R) - n(F \cap G \cap R) \\ &= 65 + 45 + 42 - 20 - 25 - 15 + 8 = 100 \end{aligned}$$

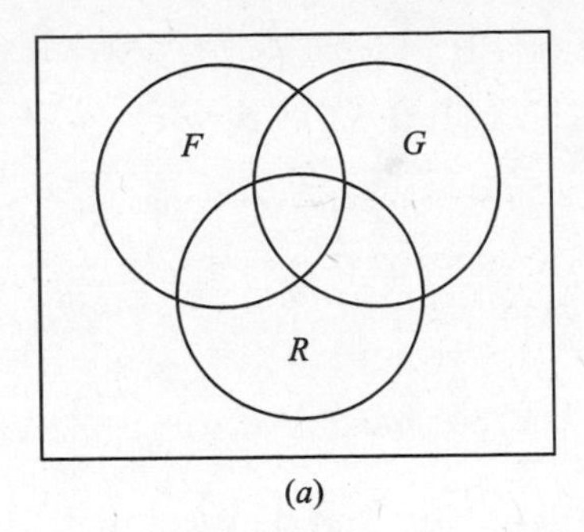

(a)

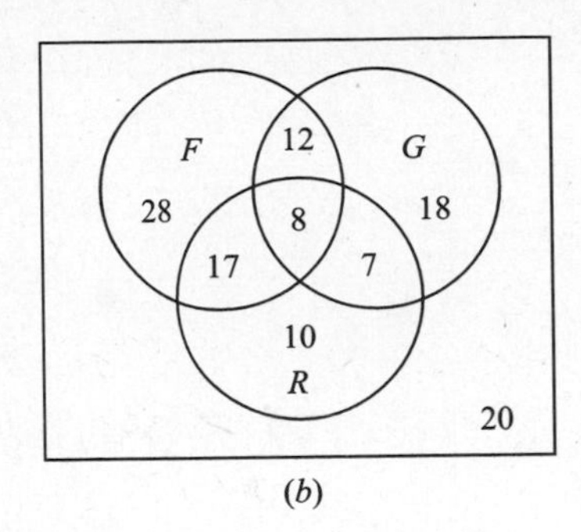

(b)

**Fig. 1-5**

(b) Using 8 study all three languages and 100 study at least one language, the remaining seven regions of the required Venn diagram Fig. 1-5(b) are obtained as follows:

$15 - 8 = 7$ study German and Russian but not French,
$25 - 8 = 17$ study French and Russian but not German,
$20 - 8 = 12$ study French and German but not Russian,
$42 - 17 - 8 - 7 = 10$ study only Russian,
$45 - 12 - 8 - 7 = 18$ study only German,
$65 - 12 - 8 - 17 = 28$ study only French,
$120 - 100 = 20$ do not study any of the languages.

(c) Use the Venn diagram of Fig. 1-5(b) to obtain:

(1) $k = 28 + 18 + 10 = 56$, (2) $k = 12 + 17 + 7 = 36$

## 1.9 CLASSES OF SETS, POWER SETS

Given a set $S$, we may wish to talk about some of its subsets. Thus we would be considering a "set of sets". Whenever such a situation arises, to avoid confusion, we will speak of a *class* of sets or a *collection* of sets. If we wish to consider some of the sets in a given class of sets, then we will use the term *subclass* or *subcollection*.

**EXAMPLE 1.8** Suppose $S = \{1, 2, 3, 4\}$. Let $\mathscr{A}$ be the class of subsets of $S$ which contain exactly three elements of $S$. Then

$$\mathscr{A} = [\{1,2,3\}, \{1,2,4\}, \{1,3,4\}, \{2,3,4\}]$$

The elements of $\mathscr{A}$ are the sets $\{1,2,3\}$, $\{1,2,4\}$, $\{1,3,4\}$, and $\{2,3,4\}$.

Let $\mathscr{B}$ be the class of subsets of $S$ which contain 2 and two other elements of $S$. Then

$$\mathscr{B} = [\{1,2,3\}, \{1,2,4\}, \{2,3,4\}]$$

The elements of $\mathscr{B}$ are $\{1,2,3\}$, $\{1,2,4\}$, and $\{2,3,4\}$. Thus $\mathscr{B}$ is a subclass of $\mathscr{A}$. (To avoid confusion, we will usually enclose the sets of a class in brackets instead of braces.)

### Power Sets

For a given set $S$, we may speak about the class of all subsets of $S$. This class is called the *power set* of $S$, and it will be denoted by $\mathscr{P}(S)$. If $S$ is finite, then so is $\mathscr{P}(S)$. In fact, the number of elements in $\mathscr{P}(S)$ is 2 raised to the power of $n(S)$; that is,

$$n(\mathscr{P}(S)) = 2^{n(S)}$$

(This is the reason $\mathscr{P}(S)$ is called the power set of $S$; it is also sometimes denoted by $2^S$.)

**EXAMPLE 1.9** Suppose $S = \{1, 2, 3\}$. Then

$$\mathscr{P}(S) = [\varnothing, \{1\}, \{2\}, \{3\}, \{1, 2\}, \{1, 3\}, \{2, 3\}, S]$$

Note that the empty set $\varnothing$ belongs to $\mathscr{P}(S)$ since $\varnothing$ is a subset of $S$. Similarly $S$ belongs to $\mathscr{P}(S)$. As expected from the above remark, $\mathscr{P}(S)$ has $2^3 = 8$ elements.

## 1.10 ARGUMENTS AND VENN DIAGRAMS

Many verbal statements are essentially statements about sets and they can therefore be described by Venn diagrams. Hence Venn diagrams can sometimes be used to determine whether or not an argument is valid. This is illustrated in the following example.

**EXAMPLE 1.10** Show that the following argument (adapted from a book on logic by Lewis Carroll, the author of *Alice in Wonderland*) is valid:

$S_1$: My saucepans are the only things I have that are made of tin.
$S_2$: I find all your presents very useful.
$S_3$: None of my saucepans is of the slightest use.

---

$S$: Your presents to me are not made of tin.

(The statements $S_1$, $S_2$, and $S_3$ above the horizontal line denote the assumptions, and the statement $S$ below the line denotes the conclusion. The argument is valid if the conclusion $S$ follows logically from the assumptions $S_1$, $S_2$, and $S_3$.)

By $S_1$ the tin objects are contained in the set of saucepans and by $S_3$ the set of saucepans and the set of useful things are disjoint: hence draw the Venn diagram of Fig. 1-6.

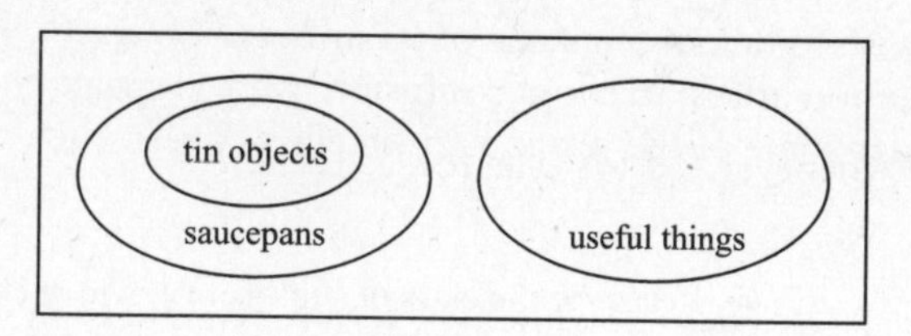


**Fig. 1-6**

By $S_2$ the set of "your presents" is a subset of the set of useful things; hence draw Fig. 1-7.

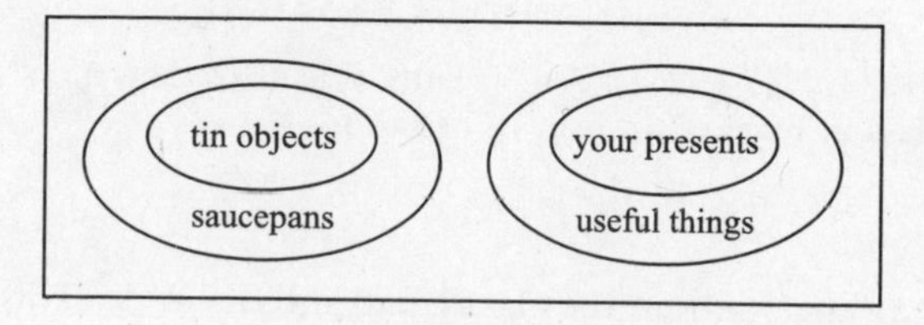


**Fig. 1-7**

The conclusion is clearly valid by the Venn diagram in Fig. 1-7 because the set of "your presents" is disjoint from the set of tin objects.

## 1.11 MATHEMATICAL INDUCTION

Consider the set $\mathbf{P} = \{1, 2, \ldots\}$ of positive integers (or counting numbers). We say that an assertion $A(n)$ is defined on $\mathbf{P}$ if $A(n)$ is true or false for each $n \in \mathbf{P}$. An essential property of $\mathbf{P}$, which is used in many proofs, follows.

**Principle of Mathematical Induction I**: Let $A(n)$ be an assertion defined on $\mathbf{P}$, that is, $A(n)$ is true or false for each integer $n \geq 1$. Suppose $A(n)$ has the following two properties:

(1) $A(1)$ is true.
(2) $A(n+1)$ is true whenever $A(n)$ is true.

Then $A(n)$ is true for every $n \geq 1$.

We shall not prove this principle. In fact, this principle is usually given as one of the axioms when $\mathbf{P}$ is developed axiomatically.

**EXAMPLE 1.11** Let $A(n)$ be the assertion that the sum of the first $n$ odd integers is $n^2$; that is,

$$A(n): \quad 1 + 3 + 5 + \cdots + (2n-1) = n^2$$

[The $n$th odd integer is $2n - 1$ and the next odd integer is $2n + 1$.] Observe that $A(n)$ is true for $n = 1$, that is,

$$A(1): \quad 1 = 1^2$$

Assuming $A(n)$ is true, we add $2n + 1$ to both sides of $A(n)$, obtaining:

$$1 + 3 + 5 + \cdots + (2n-1) + (2n+1) = n^2 + (2n+1) = (n+1)^2$$

However, this is $A(n+1)$. That is, $A(n+1)$ is true whenever $A(n)$ is true. By the principle of mathematical induction, $A(n)$ is true for all $n \geq 1$.

There is another form of the principle of mathematical induction which is sometimes more convenient to use. Although it appears different, it is really equivalent to the above principle of induction.

**Principle of Mathematical Induction II**: Let $A(n)$ be an assertion defined on the set $\mathbf{P}$ of positive integers which satisfies the following two conditions:

(1) $A(1)$ is true.
(2) $A(n)$ is true whenever $A(k)$ is true for $1 \leq k < n$.

Then $A(n)$ is true for every $n \geq 1$.

The above two principles may also be stated in terms of subsets of $\mathbf{P}$ rather than in terms of assertions defined on $\mathbf{P}$. (See Problem 1.40.) Although the languages are different, they are logically equivalent.

**Remark**: Sometimes one wants to prove that an assertion $A$ is true for a set of integers of the form

$$\{a, a+1, a+2, \ldots\}$$

where $a$ is any integer, possibly 0. This can be done by simply replacing 1 by the integer $a$ in either of the above principles of mathematical induction.

## 1.12 AXIOMATIC DEVELOPMENT OF SET THEORY

Any axiomatic development of a branch of mathematics begins with the following:

(1) undefined terms,
(2) undefined relations,
(3) axioms relating the undefined terms and undefined relations.

Then, one develops theorems based upon the axioms and definitions.

Consider, for example, the axiomatic development of plane Euclidean geometry. It begins with the following:

(1) "points" and "lines" are undefined terms;
(2) "point on a line" or, equivalently, "line contains a point" is an undefined relation.

Two of the many axioms of Euclidean geometry follow:

**Axiom 1**: Two distinct points are on one and only one line.

**Axiom 2**: Two distinct lines cannot contain more than one point in common.

The axiomatic development of set theory begins with the following:

(1) "element" and "set" are undefined terms;
(2) "element belongs to a set" is the undefined relation.

Two of the axioms (called principles) of set theory follow:

**Principle of Extension**: Two sets $A$ and $B$ are equal if and only if they have the same elements, that is, if every element in $A$ belongs to $B$ and every element in $B$ belongs to $A$.

**Principle of Abstraction**: Given any set $\mathbf{U}$ and any property $P$, there is a set $A$ such that the elements of $A$ are exactly those elements in $\mathbf{U}$ which have the property $P$; that is,

$$A = \{x : x \in \mathbf{U}, P(x) \text{ is true}\}$$

There are other axioms which are not listed. As our treatment of set theory is mainly intuitive, especially Part I, we will refrain from any further discussion of the axiomatic development of set theory.

## Solved Problems

### SETS AND SUBSETS

**1.1.** Which of these sets are equal: $\{r, t, s\}$, $\{s, t, r, s\}$, $\{t, s, t, r\}$, $\{s, r, s, t\}$?

They are all equal. Order and repetition do not change a set.

**1.2.** List the elements of the following sets where $\mathbf{P} = \{1, 2, 3, \ldots\}$.

(*a*) $A = \{x : x \in \mathbf{P},\ 3 < x < 12\}$
(*b*) $B = \{x : x \in \mathbf{P},\ x \text{ is even},\ x < 15\}$
(*c*) $C = \{x : x \in \mathbf{P},\ 4 + x = 3\}$
(*d*) $D = \{x : x \in \mathbf{P},\ x \text{ is a multiple of } 5\}$.

(*a*) A consists of the positive integers between 3 and 12; hence

$$A = \{4, 5, 6, 7, 8, 9, 10, 11\}$$

(*b*) $B$ consists of the even positive integers less than 15; hence

$$B = \{2, 4, 6, 8, 10, 12, 14\}$$

(*c*) There are no positive integers which satisfy the condition $4 + x = 3$; hence $C$ contains no elements. In other words, $C = \varnothing$, the empty set.

(*d*) $D$ is infinite, so we cannot list all its elements. However, sometimes we write

$$D = \{5, 10, 15, \ldots, 5n, \ldots\} \qquad \text{or simply} \qquad D = \{5, 10, 15, \ldots\}$$

where we assume everyone understands that we mean the multiples of 5.

**1.3.** Consider the following sets:

$$\varnothing, \quad A = \{1\}, \quad B = \{1,3\}, \quad C = \{1,5,9\}, \quad D = \{1,2,3,4,5\},$$
$$E = \{1,3,5,7,9\}, \quad \mathbf{U} = \{1,2,\ldots,8,9\}$$

Insert the correct symbol $\subseteq$ or $\not\subseteq$ between each pair of sets:

(*a*) $\varnothing, A$ (*c*) $B, C$ (*e*) $C, D$ (*g*) $D, E$
(*b*) $A, B$ (*d*) $B, E$ (*f*) $C, E$ (*h*) $D, \mathbf{U}$

(*a*) $\varnothing \subseteq A$ because $\varnothing$ is a subset of every set.
(*b*) $A \subseteq B$ because 1 is the only element of $A$ and it belongs to $B$.
(*c*) $B \not\subseteq C$ because $3 \in B$ but $3 \notin C$.
(*d*) $B \subseteq E$ because the elements of $B$ also belong to $E$.
(*e*) $C \not\subseteq D$ because $9 \in C$ but $9 \notin D$.
(*f*) $C \subseteq E$ because the elements of $C$ also belong to $E$.
(*g*) $D \not\subseteq E$ because $2 \in D$ but $2 \notin E$.
(*h*) $D \subseteq U$ because the elements of $D$ also belong to $\mathbf{U}$.

**1.4.** Show that $A = \{2,3,4,5\}$ is not a subset of $B = \{x : x \in P,\ x \text{ is even}\}$.

It is necessary to show that at least one element in $A$ does not belong to $B$. Now $3 \in A$ and, since $B$ consists of even numbers, $3 \notin B$; hence $A$ is not a subset of $B$.

**1.5.** Show that $A = \{2,3,4,5\}$ is a proper subset of $C = \{1,2,3,\ldots,8,9\}$.

Each element of $A$ belongs to $C$ so $A \subseteq C$. On the other hand, $1 \in C$ but $1 \in A$. Hence $A \neq C$. Therefore $A$ is a proper subset of $C$.

**1.6.** Determine whether or not each set is the null set:

(*a*) $X = \{x : x^2 = 9,\ 2x = 4\}$, (*b*) $Y = \{x : x \neq x\}$, (*c*) $Z = \{x : x + 8 = 8\}$.

(*a*) No number satisfies both $x^2 = 9$ and $2x = 4$; hence $X$ is the empty set; i.e., $X = \varnothing$.
(*b*) We interpret "=" to mean "is identical with" and so $Y$ is empty. In fact, some texts define the empty set as follows:

$$\varnothing \equiv \{x : x \neq x\}$$

(*c*) The number zero satisfies $x + 8 = 8$ and zero is the only solution; hence $Z = \{0\}$. Thus $Z$ is not the empty set since it contains 0. That is, $Z \neq \varnothing$.

## SET OPERATIONS

Problems 1-7 to 1-10 refer to the universal set $\mathbf{U} = \{1, 2, \ldots, 9\}$ and the sets:

$$A = \{1,2,3,4,5\} \qquad C = \{5,6,7,8,9\} \qquad E = \{2,4,6,8\}$$
$$B = \{4,5,6,7\} \qquad D = \{1,3,5,7,9\} \qquad F = \{1,5,9\}$$

**1.7.** Find:

(*a*) $A \cup B$ and $A \cap B$, (*c*) $A \cup C$ and $A \cap C$, (*e*) $E \cup E$ and $E \cap E$
(*b*) $B \cup D$ and $B \cap D$, (*d*) $D \cup E$ and $D \cap E$, (*f*) $D \cup F$ and $D \cap F$

Recall that the union $X \cup Y$ consists of those elements in either $X$ or $Y$ (or both), and that the intersection $X \cap Y$ consists of those elements in both X and Y.

(*a*) $A \cup B = \{1,2,3,4,5,6,7\}$, $A \cap B = \{4,5\}$
(*b*) $B \cup D = \{1,3,4,5,6,7,9\}$, $B \cap D = \{5,7\}$
(*c*) $A \cup C = \{1,2,3,4,5,6,7,8,9\} = \mathbf{U}$, $A \cap C = \varnothing$
(*d*) $D \cup E = \{1,2,3,4,5,6,7,8,9\} = \mathbf{U}$, $D \cap E = \varnothing$
(*e*) $E \cup E = \{2,4,6,8\} = E$, $E \cap E = \{2,4,6,8\} = E$
(*f*) $D \cup F = \{1,3,5,7,9\} = D$, $D \cap F = \{1,5,9\} = F$

Observe that $F \subseteq D$; so by Theorem 1.3 we must have $D \cup F = D$ and $D \cap F = F$.

**1.8.** Find: (*a*) $A^c$, $B^c$, $D^c$, $E^c$; (*b*) $\mathbf{U}^c$, $\varnothing^c$.

(*a*) The complement $X^c$ consists of those elements in the universal set $\mathbf{U}$ which do not belong to $X$. Thus:

$$A^c = \{6,7,8,9\}, \qquad B^c = \{1,2,3,8,9\}, \qquad D^c = \{2,4,6,8\} = E, \qquad E^c = \{1,3,5,7,9\} = D$$

(*Note*: Since $D^c = E$, we must have $E^c = D$.)

(*b*) Here $\mathbf{U}^c = \varnothing$, and $\varnothing^c = \mathbf{U}$, and this is always true.

**1.9.** Find: (*a*) $A \backslash B$, $B \backslash A$, $D \backslash E$, $F \backslash D$; (*b*) $A \oplus B$, $C \oplus D$, $E \oplus F$.

(*a*) The difference $X \backslash Y$ consist of the elements in $X$ which do not belong to $Y$. Thus:

$$A \backslash B = \{1,2,3\}, \qquad B \backslash A = \{6,7\}, \qquad D \backslash E = \{1,3,5,7,9\} = D, \qquad F \backslash D = \varnothing.$$

(*Note*: Since $D$ and $E$ are disjoint, we must have $D \backslash E = D$; and since $F \subseteq D$, we must have $F \backslash D = \varnothing$.)

(*b*) The symmetric difference $X \oplus Y$ consists of the elements in $X$ or in $Y$ but not in both $X$ and $Y$. Thus:

$$A \oplus B = \{1,2,3,6,7\}, \qquad C \oplus D = \{1,3,8,9\}, \qquad E \oplus F = \{2,4,6,8,1,5,9\} = E \cup F$$

(*Note*: Since $E$ and $F$ are disjoint, we must have $E \oplus F = E \cup F$.)

**1.10.** Find: (*a*) $A \cap (B \cup E)$, (*b*) $(A \backslash B)^c$, (*c*) $(A \cap D) \backslash B$, (*d*) $(B \cap F) \cup (C \cap E)$.

(*a*) First compute $B \cup E = \{2,4,5,6,7,8\}$. Then $A \cap (B \cup E) = \{2,4,5\}$.
(*b*) $A \backslash E = \{1,3,5\}$. Then $(A \backslash E)^c = \{2,4,6,7,8,9)$.
(*c*) $A \cap D = \{1,3,5\}$. Now $(A \cap D) \backslash B = \{1,3\}$.
(*d*) $B \cap F = \{5\}$ and $C \cap E = \{6,8)$. So $(B \cap F) \cup (C \cap E) = \{5,6,8)$.

**1.11.** Show that we can have $A \cap B = A \cap C$ without $B = C$.

Let $A = \{1, 2\}$, $B = \{2, 3\}$, and $C = \{2, 4\}$. Then $A \cap B = (2\}$ and $A \cap C = \{2)$. Thus $A \cap B = A \cap C$ but $B \neq C$.

**1.12.** Prove: $B \backslash A = B \cap A$. Thus the set operation of difference can be written in terms of the operations of intersection and complementation.

$$B \backslash A = \{x : x \in B,\ x \notin A) = \{x : x \in B,\ x \in A^c) = B \cap A^c$$

**1.13.** Prove Theorem 1.3: The following are equivalent: $A \subseteq B$, $A \cap B = A$, and $A \cup B = B$.

Suppose $A \subseteq B$. Let $x \in A$. Then $x \in B$, hence $x \in A \cap B$ and so $A \subseteq A \cap B$. By Theorem 1.2, $(A \cap B) \subseteq A$. Therefore $A \cap B = A$. On the other hand, suppose $A \cap B = A$. Let $x \in A$. Then $x \in A \cap B$, hence $x \in B$. Therefore, $A \subseteq B$. Both results show that $A \subseteq B$ is equivalent to $A \cap B = A$.

Suppose again that $A \subseteq B$. Let $x \in A \cup B$. Then $x \in A$ or $x \in B$. If $x \in A$, then $x \in B$ because $A \subseteq B$. In either case, $x \in B$. Therefore $A \cup B \subseteq B$. By Theorem 1.2, $B \subseteq A \cup B$. Therefore $A \cup B = B$. Now suppose $A \cup B = B$. Let $x \in A$. Then $x \in A \cup B$ by definition of union of sets. Hence $x \in B = A \cup B$. Therefore $A \subseteq B$. Both results show that $A \subseteq B$ is equivalent to $A \cup B = B$.

Thus $A \subseteq B$, $A \cap B = A$ and $A \cup B = B$ are equivalent.

## VENN DIAGRAMS, ALGEBRA OF SETS, DUALITY

**1.14.** Illustrate DeMorgan's law $(A \cup B)^c = A^c \cap B^c$ (proved in Section 1.7) using Venn diagrams.

Shade the area outside $A \cup B$ in a Venn diagram of sets $A$ and $B$. This is shown in Fig. 1-8($a$); hence the shaded area represents $(A \cup B)^c$. Now shade the area outside $A$ in a Venn diagram of $A$ and $B$ with strokes in one direction (///), and then shade the area outside $B$ with strokes in another direction (\\\\\\). This is shown in Fig. 1-8($b$). Thus the cross-hatched area (area where both lines are present) represents the intersection of $A^c$ and $B^c$, that is, $A^c \cap B^c$. Both $(A \cup B)^c$ and $A^c \cap B^c$ are represented by the same area; hence the Venn diagrams indicate $(A \cup B)^c = A^c \cap B^c$. (We emphasize that a Venn diagram is not a formal proof but it can indicate relationships between sets.)

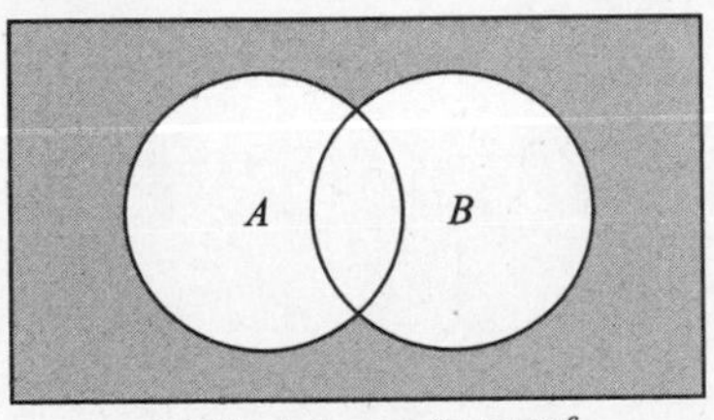

($a$) Shaded area: $(A \cup B)^c$

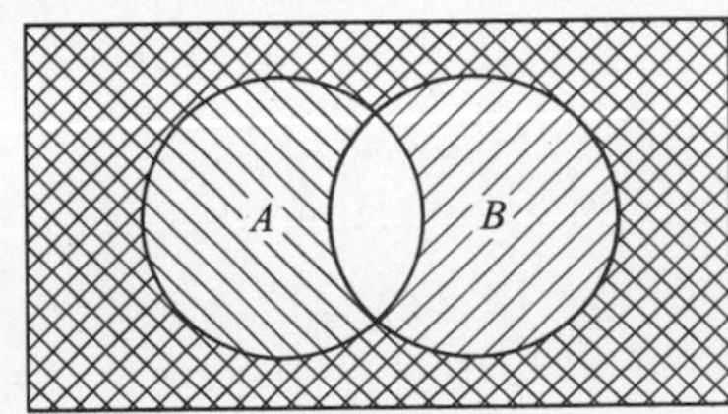

($b$) Cross-hatched area: $A^c \cap B^c$

Fig. 1-8

**1.15.** Consider the Venn diagram of two arbitrary sets $A$ and $B$ as pictured in Fig. 1-1($c$). Shade the sets: ($a$) $A \cap B^c$, ($b$) $(B \backslash A)^c$.

($a$) First shade the area represented by $A$ with strokes in one direction (///), and then shade the area represented by $B^c$ (the area outside $B$), with strokes in another direction (\\\\\\). This is shown in Fig. 1-9($a$). The cross-hatched area is the intersection of these two sets and represents $A \cap B^c$; and this is shown in Fig. 1-9($b$). Observe that $A \cap B^c = A \backslash B$. In fact, $A \backslash B$ is sometimes defined to be $A \cap B^c$.

(*b*) First shade the area represented by $B\backslash A$ (the area of $B$ which does not lie in $A$) as in Fig. 1-10(*a*). Then the area outside this shaded region, which is shown in Fig. 1-10(*b*), represents $(B\backslash A)^c$.

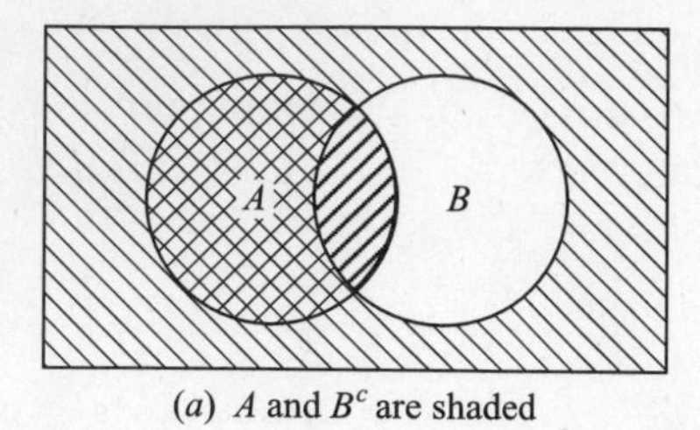

(*a*) $A$ and $B^c$ are shaded

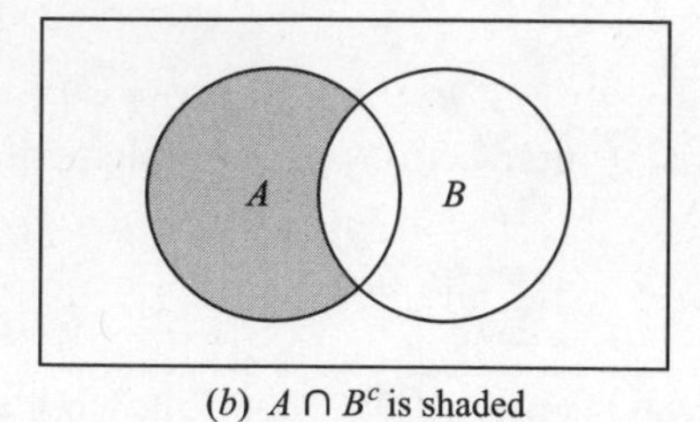

(*b*) $A \cap B^c$ is shaded

**Fig. 1-9**

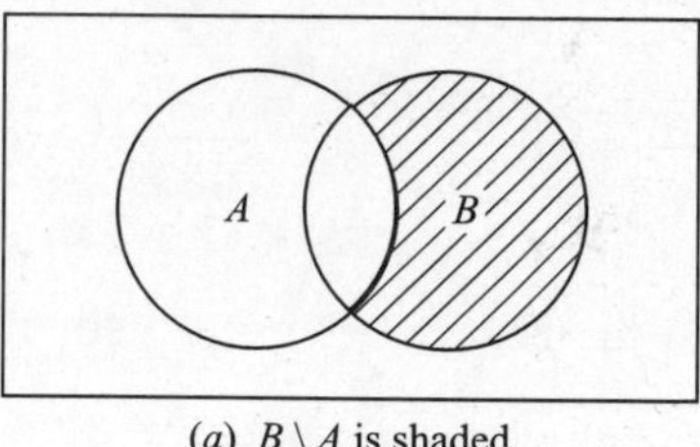

(*a*) $B \backslash A$ is shaded

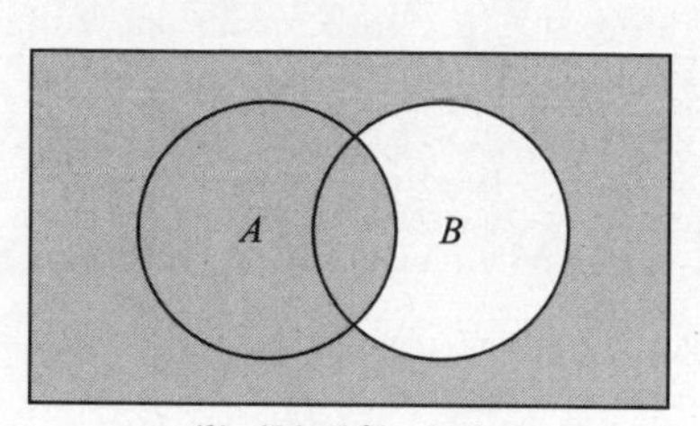

(*b*) $(B \backslash A)^c$ is shaded

**Fig. 1-10**

**1.16.** Prove Theorem 1.4: Distributive law (4*b*)

$$A \cap (B \cup C) = (A \cap B) \cup (A \cap C)$$

Illustrate the law using Venn diagrams.

By definition of union and intersection,

$$\begin{aligned} A \cap (B \cup C) &= \{x : x \in A,\ X \in B \cup C\} \\ &= \{x : x \in A,\ x \in B \text{ or } x \in A,\ x \in C) = (A \cap B) \cup (A \cap C) \end{aligned}$$

Here we use the analogous logical law

$$p \wedge (q \vee r) = (p \wedge q) \vee (p \wedge r)$$

where $\wedge$ denotes "and" and $\vee$ denotes "or".

**Venn Diagram**

Draw three intersecting circles labeled $A, B, C$, as in Fig. 1-11(*a*). Now, as in Fig. 1-11(*b*) shade $A$ with strokes in one direction (///) and shade $B \cup C$ with strokes in another direction (\\\). Then the cross-hatched area is $A \cap (B \cup C)$, as shaded in Fig. 1-11(*c*). Next shade $A \cap B$ and then $A \cap C$, as in Fig. 1-11(*d*). The total area shaded is $(A \cap B) \cup (A \cap C)$, as shaded in Fig. 1-11(*e*). As expected by the distributive law, $A \cap (B \cup C)$ and $(A \cap B) \cup (A \cap C)$ are both represented by the same set of points.

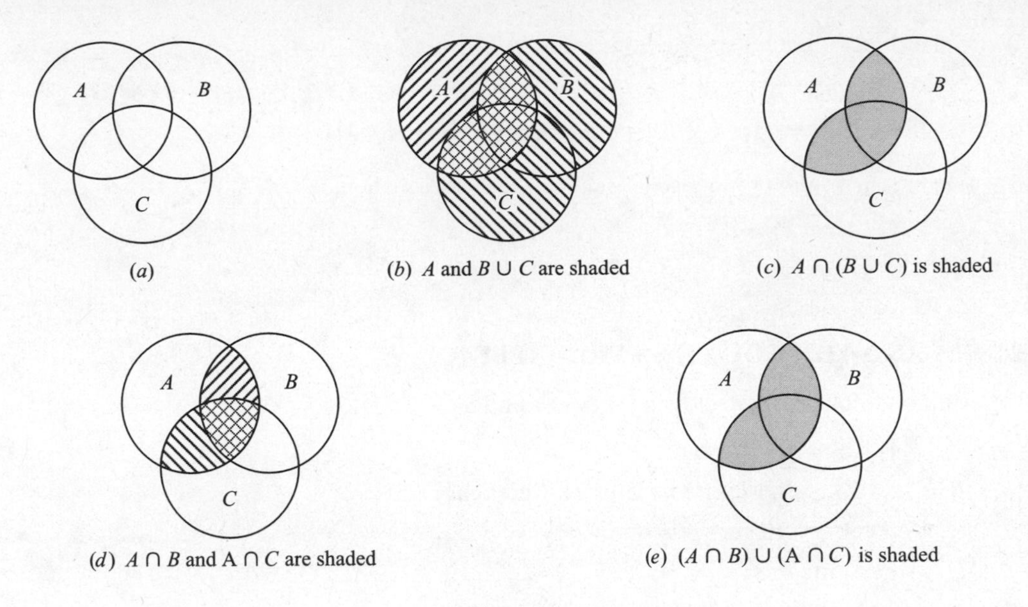


**Fig. 1-11**

**1.17.** Prove the commutative laws: (*a*) $A \cup B = B \cup A$, (*b*) $A \cap B = B \cap A$.

(*a*) $A \cup B = \{x : x \in A \text{ or } x \in B\} = \{x : x \in B \text{ or } x \in A\} = B \cup A$.
(*b*) $A \cap B = \{x : x \in A \text{ and } x \in B\} = \{x : x \in B \text{ and } x \in A\} = B \cap A$.

**1.18.** Prove: $(A \cup B) \backslash (A \cap B) = (A \backslash B) \cup (B \backslash A)$. (Thus either one may be used to define the symmetric difference $A \oplus B$.)

Using $X \backslash Y = X \cap Y^c$ and the laws in Table 1-1, including DeMorgan's laws, we obtain

$$\begin{aligned}(A \cup B) \backslash (A \cap B) &= (A \cup B) \cap (A \cap B)^c = (A \cup B) \cap (A^c \cup B^c) \\ &= (A \cap A^c) \cup (A \cap B^c) \cup (B \cap A^c) \cup (B \cap B^c) \\ &= \varnothing \cup (A \cap B^c) \cup (B \cap A^c) \cup \varnothing \\ &= (A \cap B^c) \cup (B \cap A^c) = (A \backslash B) \cup (B \backslash A)\end{aligned}$$

**1.19.** Prove the following identity: $(A \cup B) \cap (A \cup B^c) = A$.

| | **Statement** | **Reason** |
|---|---|---|
| 1. | $(A \cup B) \cap (A \cup B^c) = A \cup (B \cap B^c)$ | Distributive law |
| 2. | $B \cap B^c = \varnothing$ | Complement law |
| 3. | $(A \cup B) \cap (A \cup B^c) = A \cup \varnothing$ | Substitution |
| 4. | $A \cup \varnothing = A$ | Identity law |
| 5. | $(A \cup B) \cap (A \cup B^c) = A$ | Substitution |

**1.20.** Write the dual of each set equation:

(*a*) $(\mathbf{U} \cap A) \cup (B \cap A) = A$ (*c*) $(A \cap \mathbf{U}) \cap (\varnothing \cup A^c) = \varnothing$
(*b*) $(A \cup B \cup C)^c = (A \cup C)^c \cap (A \cup B)^c$ (*d*) $(A \cap \mathbf{U})^c \cap A = \varnothing$

Interchange $\cup$ and $\cap$ and also **U** and $\varnothing$ in each set equation:

(*a*) $(\varnothing \cup A) \cap (B \cup A) = A$ (*c*) $(A \cup \varnothing) \cup (\mathbf{U} \cap A^c) = \mathbf{U}$
(*b*) $(A \cap B \cap C)^c = (A \cap C)^c \cup (A \cap B)^c$ (*d*) $(A \cup \varnothing)^c \cup A = \mathbf{U}$

## FINITE SETS AND THE COUNTING PRINCIPLE

**1.21.** Determine which of the following sets are finite.

(*a*) $A =$ {seasons in the year},
(*b*) $B =$ {states in the United States of America},
(*c*) $C =$ {positive integers less than 1},
(*d*) $D =$ {odd integers},
(*e*) $E =$ {positive integral divisors of 12},
(*f*) $F =$ {cats living in the United States}.

(*a*) $A$ is finite since there are four seasons in the year, i.e., $n(A) = 4$.
(*b*) $B$ is finite because there are 50 states in the United States, i.e., $n(B) = 50$.
(*c*) There are no positive integers less than 1; hence $C$ is empty. Thus $C$ is finite and $n(C) = \varnothing$.
(*d*) $D$ is infinite.
(*e*) The positive integer divisors of 12 are 1, 2, 3, 4, 6, 12. Hence $E$ is finite and $n(E) = 6$.
(*f*) Although it may be difficult to find the number of cats living in the United States, there is still a finite number of them at any point in time. Hence $F$ is finite.

**1.22.** Suppose 50 science students are polled to see whether or not they have studied French (F) or German (G) yielding the following data:

25 studied French, 20 studied German, 5 studied both.

Find the number of the students who: (*a*) studied only French, (*b*) did not study German, (*c*) studied French or German, (*d*) studied neither language.

(*a*) Here 25 studied French, and 5 of them also studied German; hence $25 - 5 = 20$ students only studied French. That is, by Theorem 1.7,

$$n(F \backslash G) = n(F) - N(F \cap G) = 25 - 5 = 20.$$

(*b*) There are 50 students of whom 20 studied German; hence $50 - 20 = 30$ did not study German. That is, by Theorem 1.6,

$$n(G^c) = n(\mathbf{U}) - n(G) = 50 - 20 = 30$$

(*c*) By the inclusion-exclusion principle, Theorem 1.8,

$$n(F \cup G) = n(F) + n(G) - n(F \cap G) = 25 + 20 - 5 = 40$$

That is, 40 students studied French or German.

(*d*) The set $F^c \cap G^c$ consists of the students who studied neither language. By DeMorgan's law, $F^c \cap G^c = (F \cup G)^c$. By (*c*), 40 studied at least one of the languages; hence

$$n(F^c \cap G^c) = n(\mathbf{U}) - n(F \cup G) = 50 - 40 = 10$$

That is, 10 students studied neither language.

**1.23.** Suppose $n(\mathbf{U}) = 70$, $n(A) = 30$, $n(B) = 45$, $n(A \cap B) = 10$. Find:
(*a*) $n(A \cup B)$, (*b*) $n(A^c)$ and $n(B^c)$, (*c*) $n(A^c \cap B^c)$, (*d*) $n(A \oplus B)$.

(*a*) By Theorem 1.9, $n(A \cup B) = n(A) + n(B) - n(A \cap B) = 30 + 45 - 10 = 65$.

(*b*) Here

$$n(A^c) = n(\mathbf{U}) - n(A) = 70 - 30 = 40 \quad \text{and} \quad n(B^c) = n(\mathbf{U}) - n(B) = 70 - 45 = 25$$

(*c*) Using DeMorgan's law,

$$n(A^c \cap B^c) = n((A \cup B)^c) = n(\mathbf{U}) - n(A \cup B) = 70 - 65 = 5$$

(*d*) First find

$$n(A \backslash B) = n(A) - n(A \cap B) = 30 - 10 = 20$$
$$n(B \backslash A) = n(B) - n(A \cap B) = 45 - 10 = 25$$

Then

$$n(A \oplus B) = n(A \backslash B) + n(B \backslash A) = 20 + 25 = 45$$

**1.24.** A small college requires its students to take at least one mathematics course and at least one science course. A survey of 140 of its sophomore students shows that:

60 completed their mathematics requirement (M),
45 completed their science requirement (S),
20 completed both requirements (M and S).

Use a Venn diagram to find the number of the students who had completed:

(*a*) exactly one of the two requirements,
(*b*) at least one of the requirements,
(*c*) neither requirement.

Translating the above data into set notation yields:

$$n(M) = 60, \qquad n(S) = 45, \qquad n(M \cap S) = 20, \qquad \text{and} \qquad n(\mathbf{U}) = 140$$

Draw a Venn diagram of sets $M$ and $S$ with four regions as in Fig. 1-12(*a*). Then, as in Fig. 1-12(*b*), assign numbers to the four regions as follows::

20 completed both $M$ and $S$, i.e. $n(M \cap S) = 20$,
$60 - 20 = 40$ completed $M$ but not $S$, i.e. $n(M \backslash S) = 40$,
$45 - 20 = 25$ completed $S$ but not $M$, i.e. $n(S \backslash M) = 25$,
$140 - 20 - 40 - 25 = 55$ completed neither $M$ nor $S$.

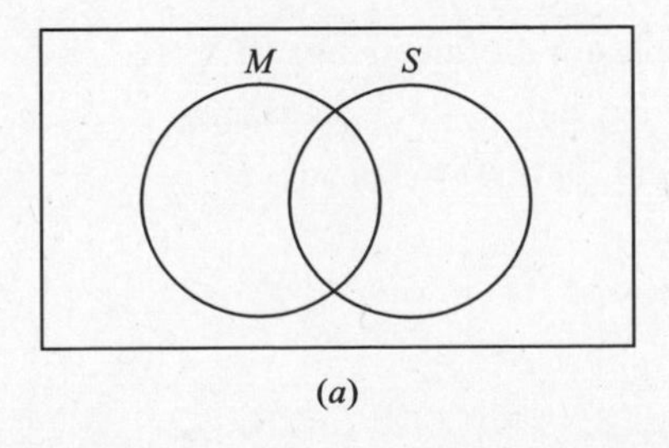


(*a*)

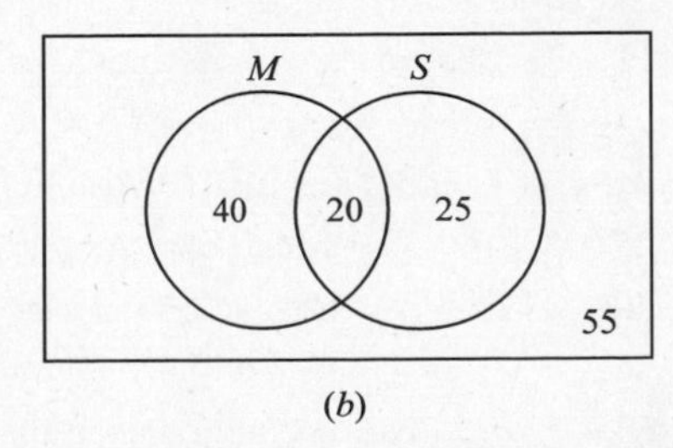


(*b*)

Fig. 1-12

By the Venn diagram:

(*a*) $40 + 25 = 65$ completed exactly one of the requirements,

(*b*) $20 + 40 + 25 = 85$ completed $M$ or $S$. Alternately, we can find $n(M \cup S)$ without the Venn diagram by using Theorem 1.7. That is,

$$n(M \cup S) = n(M) + n(S) - n(M \cap S) = 60 + 45 - 20 = 85$$

(*c*) 55 completed neither requirement.

**1.25.** In a survey of 60 people, it was found that:

25 read *Newsweek* magazine
26 read *Time*
26 read *Fortune*
9 read both *Newsweek* and *Fortune*
11 read both *Newsweek* and *Time*
8 read both *Time* and *Fortune*
3 read all three magazines

(*a*) Find the number of people who read at least one of the three magazines.

(*b*) Fill in the correct number of people in each of the eight regions of the Venn diagram in Fig. 1-13(*a*) where $N$, $T$, and $F$ denote the set of people who read *Newsweek*, *Time*, and *Fortune*, respectively.

(*c*) Find the number of people who read exactly one magazine.

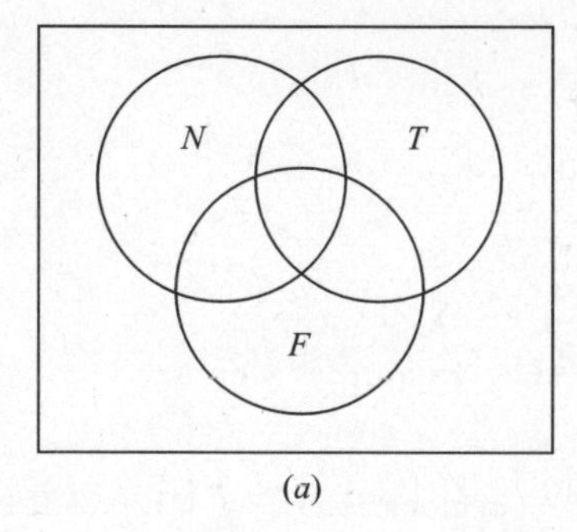


(*a*)

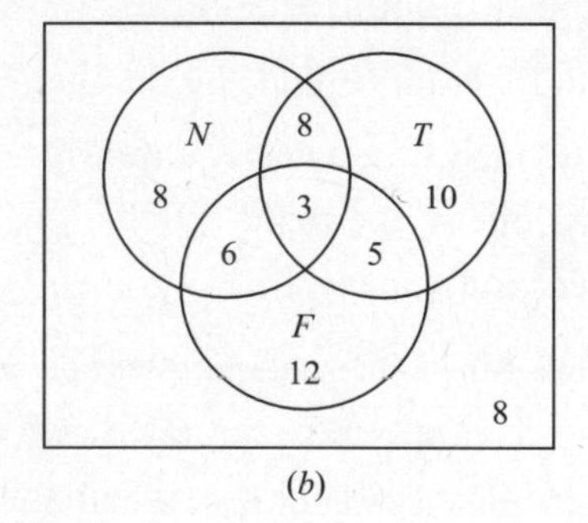


(*b*)

**Fig. 1-13**

(*a*) We want $n(N \cup T \cup F)$. By Corollary 1.9,

$$n(N \cup T \cup F) - n(N) + N(T) + n(F) - n(N \cap F) - n(N \cap T) - n(T \cap F) + n(N \cap T \cap F)$$
$$= 25 + 26 + 26 - 11 - 9 - 8 + 3 = 52$$

(*b*) The required Venn diagram in Fig 1-13(*b*) is obtained as follows:

3 read all three magazines
$11 - 3 = 8$ read *Newsweek* and *Time* but not all three magazines
$9 - 3 = 6$ read *Newsweek* and *Fortune* but not all three magazines
$8 - 3 = 5$ read *Time* and *Fortune* but not all three magazines
$25 - 8 - 6 - 3 = 8$ read only *Newsweek*
$26 - 8 - 5 - 3 = 10$ read only *Time*
$26 - 6 - 5 - 3 = 12$ read only *Fortune*
$60 - 52 = 8$ read no magazine at all

(*c*) $8 + 10 + 12 = 30$ read only one magazine.

**1.26.** Prove Theorem 1.8: If $A$ and $B$ are finite sets, then $A \cup B$ and $A \cap B$ are finite and $n(A \cup B) = n(A) + n(B) - n(A \cap B)$.

If $A$ and $B$ are finite, then clearly $A \cap B$ and $A \cup B$ are finite.

Suppose we count the elements of $A$ and then count the elements of $B$. Then every element in $A \cap B$ would be counted twice, once in $A$ and once in $B$. Hence

$$n(A \cup B) = n(A) + n(B) - n(A \cap B)$$

Alternatively (Problems 1.37 and 1.50),

(i) $A$ is the disjoint union of $A \backslash B$ and $A \cap B$,

(ii) $B$ is the disjoint union of $B \backslash A$ and $A \cap B$,

(iii) $A \cup B$ is the disjoint union of $A \backslash B$, $A \cap B$, and $B \backslash A$.

Therefore, by Lemma 1.5 and Theorem 1.7,

$$\begin{aligned} n(A \cup B) &= n(A \backslash B) + n(A \cap B) + n(B \backslash A) \\ &= n(A) - n(A \cap B) + n(A \cap B) + n(B \backslash A) - n(A \cap B) \\ &= n(A) + n(B) - n(A \cap B) \end{aligned}$$

## CLASSES OF SETS

**1.27.** Find the elements of the set $A = [\{1,2,3\}, \{4,5\}, \{6,7,8\}]$, and determine whether each of the following is true or false:

(*a*) $1 \in A$ (*c*) $\{6,7,8\} \in A$ (*e*) $\varnothing \in A$

(*b*) $\{1,2,3\} \subseteq A$ (*d*) $\{\{4,5\}\} \subseteq A$ (*f*) $\varnothing \subseteq A$

$A$ is a collection (class) of sets; its elements are the sets $\{1,2,3\}$, $\{4,5\}$, and $\{6,7,8\}$.

(*a*) *False.* 1 is not one of the elements of $A$.

(*b*) *False.* $\{1,2,3\}$ is not a subset of $A$; it is one of the elements of $A$.

(*c*) *True.* $\{6,7,8\}$ is one of the elements of $A$.

(*d*) *True.* $\{\{4,5\}\}$, the set consisting of the element $\{4,5\}$ is a subset of $A$.

(*e*) *False.* The empty set $\varnothing$ is not an element of $A$, i.e., it is not one of the three elements of $A$.

(*f*) *True.* The empty set $\varnothing$ is a subset of every set; even a collection of sets.

**1.28.** Consider that class $A$ of sets in Problem 1.27. Find the subclass $B$ of $A$ where $B$ consists of the sets in $A$ with exactly: (*a*) three elements, (*b*) four elements.

(*a*) There are two sets in $A$ with three elements, $\{1,2,3\}$ and $\{6,7,8\}$. Hence $B = [\{1,2,3\}, \{6,7,8\}]$.

(*b*) There are no sets in $A$ with four elements; hence $B$ is empty, that is, $B = \varnothing$.

**1.29.** Determine the power set $\mathscr{P}(A)$ of $A = \{a,b,c,d\}$.

The elements of $\mathscr{P}(A)$ are the subsets of $A$. Hence

$$\begin{aligned} \mathscr{P}(A) = [&A, \{a,b,c\}, \{a,b,d\}, \{a,c,d\}, \{b,c,d\}, \{a,b\}, \{a,c\}, \\ &\{a,d\}, \{b,c\}, \{b,d\}, \{c,d\}, \{a\}, \{b\}, \{c\}, \{d\}, \varnothing] \end{aligned}$$

As expected, $\mathscr{P}(A)$ has $2^4 = 16$ elements.

**1.30.** Find the number of elements in the power set of each of the following sets:

(*a*) {days of the week}, (*c*) {seasons of the year},
(*b*) {positive divisors of 12}, (*d*) {letters in the word "yes"}.

Recall that $\mathscr{P}(A)$ contains $2^{|A|}$ elements. Hence:

(*a*) $2^7 = 128$.
(*b*) $2^6 = 64$ since there are six divisors, 1, 2, 3, 4, 6, 12, of 12.
(*c*) $2^4 = 16$ since there are four seasons.
(*d*) $2^3 = 8$.

## ARGUMENTS AND VENN DIAGRAMS

**1.31.** Determine the validity of the following argument:

$S_1$: All my friends are musicians
$S_2$: John is my friend.
$S_3$: None of my neighbors are musicians.

---

$S$: John is not my neighbor.

The premises $S_1$ and $S_3$ lead to the Venn diagram in Fig. 1-14. By $S_2$, John belongs to the set of friends which is disjoint from the set of neighbors. Thus $S$ is a valid conclusion and so the argument is valid.

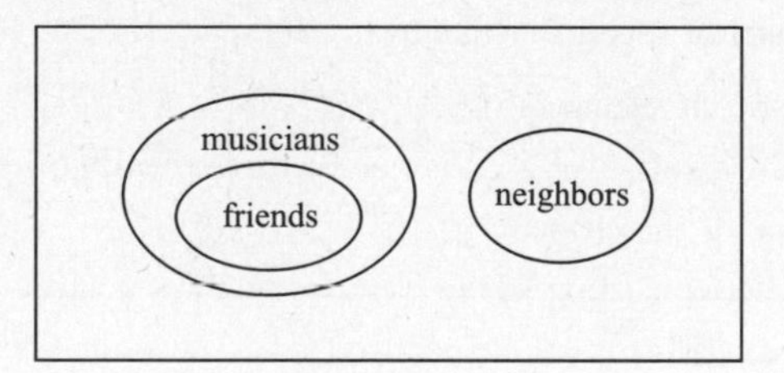


**Fig. 1-14**

**1.32.** Consider the following assumptions:

$S_1$: Poets are happy people.
$S_2$: Every doctor is wealthy.
$S_3$: No happy person is wealthy.

Determine the validity of each of the following conclusions:

(*a*) No poet is wealthy. (*b*) Doctors are happy people.
(*c*) No person can be both a poet and a doctor.

The three premises lead to the Venn diagram in Fig. 1-15. From the diagram it follows that (*a*) and (*c*) are valid conclusions whereas (*b*) is not valid.

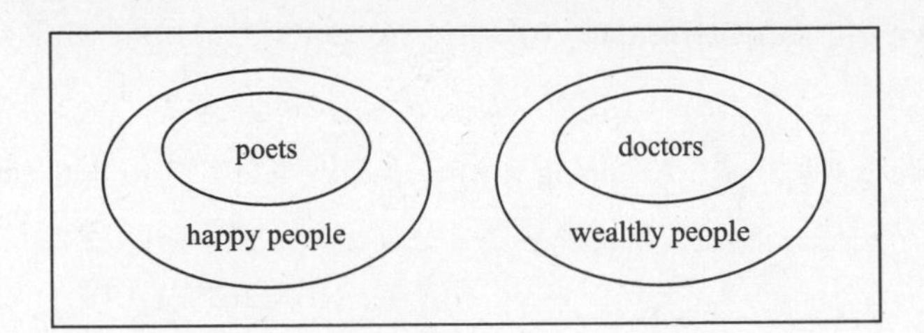


**Fig. 1-15**

**1.33.** Determine the validity of the following argument:

$S_1$: Babies are illogical.
$S_2$: Nobody is despised who can manage a crocodile.
$S_3$: Illogical people are despised.

---

$S$: Babies cannot manage crocodiles.

(The above argument is adapted from Lewis Carroll, *Symbolic Logic*; he is the author of *Alice in Wonderland.*)

The three premises lead to the Venn diagram in Fig. 1-16. Since the set of babies and the set of people who can manage crocodiles are disjoint, "Babies cannot manage crocodiles" is a valid conclusion.

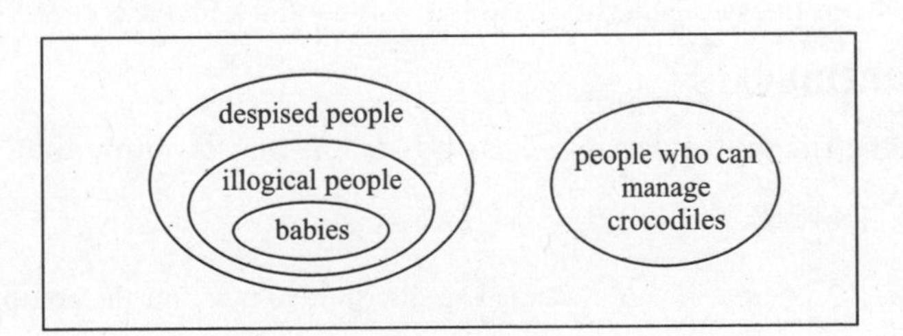


**Fig. 1-16**

## MATHEMATICAL INDUCTION

**1.34.** Prove the assertion $A(n)$ that the sum of the first $n$ positive integers is $\frac{1}{2}n(n+1)$; that is,

$$A(n)\colon\ 1+2+3+\cdots+n=\tfrac{1}{2}n(n+1)$$

The assertion holds for $n = 1$ since

$$A(1)\colon\ 1=\tfrac{1}{2}(1)(1+1)$$

Assuming $A(n)$ is true, we add $n+1$ to both sides of $A(n)$, obtaining

$$\begin{aligned}1+2+3+\cdots+n+(n+1)&=\tfrac{1}{2}n(n+1)+(n+1)\\&=\tfrac{1}{2}[n(n+1)+2(n+1)]\\&=\tfrac{1}{2}[(n+1)(n+2)]\end{aligned}$$

which is $A(n+1)$. That is, $A(n+1)$ is true whenever $A(n)$ is true. By the principle of induction, $A(n)$ is true for all $n \geq 1$.

**1.35.** Prove the following assertion (for $n \geq 0$):

$$A(n): \quad 1 + 2 + 2^2 + 2^3 + \cdots + 2^n = 2^{n+1} - 1$$

$A(0)$ is true since $1 = 2^1 - 1$. Assuming $A(n)$ is true, we add $2^{n+1}$ to both sides of $A(n)$, obtaining

$$\begin{aligned} 1 + 2 + 2^2 + 2^3 + \cdots + 2^n + 2^{n+1} &= 2^{n+1} - 1 + 2^{n+1} \\ &= 2(2^{n+1}) - 1 \\ &= 2^{n+2} - 1 \end{aligned}$$

which is $A(n+1)$. Thus $A(n+1)$ is true whenever $A(n)$ is true. By the principle of induction, $A(n)$ is true for all $n \geq 0$.

**1.36.** Prove: (*a*) $n^2 \geq 2n + 1$ for $n \geq 3$, (*b*) $n! \geq 2^n$ for $n \geq 4$.

(*a*) Since $3^2 = 9$ and $2(3) + 1 = 7$, the formula is true for $n = 3$. Using $n^2 \geq 2n + 1$ in the second step and $2n \geq 1$ in the fourth step, we have

$$(n+1)^2 = n^2 + 2n + 1 \geq (2n+1) + 2n + 1 = 2n + 2 + 2n \geq 2n + 2 + 1 = 2(n+1) + 1$$

Thus the formula is true for $n + 1$. By induction, the formula is true for all $n \geq 3$.

(*b*) Since $4! = 1 \cdot 2 \cdot 3 \cdot 4 = 24$ and $2^4 = 16$, the formula is true for $n = 4$. Assuming $n! \geq 2^n$ we have

$$(n+1)! = n!(n+1) \geq 2^n(n+1) \geq 2^n(2) = 2^{n+1}$$

Thus the formula is true for $n + 1$. By induction, the formula is true for all $n \geq 4$.

## MISCELLANEOUS PROBLEMS

**1.37.** Show that $A$ is the disjoint union of $A \backslash B$ and $A \cap B$; that is, show that:

(*a*) $A = (A \backslash B) \cup (A \cap B)$, (*b*) $(A \backslash B) \cap (A \cap B) = \varnothing$.

(*a*) By Problem 1.12, $A \backslash B = A \cap B^c$. Using the distributive law and the complement law, we get

$$(A \backslash B) \cup (A \cap B) = (A \cap B^c) \cup (A \cap B) = A \cap (B^c \cup B) = A \cap U = A$$

(*b*) Also,

$$(A \backslash B) \cap (A \cap B) = (A \cap B^c) \cap (A \cap B) = A \cap (B^c \cap B) = A \cap \varnothing = \varnothing.$$

**1.38.** Prove Corollary 1.9. Suppose $A$, $B$, $C$ are finite sets. Then $A \cup B \cup C$ is finite and

$$n(A \cup B \cup C) = n(A) + n(B) + n(C) - n(A \cap B) - n(A \cap C) - n(B \cap C) + n(A \cap B \cap C)$$

Clearly $A \cup B \cup C$ is finite when $A$, $B$, $C$ are finite. Using

$$(A \cup B) \cap C = (A \cap C) \cup (B \cap C) \qquad \text{and} \qquad (A \cap B) \cap (B \cap C) = A \cap B \cap C$$

and using Theorem 1.8 repeatedly, we have

$$\begin{aligned} n(A \cup B \cup C) &= n(A \cup B) + n(C) - n[(A \cap C) \cup (B \cap C)] \\ &= [n(A) + n(B) - n(A \cap B)] + n(C) - [n(A \cap C) + n(B \cap C) - n(A \cap B \cap C)] \\ &= n(A) + n(B) + n(C) - n(A \cap B) - n(A \cap C) - n(B \cap C) + n(A \cap B \cap C) \end{aligned}$$

as required.

**1.39.** A set $A$ of real numbers is said to be *bounded from above* if there exists a number $M$ such that $x \leq M$ for every $x$ in $A$. (Such a number $M$ is called an *upper bound* of $M$.)

(*a*) Suppose $A$ and $B$ are sets which are bounded from above with respective upper bounds $M_1$ and $M_2$. What can be said about the union and intersection of $A$ and $B$?

(*b*) Suppose $C$ and $D$ are sets of real numbers which are unbounded. What can be said about the union and intersection of $C$ and $D$?

(*a*) Both the union and intersection are bounded from above. In fact, the larger of $M_1$ and $M_2$ is always an upper bound for $A \cup B$, and the smaller of $M_1$ and $M_2$ is always an upper bound for $A \cap B$.

(*b*) The union of $C$ and $D$ must be unbounded, but the intersection could be either bounded or unbounded.

**1.40.** Restate the Principle of Mathematical Induction I and II in terms of sets, rather than assertions.

(*a*) *Principle of Mathematical Induction I*: Let $S$ be a subset of $\mathbf{P} = \{1, 2, \ldots\}$ with two properties:

$$(1)\quad 1 \in S. \qquad (2)\quad \text{If } n \in S, \text{ then } n + 1 \in S.$$

Then $S = \mathbf{P}$.

(*b*) *Principle of Mathematical Induction II*: Let $S$ be a subset of $\mathbf{P} = \{1, 2, \ldots\}$ with two properties:

$$(1)\quad 1 \in S. \qquad (2)\quad \text{If } \{1, 2, \ldots, n-1\} \subseteq S, \text{ then } n \in S.$$

Then $S = \mathbf{P}$.

## Supplementary Problems

### SETS AND SUBSETS

**1.41.** Which of the following sets are equal?

$A = \{x : x^2 - 4x + 3 = 0\}$ $\quad C = \{x : x \in \mathbf{P}, x < 3\}$ $\quad E = \{1, 2\}$ $\quad G = \{3, 1\}$

$B = \{x : x^2 - 3x + 2 = 0\}$ $\quad D = \{x : x \in \mathbf{P}, x \text{ is odd}, x < 5\}$ $\quad F = \{1, 2, 1\}$ $\quad H = \{1, 1, 3\}$

**1.42.** List the elements of the following sets if the universal set is $\mathbf{U} = \{\text{a}, \text{b}, \text{c}, \ldots, \text{y}, \text{z}\}$. Furthermore, identify which of the sets, if any, are equal.

$A = \{x : x \text{ is a vowel}\}$ $\quad C = \{x : x \text{ precedes f in the alphabet}\}$

$B = \{x : x \text{ is a letter in the word "little"}\}$ $\quad D = \{x : x \text{ is a letter in the word "title"}\}$

**1.43.** Let

$$A = \{1, 2, \ldots, 8, 9\}, \quad B = \{2, 4, 6, 8\}, \quad C = \{1, 3, 5, 7, 9\}, \quad D = \{3, 4, 5\}, \quad E = \{3, 5\}$$

Which of the above sets can equal a set $X$ under each of the following conditions?

(*a*) $X$ and $B$ are disjoint.

(*b*) $X \subseteq D$ but $X \not\subseteq B$.

(*c*) $X \subseteq A$ but $X \not\subseteq C$.

(*d*) $X \subseteq C$ but $X \not\subseteq A$.

**1.44.** Consider the following sets:

$$\varnothing, \quad A = \{a\}, \quad B = \{c, d\}, \quad C = \{a, b, c, d\}, \quad D = \{a, b\}, \quad E = \{a, b, c, d, e\}.$$

Insert the correct symbol, $\subseteq$ or $\not\subseteq$, between each pair of sets:

(*a*) $\varnothing, A$ (*c*) $A, B$ (*e*) $B, C$ (*g*) $C, D$
(*b*) $D, E$ (*d*) $D, A$ (*f*) $D, C$ (*h*) $B, D$

## SET OPERATIONS

**1.45.** Let $\mathbf{U} = \{1, 2, 3, \ldots, 8, 9\}$ be the universal set and let:

$$A = \{1, 2, 5, 6\}, \quad B = \{2, 5, 7\}, \quad C = \{1, 3, 5, 7, 9\}$$

Find: (*a*) $A \cap B$ and $A \cap C$, (*b*) $A \cup B$ and $A \cup C$, (*c*) $A^c$ and $C^c$.

**1.46.** For the sets in Problem 1.45, find: (*a*) $A \backslash B$ and $A \backslash C$, (*b*) $A \oplus B$ and $A \oplus C$.

**1.47.** For the sets in Problem 1.45, find: (*a*) $(A \cup C) \backslash B$, (*b*) $(A \cup B)^c$, (*c*) $(B \oplus C) \backslash A$.

**1.48.** Let $A = \{a, b, c, d, e\}$, $B = \{a, b, d, f, g\}$, $C = \{b, c, e, g, h\}$, $D = \{d, e, f, g, h\}$. Find:

(*a*) $A \cup B$ (*c*) $B \cap C$ (*e*) $C \backslash D$ (*g*) $A \oplus B$
(*b*) $C \cap D$ (*d*) $A \cap D$ (*f*) $D \backslash A$ (*h*) $A \oplus C$

**1.49.** For the sets in Problem 1.48, find:

(*a*) $A \cap (B \cup D)$ (*c*) $(A \cup D) \backslash C$ (*e*) $(C \backslash A) \backslash D$ (*g*) $(A \cap D) \backslash (B \cup C)$
(*b*) $B \backslash (C \cup D)$ (*d*) $B \cap C \cap D$ (*f*) $(A \oplus D) \backslash B$ (*h*) $(A \backslash C) \cap (B \cap D)$

**1.50.** Let $A$ and $B$ be any sets. Prove $A \cup B$ is the disjoint union of $A \backslash B$, $A \cap B$, and $B \backslash A$.

**1.51.** Prove the following:

(*a*) $A \subseteq B$ if and only if $A \cap B^c = \varnothing$ (*c*) $A \subseteq B$ if and only if $B^c \subseteq A^c$
(*b*) $A \subseteq B$ if and only if $A^c \cup B = U$ (*d*) $A \subseteq B$ if and only if $A \backslash B = \varnothing$

(Compare with Theorem 1.3.)

**1.52.** Prove the absorption laws: (*a*) $A \cup (A \cap B) = A$, (*b*) $A \cap (A \cup B) = A$.

**1.53.** The formula $A \backslash B = A \cap B^c$ defines the difference operation in terms of the operations of intersection and complement. Find a formula that defines the union $A \cup B$ in terms of the operations of intersection and complement.

**1.54.** (*a*) Prove: $A \cap (B \backslash C) = (A \cap B) \backslash (A \cap C)$.
(*b*) Give an example to show that $A \cup (B \backslash C) \neq (A \cup B) \backslash (A \cup C)$.

**1.55.** Prove the following properties of the symmetric difference:

(*a*) $A \oplus (B \oplus C) = (A \oplus B) \oplus C$ (Associative law)
(*b*) $A \oplus B = B \oplus A$ (Commutative law)
(*c*) If $A \oplus B = A \oplus C$, then $B = C$ (Cancellation law)
(*d*) $A \cap (B \oplus C) = (A \cap B) \oplus (A \cap C)$ (Distributive law)

### VENN DIAGRAMS, ALGEBRA OF SETS, DUALITY

**1.56.** The Venn diagram in Fig. 1-17 shows sets $A$, $B$, $C$. Shade the following sets:

(a) $A\backslash(B \cup C)$, (b) $A^c \cap (B \cap C)$, (c) $(A \cup C) \cap (B \cup C)$.

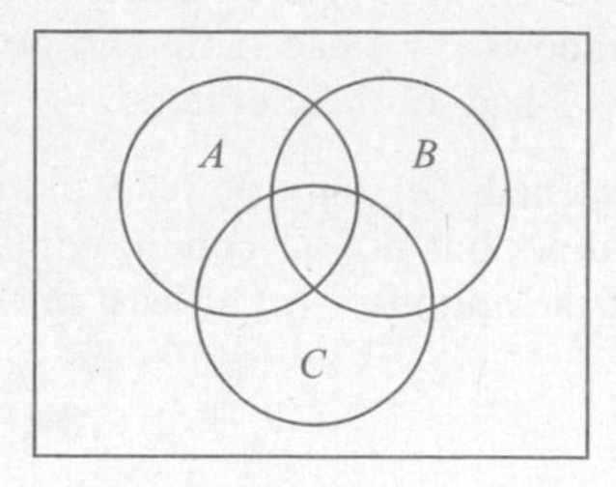


**Fig. 1-17**

**1.57.** Write the dual of each equation:

(a) $A = (B^c \cap A) \cup (A \cap B)$, (b) $(A \cap B) \cup (A \cap B^c) \cup (A^c \cap B) \cup (A^c \cap B^c) = U$

**1.58.** Use the laws in Table 1-1 to prove:

(a) $(A \cap B) \cup (A \cap B^c) = A$, (b) $A \cup B = (A \cap B^c) \cup (A^c \cap B) \cup (A \cap B)$

### FINITE SETS AND THE COUNTING PRINCIPLE

**1.59.** Determine which of the following sets are finite:

(a) lines parallel to the $x$ axis,
(b) letters in the English alphabet,
(c) months in the year,
(d) animals living on the earth,
(e) circles through the origin (0, 0),
(f) positive multiple of 5.

**1.60.** Given $n(U) = 20$, $n(A) = 12$, $n(B) = 9$, $n(A \cap B) = 4$. Find:

(a) $n(A \cup B)$, (b) $n(A^c)$, (c) $n(B^c)$, (d) $n(A\backslash B)$, (e) $n(\varnothing)$.

**1.61.** Among the 90 students in a dormitory, 35 own an automobile, 40 own a bicycle, and 10 have both an automobile and a bicycle. Find the number of the students who:

(a) do not have an automobile.
(b) have an automobile or a bicycle;
(c) have neither an automobile nor a bicycle;
(d) have an automobile or a bicycle, but not both.

**1.62.** Among 120 Freshmen at a college, 40 take mathematics, 50 take English, and 15 take both mathematics and English. Find the number of the Freshmen who:

(a) do not take mathematics;
(b) take mathematics or English;
(c) take mathematics, but not English;
(d) take English, but not mathematics;
(e) take exactly one of the two subjects;
(f) take neither mathematics nor English.

**1.63.** A survey on a sample of 25 new cars being sold at a local auto dealer was conducted to see which of three popular options, air-conditioning ($A$), radio ($R$), and power windows ($W$), were already installed. The survey found:

15 had air-conditioning 5 had air-conditioning and power windows
12 had radio 9 had air-conditioning and radio
11 had power windows 4 had radio and power windows
3 had all three options

Find the number of cars that had: (*a*) only power windows, (*b*) only air-conditioning, (*c*) only radio, (*d*) radio and power windows but not air-conditioning, (*e*) air-conditioning and radio, but not power windows, (*f*) only one of the options, (*g*) at least one option, (*h*) none of the options.

## CLASSES OF SETS, POWER SETS

**1.64.** Let $A = [\{a, b\}, \{c\}, \{d, e, f\}]$. List the elements of $A$ and determine whether each of the following statements is true or false:

(*a*) $a \in A$ (*c*) $c \in A$ (*e*) $\{d, e, f\} \subseteq A$ (*g*) $\varnothing \in A$
(*b*) $\{a\} \subseteq A$ (*d*) $\{c\} \in A$ (*f*) $\{\{a, b\}\} \subseteq A$ (*h*) $\varnothing \subseteq A$

**1.65.** Let $B = [\varnothing, \{1\}, \{2, 3\}, \{3, 4\}]$. List the elements of $B$ and determine whether each of the following statements is true or false:

(*a*) $1 \in B$ (*c*) $\{1\} \in B$ (*e*) $\{\{2, 3\}\} \subseteq B$ (*g*) $\varnothing \subseteq A$
(*b*) $\{1\} \subseteq B$ (*d*) $\{2, 3\} \subseteq B$ (*f*) $\varnothing \in A$ (*h*) $\{\varnothing\} \subseteq A$

**1.66.** Let $A = \{1, 2, 3, 4, 5\}$. (*a*) Find the power set $\mathscr{P}(A)$ of $A$. (*b*) Find the subcollection $\mathscr{B}$ of $\mathscr{P}(A)$ where each element of $\mathscr{B}$ consists of 1 and two other elements of $A$.

**1.67.** Find the power set $\mathscr{P}(A)$ of the set $A$ in Probiem 1.64.

**1.68.** Suppose $A$ is a finite set and $n(A) = m$. Prove that $\mathscr{P}(A)$ has $2^m$ elements.

## ARGUMENTS AND VENN DIAGRAMS

**1.69.** Draw a Venn diagram for the following assumptions:

$S_1$: No practical car is expensive.
$S_2$: Cars with sunroofs are expensive.
$S_3$: All wagons are practical.

Use the Venn diagram to determine whether or not each of the following is a valid conclusion:

(*a*) No practical car has a sunroof. (*c*) No wagon has a sunroof.
(*b*) All practical cars are wagons. (*d*) Cars with sunroofs are not practical.

**1.70.** Draw a Venn diagram for the following assumptions:

$S_1$: I planted all my expensive trees last year.
$S_2$: All my fruit trees are in my orchard.
$S_3$: No tree in my orchard was planted last year.

Use the Venn diagram to determine whether or not each of the following is a valid conclusion:

(*a*) The fruit trees were planted last year. (*c*) No fruit tree is expensive.
(*b*) No expensive tree is in the orchard. (*d*) Only fruit trees are in the orchard.

**1.71.** Draw a Venn diagram for the following assumptions:

$S_1$: All poets are poor.

$S_2$: In order to be a teacher, one must graduate from college.

$S_3$: No college graduate is poor.

Use the Venn diagram to determine whether or not each of the following is a valid conclusion:

(*a*) Teachers are not poor.
(*b*) Poets are not teachers.
(*c*) College graduates do not become poets.
(*d*) Every poor person becomes a poet.

**1.72.** Draw a Venn diagram for the following assumptions:

$S_1$: All mathematicians are interesting people.

$S_2$: Only uninteresting people become insurance salespersons.

$S_3$: Every genius is a mathematician.

Use the Venn diagram to determine whether or not each of the following is a valid conclusion:

(*a*) No genius is an insurance salesperson.
(*b*) Insurance salespersons are not mathematicians.
(*c*) Every interesting person is a genius.

## MATHEMATICAL INDUCTION

**1.73.** Prove: $2+4+6+\cdots+2n=n(n+1)$.

**1.74.** Prove: $1+4+7+\cdots+(3n-2)=2n(3n-1)$.

**1.75.** Prove: $\dfrac{1}{1\cdot 3}+\dfrac{1}{3\cdot 5}+\dfrac{1}{5\cdot 7}+\cdots+\dfrac{1}{(2n-1)(2n+1)}=\dfrac{1}{2n+1}$.

**1.76.** Prove: $1^2+2^2+3^2+\cdots+n^2=\dfrac{n(n+1)(2n+1)}{6}$.

**1.77.** Prove: Given $a^0=1$ and $a^n=a^{n-1}a$ for $n>0$. Prove: (*a*) $a^m a^n=a^{m+n}$, (*b*) $(a^m)^n=a^{mn}$.

# Answers to Supplementary Problems

**1.41.** $B=C=E=F$; $A=D=G=H$

**1.42.** $A=\{\text{a,e,i,o,u}\}$; $B=D=\{\text{l,i,t,e}\}$; $C=\{\text{a,b,c,d,e}\}$

**1.43.** (*a*) $C$ and $E$; (*b*) $D$ and $E$; (*c*) $A$, $B$, $D$; (*d*) none

**1.44.** (*a*) $\subseteq$; (*b*) $\subseteq$; (*c*) $\nsubseteq$; (*d*) $\nsubseteq$; (*e*) $\subseteq$; (*f*) $\subseteq$; (*g*) $\nsubseteq$; (*h*) $\nsubseteq$

**1.45.** (*a*) $A\cap B=\{2,5\}$, $A\cap C=\{1,5\}$; (*b*) $A\cup B=\{1,2,5,6,7\}$, $A\cup C=\{1,2,3,5,6,7,9\}$; (*c*) $A^c=\{3,4,7,8,9\}$, $C^c=\{2,4,6,8\}$

**1.46.** (*a*) $A\backslash B=\{1,6\}$, $A\backslash C=\{2,6\}$; (*b*) $A\oplus B=\{1,6,7\}$, $A\oplus C=\{2,3,6,7,9\}$

**1.47.** (*a*) $\{1,3,6,7,9\}$, (*b*) $\{3,4,8,9\}$, (*c*) $\{3,9\}$

**1.48.** (*a*) $\{a,b,c,d,e,f,g\}$; (*b*) $\{e,g,h\}$; (*c*) $\{b,g\}$; (*d*) $\{d,e\}$; (*e*) $\{b,c\}$; (*f*) $\{f,g,h\}$; (*g*) $\{c,e,f,g\}$; (*h*) $\{a,d,g,h\}$

**1.49.** (*a*) $\{a,b,d,e\}$; (*b*) $\{a\}$; (*c*) $\{a,d,f\}$; (*d*) $\{g\}$; (*e*) $\varnothing$; (*f*) $\{c,h\}$; (*g*) $\varnothing$; (*h*) $\{a,d\}$

**1.53.** $A \cup B = (A^c \cap B^c)^c$

**1.54.** (*b*) $A=\{a\}$; $B=\{b\}$; $C=\{c\}$, $A\cup(B\backslash C)=\{a\}$, $(A\cup B)\backslash(A\cup C)=\{b\}$

**1.56.** See Fig. 1-18.

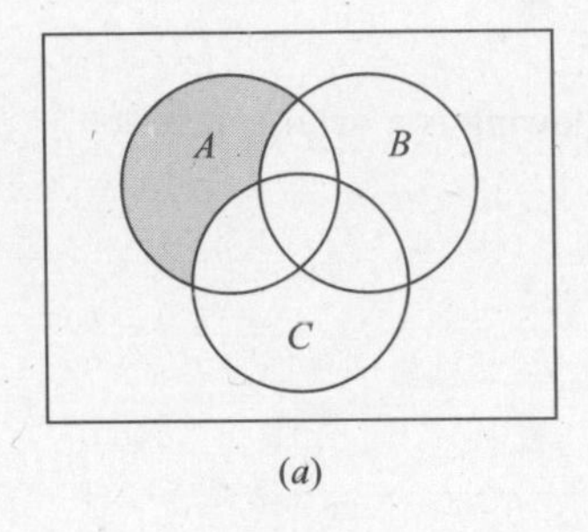

(*a*)

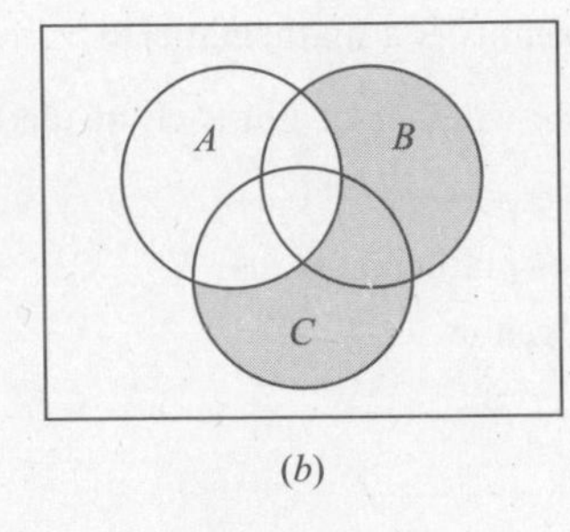

(*b*)

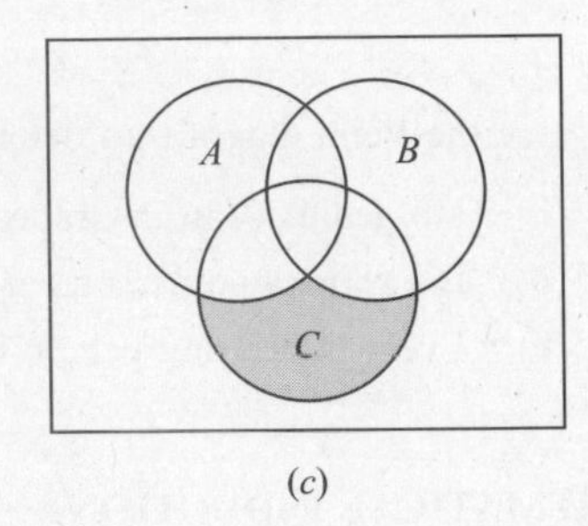

(*c*)

**Fig. 1-18**

**1.57.** (*a*) $A = (B^c \cup A) \cap (A \cup B)$
(*b*) $(A\cup B)\cap(A\cup B^c)\cap(A^c\cup B)\cap(A^c\cup B^c)=\varnothing$

**1.59.** (*b*), (*c*), and (*d*)

**1.60.** (*a*) 17; (*b*) 8; (*c*) 11; (*d*) 8; (*e*) 0

**1.61.** (*a*) 55; (*b*) 75; (*c*) 15, (*d*) 55

**1.62.** (*a*) 80; (*b*) 75; (*c*) 25; (*d*) 35; (*e*) 60; (*f*) 45

**1.63.** Use the data to first fill in the Venn diagram of $A$ (air-conditioning), $R$ (radio), $W$ (power windows) in Fig. 1-19. Then: (*a*) 5; (*b*) 4; (*c*) 2; (*d*) 4; (*e*) 6; (*f*) 11; (*g*) 23; (*h*) 2.

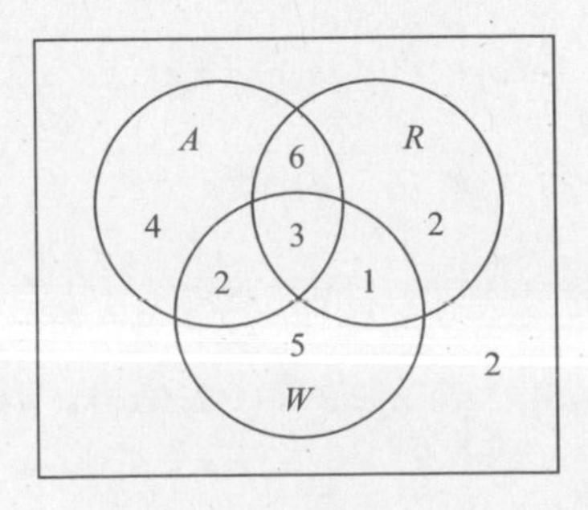


**Fig. 1-19**

**1.64.** Three elements: $\{a, b\}$, $\{c\}$, and $\{d, e, f\}$. (*a*) F; (*b*) F; (*c*) F; (*d*) T; (*e*) F; (*f*) T; (*g*) F; (*h*) T

**1.65.** Four elements: $\varnothing$, $\{1\}$, $\{2, 3\}$, and $\{3, 4\}$. (*a*) F; (*b*) F; (*c*) T; (*d*) F; (*e*) T; (*f*) T; (*g*) T; (*h*) T

**1.66.** (*a*) $\mathscr{P}(A)$ has $2^5 = 32$ elements as follows (where $135 = \{1, 3, 5\}$):

$[\varnothing, 1, 2, 3, 4, 5, 12, 13, 14, 15, 23, 24, 25, 34, 35, 45, 123, 124, 125, 134, 135, 145, 234, 235, 245, 345, 1234, 1235, 1245, 1345, 2345, A]$

(*b*) $\mathscr{B}$ has 6 elements: [123, 124, 125, 134, 135, 145].

**1.67.** $A$ has 3 elements, so $\mathscr{P}(A)$ has $2^3 = 8$ elements as follows (where $[ab, c] = [\{ab\}, \{c\}]$):

$$\{\varnothing,\ [ab],\ [c],\ [def],\ [ab, c],\ [ab, def], [c, def],\ A\}$$

Note that $\mathscr{P}(A)$ is a collection of collections of sets.

**1.68.** Let $X$ be an arbitrary element of $\mathscr{P}(A)$. For each $a \in A$, there are two possibilities, $a \in X$ or $a \not\subseteq X$. Since there are $m$ elements in $A$, there are $2 \cdot 2 \cdot \ldots \cdot 2$ ($m$ factors) $= 2^m$ different sets $X$. That is, $\mathscr{P}(A)$ has $2^m$ elements.

**1.69.** See Fig. 1-20. (*a*) Yes; (*b*) no; (*c*) yes; (*d*) yes

**1.70.** See Fig. 1-21. (*a*) No; (*b*) yes; (*c*) yes; (*d*) no

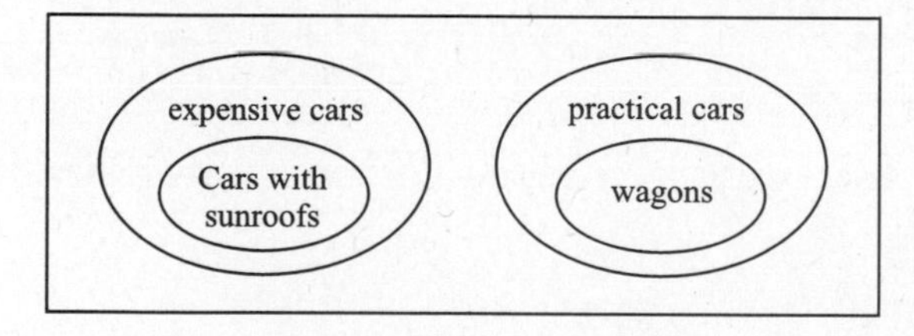


Fig. 1-20

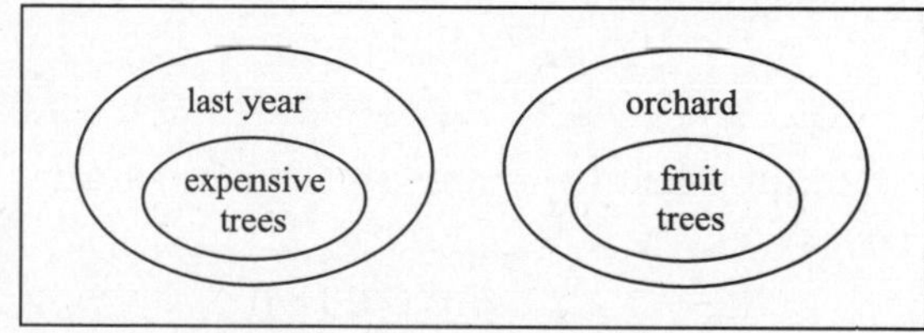


Fig. 1-21

**1.71.** See Fig. 1-22. (*a*) Yes; (*b*) yes; (*c*) yes; (*d*) no

**1.72.** See Fig. 1-23. (*a*) Yes; (*b*) yes; (*c*) no

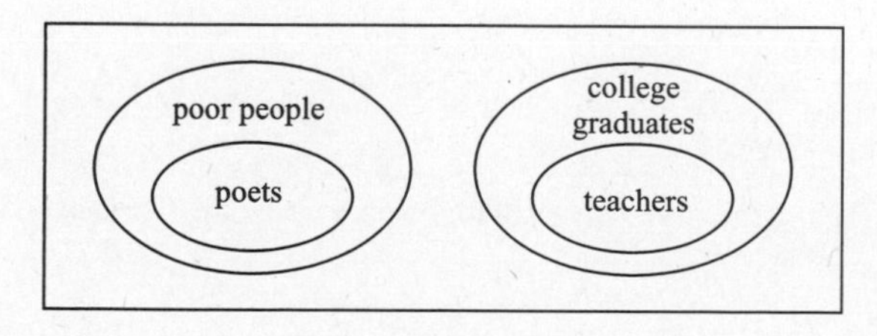


Fig. 1-22

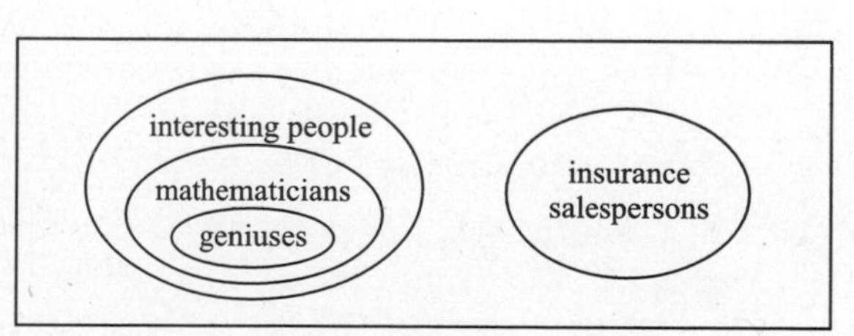


Fig. 1-23

# Chapter 2

# Sets and Elementary Properties of the Real Numbers

## 2.1 INTRODUCTION

This chapter investigates some sets and basic properties of the real numbers **R** and the integers

$$\mathbf{Z} = \{\ldots, -2, -1, 0, 1, 2, \ldots\}$$

(The letter **Z** comes from the word *Zahlen* which means number in German.)

The following simple rules concerning the addition and multiplication of these numbers are assumed:

(*a*) Associative law for multiplication and addition:

$$(a+b)+c = a+(b+c) \qquad \text{and} \qquad (ab)c = a(bc)$$

(*b*) Commutative law for multiplication and addition:

$$a+b = b+a \qquad \text{and} \qquad ab = ba$$

(*c*) Distributive law:

$$a(b+c) = ab + ac$$

(*d*) Additive identity and multiplicative identity: There exists a zero element 0 and a unity element 1 such that, for any number $a$,

$$a+0 = 0+a = a \qquad \text{and} \qquad a \cdot 1 = 1 \cdot a = a$$

(*e*) Additive inverse (negative): For any number $a$, there exists its negative $-a$ such that

$$a + (-a) = (-a) + a = 0$$

(*f*) Multiplicative inverse: For any number $a \neq 0$, there exists an inverse $a^{-1}$ such that

$$a \cdot a^{-1} = a^{-1} \cdot a = 1$$

Subtraction and division (except by 0) are defined in **R** by

$$a - b \equiv a + (-b) \qquad \text{and} \qquad a \cdot b^{-1}$$

Observe that subtraction uses property (*e*) of negatives, and division uses property (*f*) of inverses.

**Warning.** The last property (*f*) holds for the real numbers **R** and the rational numbers **Q**, but does not hold for the integers **Z**. That is, one can add, subtract, multiply, and divide (except by 0) in **R** and **Q**, but only add, subtract, and multiply in **Z**.

## 2.2 REAL NUMBER SYSTEM R

The notation **R** will be used to denote the real numbers. These are the numbers one uses in basic arithmetic and algebra. **R** together with its properties is called the *real number system.*

The set **R** of real numbers includes the following sets of numbers:

$\mathbf{Z} = \{\ldots, -2, -1, 0, 1, 2, \ldots\} =$ set of *integers* (signed whole numbers)
$\mathbf{P} = \{1, 2, 3, \ldots\} =$ set of positive integers (counting numbers)
$\mathbf{N} = \{0, 1, 2, \ldots\} =$ set of nonnegative integers (natural numbers)
$\mathbf{Q} =$ set of *rational numbers*, i.e. numbers which are ratios of integers

Examples of rational numbers are $2/3$ and $-3/4$. Those real numbers that are not rational, such as $\pi$ and $\sqrt{2}$, i.e., real numbers which cannot be represented as the ratio of integers, are called *irrational numbers*. The integer 0 is also a real number. Furthermore, for each positive real number, there is a corresponding negative real number.

### Real Line R, Decimal Expansion

One of the most important properties of the real numbers is that they can be represented graphically by points on a straight line. Specifically, as pictured in Fig. 2-1, a point, called the *origin*, is chosen to represent 0, and another point, usually to the right of 0, is chosen to represent 1. The direction from 0 to 1 is the *positive direction* and is sometimes indicated by an arrowhead at the end of the line. The distance between 0 and 1 is the *unit length*. Now there is a natural way to pair off the points on the line and the real numbers, that is, where each point on the line corresponds to a unique real number and vice versa. The positive real numbers are those to the right of 0 (on the same side as 1) and the negative numbers are those to the left of 0. The points representing the rational numbers $5/4$ and $-3/2$ are indicated in Fig. 2-1. We refer to such a line as the *real line* or the *real line* **R**.

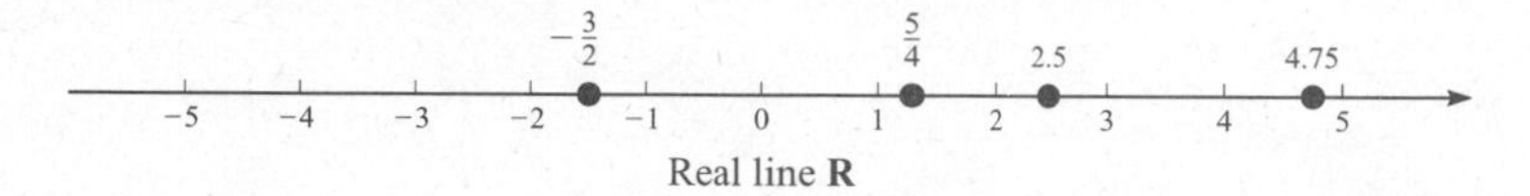


**Fig. 2-1**

Real numbers can also be represented by decimals. The decimal expansion of a rational number will either stop as in $\frac{3}{4} = 0.75$ or will have a pattern that repeats indefinitely, such as $\frac{17}{11} = 1.545454\ldots$. Even when the decimal expansion stops, it can be rewritten using repeated 9's, for example, $\frac{3}{4} = 0.74999\ldots$. The decimal expansion of an irrational number never stops nor does it have a repeating pattern. The points representing the decimal 2.5 and 4.75 are indicated in Fig. 2-1.

## 2.3 ORDER AND INEQUALITIES

Let $a$ and $b$ be real numbers. We say $a$ is *less* than $b$, written

$$a < b$$

if the difference $b - a$ is positive. Geometrically, $a < b$ if and only if the point $a$ lies to the left of the point $b$ on the real line **R**.

Observe that we define order in **R** in terms of the positive real numbers denoted by $\mathbf{R}^+$. All the usual properties of this order relation are a consequence of the following two properties of the positive real numbers $\mathbf{R}^+$:

[$\mathbf{P}_1$] If $a$ and $b$ are positive, then $a + b$ and $ab$ are positive.
[$\mathbf{P}_2$] For any real number $a$, either $a$ is positive, $a = 0$, or $-a$ is positive.

The following additional notation and terminology are used:

$a > b$, means $b < a$; read: *a is greater than b*
$a \leq b$, means $a < b$ or $a = b$; read: *a is less than or equal to b*
$a \geq b$, means $b \leq a$; read: *a is greater than or equal to b*

Any statement of the form $a < b$, $a \le b$, $a > b$, or $a \ge b$ is called an *inequality*; and any statement of the form $a < b$ or $a > b$ is sometimes called a *strict inequality*.

**EXAMPLE 2.1**

(*a*) $2 < 5$; $-6 < -3$; $4 \le 4$; $5 > -8$; $6 \ge 0$; $-7 \le 0$.

(*b*) Sorting the numbers 4, −7, 9, −2, 6, 0, −11, 13, −1, −5 in increasing order we obtain:

$$-11, -7, -5, -2, -1, 0, 3, 4, 6, 9, 13$$

(*c*) A real number $a$ is positive iff $a > 0$, and $a$ is negative iff $a < 0$. (Recall that "iff" is short for "if and only if.")

(*d*) The statement $2 < x < 7$ means $2 < x$ and $x < 7$; hence $x$ will lie between 2 and 7 on the real line **R**.

Basic properties of the inequality relations follow.

**Proposition 2.1**: Let $a, b, c$ be real numbers. Then:

(i) $a \le a$.
(ii) If $a \le b$ and $b \le a$, then $a = b$.
(iii) If $a \le b$ and $b \le c$, then $a \le c$.

**Proposition 2.2 (Law of Trichotomy)**: For any real numbers $a$ and $b$, exactly one of the following holds:

$$a < b, \qquad a = b, \qquad \text{or} \qquad a > b$$

**Proposition 2.3**: Let $a, b, c$ be real numbers such that $a \le b$. Then:

(i) $a + c \le b + c$.
(ii) $ac \le bc$ when $c > 0$; but $ac \ge bc$ when $c < 0$.

**Remark**: Observe that the above two properties $[\mathbf{P}_1]$ and $[\mathbf{P}_2]$ of the positive real numbers $\mathbf{R}^+$ are also true for the positive rational numbers $\mathbf{Q}^+$ viewed as a subset of the rational numbers **Q**, and the positive integers $\mathbf{P} = \mathbf{Z}^+$ viewed as a subset of the integers **Z**. Accordingly, Propositions 2.1, 2.2, and 2.3 also hold for the rational numbers **Q** and the integers **Z**.

## 2.4 ABSOLUTE VALUE, DISTANCE

The *absolute value* of a real number $a$, denoted by $|a|$, may be viewed as the distance between $a$ and the origin 0 on the real line **R**. Formally, $|a| = a$ or $-a$ according as $a$ is positive or negative, and $|0| = 0$. That is:

$$|a| = \begin{cases} a, & \text{if } a \ge 0 \\ -a, & \text{if } a < 0 \end{cases}$$

Accordingly, $|a|$ is always positive when $a \ne 0$. Intuitively, $|a|$ may be viewed as the magnitude of $a$ without regard to sign.

The distance $d$ between two points (real numbers) $a$ and $b$ is denoted by $d(a, b)$ and is obtained from the formula

$$d = d(a, b) = |a - b| = |b - a|$$

Alternatively:

$$d = \begin{cases} |a| + |b|, & \text{if } a \text{ and } b \text{ have different signs} \\ |a| - |b|, & \text{if } a \text{ and } b \text{ have the same sign and } |a| \ge |b| \end{cases}$$

These two cases are pictured in Fig. 2-2.

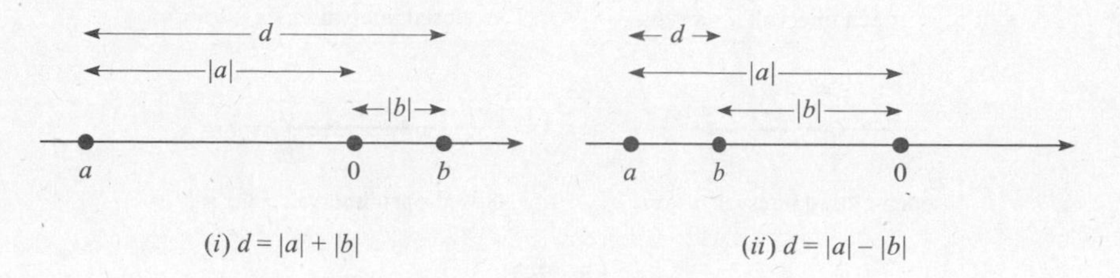


**Fig. 2-2**

**EXAMPLE 2.2**

(*a*) $|-3| = 3$, $|7| = 7$, $|-13| = 13$, $|4.25| = 4.25$, $|-0.75| = 0.75$.

(*b*) $|2-7| = |-5| = 5$, $|7-2| = |5| = 5$, $|-3-8| = |-11| = 11$.

(*c*) Using Fig. 2-2,

$$d(-2, 9) = 2 + 9 = 11, \qquad d(5, 8) = 8 - 5 = 3, \qquad d(-4, -11) = 11 - 4 = 7$$

The following proposition gives some properties of the absolute value function. [Problems 2.14 and 2.15 prove (iii) and (iv).]

**Proposition 2.4**: Let $a$ and $b$ be any real numbers.

(i) $|a| \geq 0$, and $|a| = 0$ iff $a = 0$.
(ii) $-|a| \leq a \leq |a|$.
(iii) $|ab| = |a|\ |b|$.
(iv) $|a \pm b| \leq |a| + |b|$.
(v) $||a| - |b|| \leq |a \pm b|$.

## 2.5 INTERVALS

Let $a$ and $b$ be distinct real numbers with, say, $a < b$. The *intervals* with *endpoints* $a$ to $b$ are denoted and defined as follows:

$(a, b) = \{x : a \leq x \leq b\}$, open interval from $a$ to $b$
$[a, b] = \{x : a \leq x \leq b\}$, closed interval from $a$ to $b$
$(a, b] = \{x : a < x \leq b\}$, open-closed interval from $a$ to $b$
$[a, b) = \{x : a \leq x < b\}$, closed-open interval from $a$ to $b$

Observe that an interval is open if it does not include its endpoints and is closed if it does include its endpoints. Also, a parenthesis "(" or ")" is used to indicate that an endpoint does not belong to the interval, and a bracket "[" or "]" is used to indicate that an endpoint does belong to the interval.

Figure 2-3 shows how we picture each of the above four intervals on the real line **R**. Notice that in each case the endpoints $a$ and $b$ are circled, the line segment between $a$ and $b$ is thickened, and the circle about the endpoint is filled if the endpoint belongs to the interval.

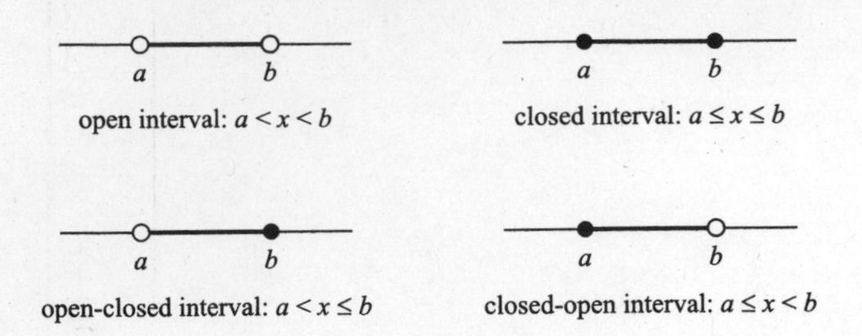


**Fig. 2-3**

**EXAMPLE 2.3**

(*a*) Find the interval satisfying each inequality, i.e., rewrite the inequality in terms of $x$ alone:

$$(1)\ 2 \le x - 5 \le 8, \quad (2)\ -1 \le x + 3 \le 4, \quad (3)\ -6 \le 3x \le 12, \quad (4)\ -6 \le -2x \le 4$$

(1) Add 5 to each side to obtain $7 \le x \le 13$.

(2) Add $-3$ to each side to obtain $-4 \le x \le 1$.

(3) Divide each side by 3 (or: multiply by $\frac{1}{3}$) to obtain $-2 \le x \le 4$.

(4) Divide each side by $-2$ (or: multiply by $-\frac{1}{2}$) and reverse inequalities to obtain $-6 \le x \le 3$.

(*b*) The inequality $|x| < 5$ may be interpreted to mean that the distance between $x$ and the origin 0 is less than 5; hence $x$ must lie between $-5$ and 5 on the real line **R**. In other words,

$$|x| < 5 \quad \text{and} \quad -5 < x < 5,$$

have the same meaning and, similarly,

$$|x| \le 5 \quad \text{and} \quad -5 \le x \le 5$$

have the same meaning.

**Definition**: A set $A$ of real numbers is said to be *dense* in **R** if every open interval contains a point of $A$ or, equivalently, if there is a point of $A$ between any two points in **R**.

The following theorem applies.

**Theorem 2.5**: The rational numbers **Q** are dense in **R**.

The proof of the above theorem lies beyond the scope of this text. It is closely related to the fact that every real number may be expressed as an infinite decimal or, equivalently, that every real number is the limit of a sequence of rational numbers.

### Infinite Intervals

Let $a$ be any real number. Then the set of real numbers $x$ satisfying $x < a$, $x \le a$, $x > a$, or $x \ge a$, is called an *infinite* interval with endpoint $a$. The interval is said to be *closed* or *open* according as the endpoint $a$ does or does not belong to the interval. The four infinite intervals may also be denoted and defined as follows:

$$(-\infty, a) = \{x : x < a\} \qquad (a, \infty) = \{x : x > a\}$$
$$(-\infty, a] = \{x : x \le a\} \qquad [a, \infty) = \{x : x \ge a\}$$

Note that the infinity symbol $\infty$ means all the numbers in the positive direction of $a$, whereas the minus infinity symbol $-\infty$ means all the numbers in the negative direction of $a$. A parenthesis is used with $\infty$ and $-\infty$ since they do not represent numbers in the interval. These infinite intervals are pictured in Fig. 2-4.

Fig. 2-4

## 2.6 BOUNDED SETS, COMPLETION PROPERTY

Let $A$ be a set of real numbers. Then $A$ is said to be:

(i) bounded, (ii) bounded from above, (iii) bounded from below

according as there exists a real number $M$ such that, for every $x \in A$:

$$\text{(i) } |x| \leq M, \quad \text{(ii) } x \leq M, \quad \text{(iii) } M \leq x$$

The number $M$ is called a *bound* in (i), an *upper bound* in (ii), and a *lower bound* in (iii). Note that $A$ is bounded if and only if $A$ is a subset of some finite interval. Specifically, $M$ is a bound of $A$ if and only if $A$ is a subset of $[-M, M]$.

If $A$ is finite then $A$ is necessarily bounded. If $A$ is infinite, then $A$ may be bounded, bounded from above (below), or unbounded.

**EXAMPLE 2.4**

(*a*) $A = \{1, 1/2, 1/3, \ldots, 1/n, \ldots\}$ is bounded since $A$ is certainly a subset of the closed *unit* interval $I = [0, 1]$.

(*b*) $B = \{2, 4, 6, \ldots\}$ is unbounded, but it is bounded from below.

(*c*) $C = \{\ldots, -5, -3, -1\}$ is unbounded, but it is bounded from above.

(*d*) $\mathbf{Z} = \{\ldots, -2, -1, 0, 1, 2, \ldots\}$ is unbounded. It has neither an upper bound nor a lower bound.

**Definition:** Let $A$ be a set of real numbers. A number $M$ is called the *least upper bound* or *supremum* of $A$, written

$$M = \sup(A)$$

if $M$ is an upper bound of $A$ but any number less than $M$ is not an upper bound of $A$, that is, for any positive number $\epsilon$, there exist $a \in A$ such that, $M - \epsilon < a$.

The following statement applies.

**Completion Property of R:** If a set $A$ of real numbers is bounded from above, then $\sup(A)$ exists.

The real numbers **R** are said to be *complete* since it satisfies the above property. We note that the rational numbers **Q** is not complete as seen by the following example.

**EXAMPLE 2.5** Let $A$ be the following subset of the rational numbers **Q**:

$$A = \{x \in \mathbf{Q} : x > 0, x^2 < 3\}$$

Observe that $A$ is bounded. However $\sup(A)$ does not exist. We cannot let $\sup(A) = \sqrt{3}$ since $\sqrt{3}$ is not a rational number.

The next two theorems (see Problems 6.17 and 6.49) follow from the completion property of **R**.

**Nested Interval Theorem**: The intersection $S = \cap_n I_n$ of a nested sequence of closed intervals is not empty. [A sequence $\{I_n\}$ of intervals is nested if $I_1 \supseteq I_2, \ldots$.]

**Heine-Borel Theorem**: Let $\mathscr{C}$ be a collection of open intervals which contain a closed interval $A = [a, b]$. Then a finite subcollection of $\mathscr{C}$ contains $A$.

## 2.7 INTEGERS Z (OPTIONAL MATERIAL)

The notation $\mathbf{Z}$ is used to denote the integers, the "signed whole numbers"; that is,

$$\mathbf{Z} = \{\ldots, -3, -2, -1, 1, 2, 3, \ldots\}$$

As noted above, $\mathbf{Z}$ satisfies all the properties in Section 2.1 except $(f)$. Accordingly, one can always add, subtract, and multiply integers obtaining integers. However, the quotient of two integers need not be an integer, hence the question of divisibility plays an important role in $\mathbf{Z}$.

One fundamental property of the integers $\mathbf{Z}$ is mathematical induction, which was discussed in Section 1.11. We give an equivalent statement below.

### Well-Ordering Principle

A property of the positive integers $\mathbf{P}$ which is equivalent to the principle of induction, although apparently very dissimilar, is the well-ordering principle (proved in Problem 2.32). Namely:

**Theorem 2.6 (Well-Ordering Principle)**: Let $S$ be a nonempty set of positive integers. Then $S$ contains a *least element*; that is, $S$ contains an element $a$ such that $a \leq s$ for every $s$ in $S$.

Generally speaking, an ordered set $S$ is said to be *well-ordered* if every subset of $S$ contains a first element. Thus Theorem 2.6 states that $\mathbf{P}$ is well-ordered.

A set $S$ of integers is said to be *bounded from below* if every element of $S$ is greater than some integer $m$ (which may be negative). (The number $m$ is called a *lower bound* of $S$.) A simple corollary of the above theorem follows:

**Corollary 2.7**: Let $S$ be a nonempty set of integers which is bounded from below. Then $S$ contains a least element.

### Division Algorithm

The following fundamental property of arithmetic (proved in Problems 2.36 and 2.37) is essentially a restatement of the result of long division.

**Theorem 2.8 (Division Algorithm)**: Let $a$ and $b$ be integers with $b \neq 0$. Then there exists integers $q$ and $r$ such that

$$a = bq + r \qquad \text{and} \qquad 0 \leq r < |b|$$

Also, the integers $q$ and $r$ are unique.

The number $q$ in the above theorem is called the *quotient*, and $r$ is called the *remainder*. We stress the fact that $r$ must be nonnegative. The theorem also states that

$$a - bq = r$$

This equation will be used subsequently.

**EXAMPLE 2.6**

(*a*) Let $a = 4461$ and $b = 16$. Dividing $a = 4461$ by $b = 16$ yields a quotient $q = 278$ and remainder $r = 13$. As expected, $a - bq + r$, that is,

$$4461 = 16(278) + 13$$

(*b*) Let $a = -262$ and $b = 3$. Here $a$ is negative. First divide $|a| = 262$ by $b = 3$ to obtain a quotient $q' = 87$ and a remainder $r' = 1$; hence

$$262 = 3(87) + 1$$

We need $a = -262$, so we multiply by $-1$ obtaining

$$-262 = 3(-87) - 1$$

However, $-1$ is negative and hence cannot be $r$. We correct this by adding and subtracting $b = 3$ as follows:

$$-262 = 3(-87) - 3 + 3 - 1 = 3(-88) + 2$$

Therefore, $q = -(q' + 1) = -88$ and $r = b - r' = 2$.

**Remark**: The result in Example 2.6(*b*) is true in general. That is, suppose $a$ is negative and suppose we want to find the quotient $q$ and remainder $r$ when $a$ is divided by $b$. First divide $|a|$ by $b$ to obtain a positive quotient $q'$ and remainder $r'$. If $r' \neq 0$, then set

$$q = -(q' + 1) \qquad \text{and} \qquad r = b - r'$$

but if $r' = 0$, then set $q = -q'$ and $r = r' = 0$.

### Divisibility

Let $a$ and $b$ be integers with $a \neq 0$. Suppose $ac = b$ for some integer $c$. We then say that $a$ *divides* $b$ or $b$ is *divisible* by $a$ and write

$$a|b$$

We may also say that $b$ is a *multiple* of $a$ or that $a$ is a *factor* or *divisor* of $b$. If $a$ does not divide $b$, we will write $a \nmid b$.

**EXAMPLE 2.7**

(*a*) $3|6$ since $3 \cdot 2 = 6$; and $-4|28$ since $(-4)(-7) = 28$.

(*b*) The divisors:

(i) of 1 are $\pm 1$ (iii) of 4 are $\pm 1, \pm 2, \pm 4$ (v) of 7 are $\pm 1, \pm 7$,

(ii) of 2 are $\pm 1, \pm 2$ (iv) of 5 are $\pm 1, \pm 5$ (vi) of 9 are $\pm 1, \pm 3, \pm 9$

(*c*) If $a \neq 0$, then $a|0$ since $a \cdot 0 = 0$.

(*d*) Every integer $a$ is divisible by $\pm 1$ and $\pm a$. These are sometimes called the *trivial divisors* of $a$.

Simple properties of divisibility follow.

(i) If $a|b$ and $b|c$, then $a|c$.

(ii) If $a|b$ then, for any integer $x$, $a|bx$.

(iii) If $a|b$ and $a|c$, then $a|(b + c)$ and $a|(b - c)$.

(iv) If $a|b$ and $b \neq 0$, then $a = \pm b$ or $|a| < |b|$.

(v) If $a|b$ and $b|a$, then $|a| = |b|$, i.e., $a = \pm b$.

(vi) If $a|1$, then $a = \pm 1$.

Putting (ii) and (iii) together, we obtain the following important result.

**Proposition 2.9**: Suppose $a|b$ and $a|c$. Then, for any integers $x$ and $y$, $a|(bx + cy)$.

The expression $bx + cy$ will be called a *linear combination* of $b$ and $c$.

### Primes

A positive integer $p > 1$ is called a *prime number* or a *prime* if its only divisors are $\pm 1$ and $\pm p$, that is, if $p$ only has trivial divisors. If $n > 1$ is not prime, then $n$ is said to be *composite*. We note (Problem 2.31) that if $n > 1$ is composite then $n = ab$ where $1 < a, b < n$.

**EXAMPLE 2.8**

(*a*) The integers 2 and 7 are primes, whereas $6 = 2 \cdot 3$ and $15 = 3 \cdot 5$ are composite.

(*b*) The primes less than 50 follow:

$$2, 3, 5, 7, 11, 13, 17, 19, 23, 29, 31, 37, 41, 43, 47$$

(*c*) Although 21, 24, and 1729 are not primes, each can be written as a product of primes:

$$21 = 3 \cdot 7, \qquad 24 = 2 \cdot 2 \cdot 2 \cdot 3 = 2^3 \cdot 3, \qquad 1729 = 7 \cdot 13 \cdot 19$$

The Fundamental Theorem of Arithmetic states that every integer $n > 1$ can be written as a product of primes in essentially one way; it is a deep and somewhat difficult theorem to prove. However, using induction, it is easy at this point to prove that such a product exists. Namely:

**Theorem 2.10**: Every integer $n > 1$ can be written as a product of primes.

Note that a product may consist of a single factor so that a prime $p$ is itself a product of primes.

We prove Theorem 2.10 here, since its proof is relatively simple.

*Proof*: The proof is by induction. Let $n = 2$. Since 2 is prime, $n$ is a product of primes. Suppose $n > 2$, and the theorem holds for positive integers less than $n$. If $n$ is prime, then $n$ is a product of primes. If $n$ is composite, then $n = ab$ where $a, b < n$. By induction, $a$ and $b$ are products of primes; hence $n = ab$ is also a product of primes.

Euclid, who proved the Fundamental Theorem of Arithmetic, also asked whether or not there was a largest prime. He answered the question thus:

**Theorem 2.11**: There is no largest prime, that is, there exists an infinite number of primes.

*Proof*: Suppose there is a finite number of primes, say $p_1, p_2, \ldots, p_m$. Consider the integer

$$n = p_1 p_2 \cdots p_m + 1$$

Since $n$ is a product of primes (Theorem 2.10), it is divisible by one of the primes, say $p_k$. Note that $p_k$ also divides the product $p_1 p_2 \cdots p_m$. Therefore $p_k$ divides

$$n - p_1 p_2 \cdots p_m = 1$$

This is impossible, and so $n$ is divisible by some other prime. This contradicts the assumption that $p_1, p_2, \ldots, p_m$ are the only primes. Thus the number of primes is infinite, and the theorem is proved.

## 2.8 GREATEST COMMON DIVISOR, EUCLIDEAN ALGORITHM

Suppose $a$ and $b$ are integers, not both 0. An integer $d$ is called a *common divisor* of $a$ and $b$ if $d$ divides both $a$ and $b$, that is, if $d|a$ and $d|b$. Note that 1 is always a positive common divisor of $a$ and $b$, and that any common divisor of $a$ and $b$ cannot be greater than $|a|$ or $|b|$. Thus there exists a largest common divisor of $a$ and $b$; it is denoted by

$$\gcd\,(a, b)$$

and it is called the *greatest common divisor* of $a$ and $b$.

**EXAMPLE 2.9**

(*a*) The common divisors of 12 and 18 are $\pm 1, \pm 2, \pm 3, \pm 6$. Thus $\gcd\,(12, 18) = 6$. Similarly,

$$\gcd\,(12, -18) = 6, \qquad \gcd\,(12, -16) = 4, \qquad \gcd\,(29, 15) = 1, \qquad \gcd\,(14, 49) = 7$$

(*b*) For any integer $a$, we have gcd $(1, a) = 1$.

(*c*) For any prime $p$, we have gcd $(p, a) = p$ or gcd $(p, a) = 1$ according as $p|a$ or $p \nmid a$.

(*d*) Suppose $a$ is positive. Then $a|b$ if and only if gcd $(a, b) = a$.

The following theorem (proved in Problem 2.43) gives an alternative characterization of the greatest common divisor.

**Theorem 2.12**: Let $d$ be the smallest positive integer of the form $ax + by$. Then $d = \gcd(a, b)$.

**Corollary 2.13**: Suppose $d = \gcd(a, b)$. Then there exists integers $x$ and $y$ such that $d = ax + by$.

Another way to characterize the greatest common divisor, without using the inequality relation, follows:

**Theorem 2.14**: A positive integer $d = \gcd(a, b)$ if and only if $d$ has the following properties:
(1) $d$ divides both $a$ and $b$;
(2) if $c$ divides both $a$ and $b$, then $c|d$.

Simple properties of the greatest common divisor follow.

(*a*) $\gcd(a, b) = \gcd(b, a)$.
(*b*) If $x > 0$, then $\gcd(ax, bx) = x \cdot \gcd(a, b)$.
(*c*) If $d = \gcd(a, b)$, then $\gcd(a/d,\ b/d) = 1$.
(*d*) For any integer $x$, $\gcd(a, b) = \gcd(a, b + ax)$.

### Euclidean Algorithm

Let $a$ and $b$ be integers, and let $d = \gcd(a, b)$. One can always find $d$ by listing all the divisors of $a$ and then all the divisors of $b$ and then choosing the largest common divisor. This procedure does not find the integers $x$ and $y$ such that

$$d = ax + by$$

This subsection gives a very efficient algorithm for finding both $d = \gcd(a, b)$ and the above integers $x$ and $y$.

This algorithm, called the Euclidean algorithm, consists of repeatedly applying the division algorithm (long division). We illustrate the algorithm with an example.

**EXAMPLE 2.10** Let $a = 540$ and $b = 168$. We find $d = \gcd(a, b)$ by dividing $a$ by $b$ and then repeatedly dividing each remainder into the divisor until obtaining a zero remainder. These steps are pictured in Fig. 2-5. The last nonzero remainder is 12. Thus

$$12 = \gcd(540, 168)$$

This follows from the fact that

$$\gcd(540, 168) = \gcd(168, 36) = \gcd(36, 24) = \gcd(24, 12) = 12$$

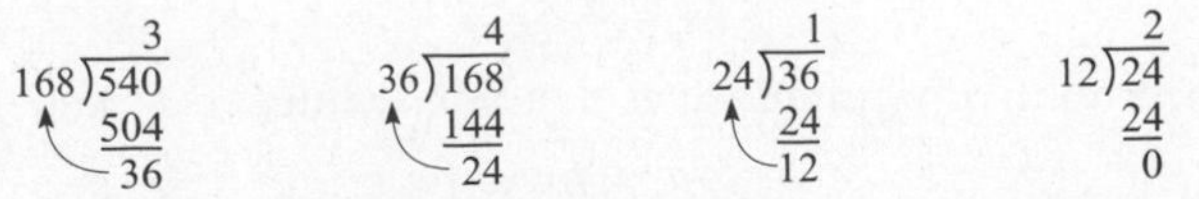

$$\begin{array}{r} 3 \\ 168\overline{)540} \\ \underline{504} \\ 36 \end{array} \qquad \begin{array}{r} 4 \\ 36\overline{)168} \\ \underline{144} \\ 24 \end{array} \qquad \begin{array}{r} 1 \\ 24\overline{)36} \\ \underline{24} \\ 12 \end{array} \qquad \begin{array}{r} 2 \\ 12\overline{)24} \\ \underline{24} \\ 0 \end{array}$$

**Fig. 2-5**

Next we find $x$ and $y$ such that

$$12 = 540x + 168y$$

The first three quotients in Fig. 2-5 yield the equations:

$$\begin{array}{llll} (1) & 540 = 3(168) + 36 & \text{or} & 36 = 540 - 3(168) \\ (2) & 168 = 4(36) + 24 & \text{or} & 24 = 168 - 4(36) \\ (3) & 36 = 1(24) + 12 & \text{or} & 12 = 36 - 1(24) \end{array}$$

Equation (3) tells us that 12 is a linear combination of 36 and 24. We use (2) to replace 24 in (3) so we can write 12 as a linear combination of 168 and 36 as follows:

$$\begin{aligned} (4)\quad 12 &= 36 - 1[168 - 4(36)] = 36 - (168) + 4(36) \\ &= 5(36) - 1(168) \end{aligned}$$

We now use (1) in (4) so we can write 12 as a linear combination of 168 and 540 as follows:

$$\begin{aligned} 12 &= 5[540 - 3(168)] - 1(168) \\ &= 5(540) - 15(168) - 1(168) \\ &= 5(540) - 16(168) \end{aligned}$$

This is our desired linear combination. Thus $x = 5$ and $y = -16$.

### Least Common Multiple

Suppose $a$ and $b$ are nonzero integers. Note that $|ab|$ is a positive common multiple of $a$ and $b$. Thus there exists a smallest positive common multiple of $a$ and $b$; it is denoted by

$$\text{lcm}\ (a, b)$$

and it is called the *least common multiple* of $a$ and $b$.

**EXAMPLE 2.11**

(*a*) $\text{lcm}\ (2, 3) = 6$, $\text{lcm}\ (4, 6) = 12$, $\text{lcm}\ (9, 10) = 90$.

(*b*) For any positive integer $a$, we have $\text{lcm}\ (1, a) = a$.

(*c*) For any prime $p$ and any positive integer $a$, $\text{lcm}\ (p, a) = a$ or $\text{lcm}\ (p, a) = ap$ according as $p|a$ or $p\nmid a$.

(*d*) Suppose $a$ and $b$ are positive integers. Then $a|b$ if and only if $\text{lcm}\ (a, b) = b$.

The next theorem gives an important relationship between the greatest common divisor and the least common multiple.

**Theorem 2.15**: Suppose $a$ and $b$ are nonzero integers. Then

$$\text{lcm}\ (a, b) = \frac{|ab|}{\text{gcd}\ (a, b)}$$

## 2.9 FUNDAMENTAL THEOREM OF ARITHMETIC

This section discusses the Fundamental Theorem of Arithmetic. First we need the notion of relatively prime integers.

Two integers $a$ and $b$ are said to be *relatively prime*, or *coprime*, if

$$\text{gcd}\ (a, b) = 1$$

Accordingly, if $a$ and $b$ are relatively prime, then there exist integers $x$ and $y$ such that

$$ax + by = 1$$

Conversely, if $ax + by = 1$, then $a$ and $b$ are relatively prime.

**EXAMPLE 2.12**

(*a*) Observe that $\gcd(12, 35) = 1$, $\gcd(49, 18) = 1$, $\gcd(21, 64) = 1$, $\gcd(-28, 45) = 1$

(*b*) If $p$ and $q$ are distinct primes, then $\gcd(p, q) = 1$.

(*c*) For any integer $a$, we have $\gcd(a, a+1) = 1$. This follows from the fact that any common divisor of $a$ and $a+1$ must divide their difference $(a+1) - a = 1$.

The relation of being relatively prime is particularly important because of the following results. We will prove the second theorem here.

**Theorem 2.16**: Suppose $\gcd(a, b) = 1$, and $a$ and $b$ both divide $c$. Then $ab$ divides $c$.

**Theorem 2.17**: Suppose $a|bc$, and $\gcd(a, b) = 1$. Then $a|c$.

*Proof*: Since $\gcd(a, b) = 1$, there exist $x$ and $y$ such that $ax + by = 1$. Multiplying by $c$ yields

$$acx + bcy = c$$

We have $a|acx$. Also, $a|bcy$ since, by hypothesis, $a|bc$. Hence $a$ divides the sum $acx + bcy = c$.

**Corollary 2.18**: Suppose a prime $p$ divides a product $ab$. Then

$$p|a \text{ or } p|b$$

This corollary dates back to Euclid. In fact, it is the basis of his proof of the fundamental theorem of arithmetic.

### Fundamental Theorem of Arithmetic

Theorem 2.10 asserts that every positive integer is a product of primes. Can different products of primes yield the same number? Clearly, we can rearrange the order of the prime factors, e.g.

$$30 = 2 \cdot 3 \cdot 5 = 5 \cdot 2 \cdot 3 = 3 \cdot 2 \cdot 5$$

The fundamental theorem of arithmetic (proved in Problem 2.49) says that this is the only way that two "different" products can give the same number. Namely:

**Theorem 2.19 (Fundamental Theorem of Arithmetic)**: Every integer $n > 1$ can be expressed uniquely (except for order) as a product of primes.

The primes in the factorization of $n$ need not be distinct. Frequently, it is useful to collect together all equal primes. Then $n$ can be expressed uniquely in the form

$$n = p_1^{m_1} p_2^{m_2} \cdots p_r^{m_r}$$

where the $m_i$ are positive and $p_1 < p_2 < \cdots < p_r$. This is called the *canonical factorization* of $n$.

**EXAMPLE 2.13** Let $a = 2^4 \cdot 3^3 \cdot 7 \cdot 11 \cdot 13$ and $b = 2^3 \cdot 3^2 \cdot 5^2 \cdot 11 \cdot 17$. Find $d = \gcd(a, b)$ and $m = \operatorname{lcm}(a, b)$.

(*a*) First we find $d = \gcd(a, b)$. Those primes $p_i$ which appear in both $a$ and $b$, i.e., 2, 3, and 11, will also appear in $d$, and the exponent of $p_i$ in $d$ will be the smaller of its exponents in $a$ and $b$. Thus

$$d = \gcd(a, b) = 2^3 \cdot 3^2 \cdot 11 = 792$$

(*b*) Next we find $m = \operatorname{lcm}(a, b)$. Those primes $p_i$ which appear in either $a$ and $b$, i.e., 2, 3, 5, 7, 11, 13 and 17 will also appear in $m$, and the exponent of $p_i$ in $m$ will be the larger of its exponents in $a$ and $b$. Thus

$$m = \operatorname{lcm}(a, b) = 2^4 \cdot 3^3 \cdot 5^2 \cdot 11 \cdot 13 \cdot 17$$

# Solved Problems

## REAL NUMBER SYSTEM R, SETS OF NUMBERS

**2.1.** Assuming **R**, **Q**, **Q**$'$, **Z**, and **P** denote respectively, the real numbers, rational numbers, irrational numbers, integers, and positive integers, state whether each of the following is true or false:

| | | | |
|---|---|---|---|
| (*a*) $-7 \in \mathbf{P}$ | (*d*) $3\pi \in \mathbf{Q}$ | (*g*) $\pi^2 \in \mathbf{R}$ | (*j*) $-6 \in \mathbf{Q}$ |
| (*b*) $\sqrt{2} \in \mathbf{Q}'$ | (*e*) $\sqrt[3]{8} \in \mathbf{P}$ | (*h*) $\sqrt{9/4} \in \mathbf{Q}'$ | (*k*) $\sqrt{-4} \in \mathbf{R}$ |
| (*c*) $4 \in \mathbf{Z}$ | (*f*) $-2 \in \mathbf{Z}$ | (*i*) $1/2 \in \mathbf{Z}$ | (*l*) $6 \in \mathbf{R}$ |

(*a*) False. **P** only contains positive integers, $-7$ is negative.

(*b*) True. $\sqrt{2}$ cannot be expressed as the ratio of two integers; hence $\sqrt{2}$ is irrational.

(*c*) True. The integers **Z** contain all the "whole" numbers, so 4 is an integer.

(*d*) False. $\pi$ is not rational and neither is $3\pi$.

(*e*) True. $\sqrt[3]{8} = 2$ is a positive integer.

(*f*) True. **Z** contains both the positive and negative "whole" numbers.

(*g*) True. $\pi$ is real and so is $\pi^2$.

(*h*) False. $\sqrt{9/4} = 3/2$ is rational, not irrational.

(*i*) False. $1/2$ is not an integer.

(*j*) True. The rational numbers include the integers.

(*k*) False. $\sqrt{-4} = 2i$ is not a real number.

(*l*) True. The real numbers include the integers.

**2.2.** Plot the numbers $\frac{5}{2}$, 3.8, $-4.5$, and $-3.3$ on the real line **R**.

The points corresponding to the numbers are shown in Fig. 2-6.

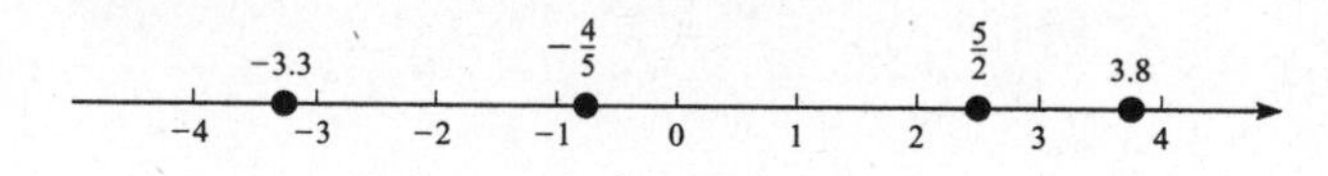


**Fig. 2-6**

**2.3.** Express each real number as an infinite decimal (that is, without ending in zeros): (*a*) $2/3$, (*b*) $4/7$, (*c*) $3/8$.

(*a*) Dividing 2 by 3 yields $2/3 = 0.6666\ldots$.

(*b*) Dividing 4 by 7 yields the following where 571428 repeats:

$$4/7 = 0.571\,428\,571\,428\ldots$$

(*c*) Dividing 3 by 8 yields $3/8 = 0.375$, which is not an infinite decimal. However, for any nonzero digit $d$, one can show that $d = d'.999\ldots$ where $d' = d - 1$. Thus replace $5000\ldots$ by $4999\ldots$ to obtain the required infinite decimal

$$3/8 = 0.3749999\ldots$$

**2.4.** Consider (*a*) the rational numbers **Q** and (*b*) the irrational numbers **Q**$'$. Determine whether or not each is closed under the operations of addition and multiplication.

A set $S$ of real numbers is closed under addition and multiplication according as the sum and product of any numbers in $S$ still belongs to $S$.

(*a*) The sum and product of rational numbers are rational; hence **Q** is closed under addition and multiplication.

(*b*) The sum and product of irrational numbers, need not be irrational. For example, $\sqrt{2}+(-\sqrt{2})=0$ is not irrational, and $\sqrt{2}\sqrt{2}=2$ is not irrational. Thus **Q**$'$ is not closed under addition and not closed under multiplication.

**2.5.** Let (*a*) $E=\{2,4,6,\ldots\}$ and (*b*) $F=\{1,3,5,\ldots\}$. Determine whether or not each is closed under the operations of addition and multiplication.

(*a*) The sum and product of positive even integers are positive and even; hence $E$ is closed under addition and multiplication.

(*b*) The sum of two odd numbers is not odd, hence $F$ is not closed under addition. However, the product of two positive odd integers is positive and odd; hence $F$ is closed under multiplication.

## ORDER AND INEQUALITIES, ABSOLUTE VALUE

**2.6.** Insert the correct symbol, <, >, or =, between each pair of real numbers:

(*a*) $4$ ____ $-7$ (*c*) $3^2$ ____ $9$ (*e*) $3^2$ ____ $5.5$

(*b*) $-2$ ____ $-9$ (*d*) $-8$ ____ $\pi$ (*f*) $6.25$ ____ $8$

For each pair of real numbers, say $a$ and $b$, determine their relative positions on the number line **R**; or, alternately, compute $b-a$, and write

$$a<b, \qquad a>b, \qquad \text{or} \qquad a=b$$

according as $b-a$ is positive, negative, or zero. Hence:

(*a*) $4>-7$, (*c*) $3^2=9$, (*e*) $3^2>5.5$,

(*b*) $-2>-9$, (*d*) $-8<\pi$, (*f*) $6.25<8$.

**2.7.** Rewrite the following geometric relationships between the given real numbers using the inequality notation: (*a*) $x$ lies to the right of 8; (*b*) $y$ lies to the left of $-2$; (*c*) $z$ lies between $-3$ and 7; (*d*) $t$ lies between 5 and 1.

Recall that $a<b$ means that $a$ lies to the left of $b$ on the real line **R**. Thus: (*a*) $x>8$ or $8<x$; (*b*) $y<-2$; (*c*) $-3<z$ and $z<7$ or, simply, $-3<z<7$; (*d*) $1<t<5$.

**2.8.** Sort the following numbers in increasing order (where $e=2.7814\ldots$):

$$5,\ -8,\ 2,\ -3,\ \pi,\ -2.8,\ 0,\ 9,\ e,\ -1.5,\ 3$$

The negative numbers will be on the left of 0, decreasing in magnitude (absolute value) from left to right, and the positive numbers will be on the right of 0 increasing in magnitude from left to right:

$$-8,\ -3,\ -2.8,\ -1.5,\ 0,\ 2,\ e,\ 3,\ \pi,\ 5,\ 9$$

**2.9.** Evaluate: (*a*) $|-4|$, $|6.2|$, $|0|$, $|-1.25|$
(*b*) $|2-5|$, $|-2+5|$, $|-2-5|$
(*c*) $|5-8|+|2-4|$, $|4-3|-|3-9|$

(*a*) The absolute value is the magnitude of the number without regard to sign. Hence:

$$|-4|=4, \quad |6.2|=6.2, \quad |0|=0, \quad |-1.25|=1.25$$

(*b*) Evaluate inside the absolute value sign first:

$$|2-5|=|-3|=3, \quad |-2+5|=|3|=3, \quad |-2-5|=|-7|=7$$

(*c*) Evaluate inside the absolute value sign first:

$$|5-8|+|2-4|=|-3|+|-2|=3+2=5$$
$$|4-3|-|3-9|=|1|-|-6|=1-6=-5$$

**2.10.** Find the distance $d$ between each pair of integers:

(*a*) 3 and $-7$ (*c*) 1 and 9 (*e*) 4 and $-4$
(*b*) $-4$ and 2 (*d*) $-8$ and $-3$ (*f*) $-5$ and $-8$

The distance $d$ between $a$ and $b$ is given by $d=|a-b|=|b-a|$. Alternatively, as shown in Fig. 2-2, $d=|a|+|b|$ when $a$ and $b$ have different signs, and $d=|a|-|b|$ when $a$ and $b$ have the same sign and $|a|\geq|b|$. Thus:

(*a*) $d=3+7=10$ (*c*) $d=9-1=8$ (*e*) $d=4+4=8$
(*b*) $d=4+2=6$ (*d*) $d=8-3=5$ (*f*) $d=8-5=3$

**2.11.** Find all integers $n$ such that: (*a*) $1<2n-6<14$, (*b*) $2<8-3n<18$.

(*a*) Add 6 to the "three sides" to get $7<2n<20$. Then divide all sides by 2 (or multiply by 1/2) to get $3.5<n<10$. Hence

$$n=4,5,6,7,8,9$$

(*b*) Add $-8$ to the three sides to get $-6<-3n<10$. Divide by $-3$ (or multiply by $-1/3$) and, since $-3$ is negative, change the direction of the inequality to get

$$2>n>-3.3 \quad \text{or} \quad -3.3<n<2$$

Hence $n=-3,-2,-1,0,1$.

**2.12.** Prove Proposition 2.1(iii): If $a\leq b$ and $b\leq c$, then $a\leq c$.

The proposition is clearly true when $a=b$ or $b=c$. Thus we need only consider the case that $a<b$ and $b<c$. Hence $b-a$ and $c-b$ are positive. Therefore, by property $[\mathbf{P}_1]$ of the positive real numbers $\mathbf{R}^+$, the sum is also positive. That is,

$$(b-a)+(c-b)=c-a$$

is positive. Thus $a<c$ and hence $a\leq c$.

**2.13.** Prove Proposition 2.3: Let $a, b, c$ be real numbers such that $a \leq b$. Then: (i) $a + c \leq b + c$, (ii) $ac \leq bc$ when $c > 0$; but $ac \geq bc$ when $c < 0$.

The proposition is certainly true when $a = b$. Hence we need only consider the case when $a < b$, that is, when $b - a$ is positive.

(i) The following difference is positive:

$$(b + c) - (a + c) = b - a$$

Hence $a + c < b + c$.

(ii) Suppose $c$ is positive. By property $[\mathbf{P}_1]$ of the positive real numbers $\mathbf{R}^+$, the following product is also positive:

$$c(b - a) = bc - ac$$

Thus $ac < bc$. Now suppose $c$ is negative. Then $-c$ is positive. Thus the following product is also positive:

$$(-c)(b - a) = ac - bc$$

Accordingly, $bc < ac$, whence $ac > bc$.

**2.14.** Prove Proposition 2.4(iii): $|ab| = |a||b|$.

The proof consists of a case-by-case analysis.

(*a*) Suppose $a = 0$ or $b = 0$.

Then $|a| = 0$ or $|b| = 0$, and so $|a||b| = 0$. Also $ab = 0$. Hence

$$|ab| = 0 = |a||b|$$

(*b*) Suppose $a > 0$ and $b > 0$.

Then $|a| = a$ and $|b| = b$. Hence

$$|ab| = ab = |a||b|$$

(*c*) Suppose $a > 0$ and $b < 0$.

Then $|a| = a$ and $|b| = -b$. Also $ab < 0$. Hence

$$|ab| = -(ab) = a(-b) = |a||b|$$

(*d*) Suppose $a < 0$ and $b > 0$.

Then $|a| = -a$ and $|b| = b$. Also $ab < 0$. Hence

$$|ab| = -(ab) = (-a)b = |a||b|$$

(*e*) Suppose $a < 0$ and $b < 0$.

Then $|a| = -a$ and $|b| = -b$. Also $ab > 0$. Hence

$$|ab| = ab = (-a)(-b) = |a||b|$$

**2.15.** Prove Proposition 2.4(iv): $|a \pm b| \leq |a| + |b|$.

Now $ab \leq |ab| = |a||b|$, and so $2ab \leq 2|a||b|$. Hence

$$(a + b)^2 = a^2 + 2ab + b^2 \leq |a|^2 + 2|a||b| + |b|^2 = (|a| + |b|)^2$$

But $\sqrt{(a + b)^2} = |a + b|$. Thus the square root of the above yields $|a + b| \leq |a| + |b|$.

Also,

$$|a - b| = |a + (-b)| \leq |a| + |-b| = |a| + |b|$$

**2.16.** Plot and describe the absolute value function $f(x) = |x|$.

For nonnegative values of $x$ we have $f(x) = x$ and hence we obtain points of the form $(a, a)$, e.g.,

$$(0,0),\ (1,1),\ (2,2),\ldots$$

For negative values of $x$ we have $f(x) = -x$ and hence we obtain points of the form $(-a, a)$, e.g.,

$$(-1,1),\ (-2,2),\ (-3,3),\ldots$$

This yields the graph in Fig. 2-7. Observe that the graph of $f(x) = |x|$ lies entirely in the upper half-plane since $f(x) \geq 0$ for every $x \in \mathbf{R}$. Also, the graph consists of the line $y = x$ in the right half-plane and the line $y = -x$ in the left half-plane.

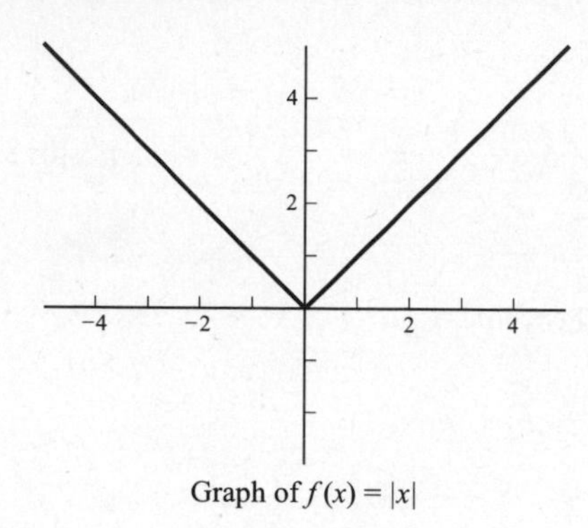


Graph of $f(x) = |x|$

**Fig. 2-7**

## INTERVALS

**2.17.** Rewrite each interval in set-builder form:

(*a*) $A = [-3, 5)$, (*b*) $B = (3, 8)$, (*c*) $C = [0, 4]$, (*d*) $D = (-7, -2]$.

Recall that a parenthesis means the endpoint does not belong to the interval, and that a bracket means that the endpoint does belong to the interval. Thus:

(*a*) $A = \{x : -3 \leq x < 5\}$ (*c*) $C = \{x : 0 \leq x \leq 4\}$

(*b*) $B = \{x : 3 < x < 8\}$ (*d*) $D = \{x : -7 < x \leq -2\}$

**2.18.** Describe and plot each interval:

(*a*) $A = (2, 4)$, (*b*) $B = [-1, 2]$, (*c*) $C = (-3, 1]$.

(*a*) $A$ consists of all numbers between 2 and 4 without the endpoints 2 and 4. See Fig. 2-8(*a*).

(*b*) $B$ consists of all points between $-1$ and 2 including both endpoints $-1$ and 2. See Fig. 2-8(*b*).

(*c*) $C$ consists of all points between $-3$ and 1 including only the endpoint 1. See Fig. 2-8(*c*).

Note that a circle about an endpoint is filled or unfilled according as the endpoint does or does not belong to the interval.

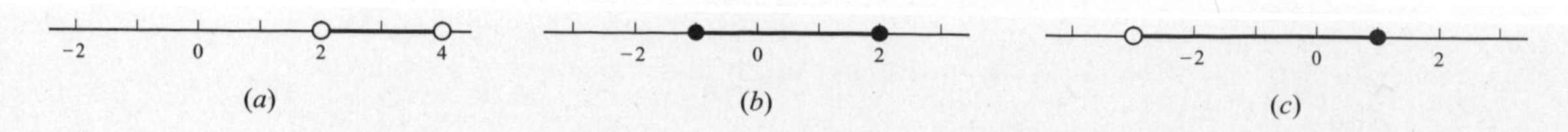


**Fig. 2-8**

**2.19.** Find the interval satisfying each inequality, i.e., rewrite the inequality in terms of $x$ alone:

(a) $3 \le x - 4 \le 9$, (b) $-2 \le x + 5 \le 3$, (c) $-8 \le 2x \le 2$, (d) $-9 \le -3x \le 15$.

(a) Add 4 to each side to obtain $7 \le x \le 13$.

(b) Add $-5$ to each side to obtain $-7 \le x \le -2$.

(c) Divide each side 2 (or multiply by 1/2) to obtain $-4 \le x \le 1$.

(d) Divide each side $-3$ and reverse inequalities to obtain $3 \ge x \ge -5$ or, in the usual form, $-5 \le x \le 3$.

**2.20.** Rewrite without the absolute value sign: (a) $|x| < 3$, (b) $|x - 2| < 5$.

(a) Here $x$ lies between $-3$ and 3; hence $-3 < x < x$.

(b) Here $-5 < x - 2 < 5$ or $-3 < x < 7$.

**2.21.** Write each open interval in the form $|x - a| < r$: (a) $2 < x < 10$, (b) $-7 < x < 3$.

Here $a$ will be the "center" and $r$ will be the "radius" of the interval, i.e., $a$ is the midpoint and $r$ is half the length of the interval. Thus find the sum $s$ of the endpoints and divide by 2 to obtain $a$, and find the distance $d$ between the endpoints and divide by 2 to obtain $r$.

(a) $s = 12$ so $a = 6$; $d = 8$ so $r = 4$; hence $|x - 6| < 4$.

(b) $s = -4$ so $a = -2$; $d = 10$ so $r = 5$; hence $|x + 2| < 5$.

**2.22.** Under what condition will the intersection of two intervals be an interval?

The intersection of two intervals will always be an interval or a singleton set $\{a\}$ or the empty set $\emptyset$. In other words, if we view

$$[a, a] = \{x : a \le x \le a\} = \{a\} \quad \text{and} \quad (a, a) = \{x : a < x < a\} = \emptyset$$

as intervals, then the intersection of any two intervals is always an interval.

**2.23.** Describe, plot, and write in interval notation each set: (a) $x > -1$, (b) $x \le 2$.

(a) All numbers greater than $-1$, and hence all numbers to the right of $-1$ as pictured in Fig. 2-9(a). The interval notation is $(-1, \infty)$ where the infinity symbol $\infty$ means all the numbers in the positive direction of $-1$.

(b) All numbers less than or equal to 2, and hence 2 and all numbers to the left of 2 as pictured in Fig. 2-9(b). The interval notation is $(-\infty, 2])$ where the minus infinity symbol $-\infty$ means all the numbers in the negative direction of 2.

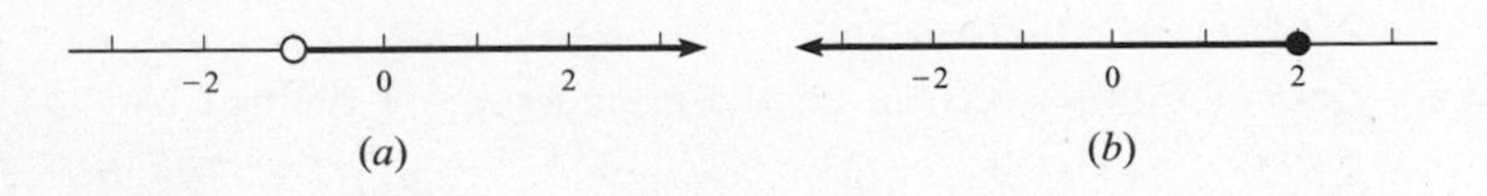


**Fig. 2-9**

**2.24.** Are the integers **Z** dense in **R**?

A set $A$ is dense in **R** if every open interval contains an element of $A$. Thus **Z** is not dense in **R** since, for example, the open interval $(1/3, 1/2)$ does not contain an integer.

## BOUNDED AND UNBOUNDED SETS

**2.25.** State whether each set of real numbers is bounded or unbounded:

(*a*) $A = \{x : x < 5\}$

(*b*) $B = \{\ldots -10, -5, 0, 5, 10, \ldots\}$

(*c*) $C = \{2^{18}, 3^{-7}, 7, 0, 8^{35}\}$

(*d*) $D = \{2, 4, 8, \ldots, 2^n, \ldots\}$

(*e*) $E = \{\frac{1}{2}, \frac{1}{4}, \frac{1}{8}, \ldots, (\frac{1}{2})^n, \ldots\}$

(*f*) $F = \{1, -1, \frac{1}{2}, -\frac{1}{2}, \frac{1}{3}, -\frac{1}{3}, \ldots\}$

(*a*) $A$ is unbounded. There are negative numbers whose absolute values are arbitrarily large.

(*b*) $B$ is unbounded.

(*c*) Although $C$ contains very large numbers, C is still bounded since $C$ is finite.

(*d*) $D$ is the set of powers of 2 and they are arbitrarily large. Thus $D$ is unbounded.

(*e*) $E$ is the set of positive powers of 1/2. Although $E$ is infinite, it is still bounded. In fact, $E$ is contained in the unit interval $I = [0, 1]$.

(*f*) Although $F$ is infinite it is still bounded. In fact, $F$ is contained in the interval $[-1, 1]$.

**2.26.** Which of the unbounded sets $A, B, D$ in Problem 2.25 are bounded from: (*a*) below, (*b*) above?

$A$ is bounded from above, $D$ is bounded from below, but $B$ has neither an upper nor a lower bound.

**2.27.** If two sets are bounded, what can be said about their union and intersection?

Both the union and intersection of bounded sets are bounded.

**2.28.** If two sets are unbounded, what can be said about their union and intersection?

The union of the sets must be unbounded, but the intersection could be either unbounded or bounded. For example, $A = (-\infty, 1]$ and $B = [-1, \infty)$ are unbounded, but $A \cap B = [-1, 1]$ is bounded. On the other hand, $C = [3, \infty)$ and $\mathbf{Z}$ are unbounded, and $C \cap \mathbf{Z} = \{3, 4, 5, \ldots\}$ is also unbounded.

## INTEGERS Z, MATHEMATICAL INDUCTION, WELL-ORDERING PRINCIPLE

The reader is referred to Section 1.11 where the principle of mathematical induction is stated and discussed.

**2.29.** Suppose $a \neq 1$. Let $A$ be the assertion on the integers $n \geq 1$ defined by

$$A(n): 1 + a + a^2 + a^3 + \cdots + a^n = \frac{a^{n+1} - 1}{a - 1}$$

Show that $A$ is true for all $n$.

$A(1)$ is true since

$$1 + a = \frac{a^2 - 1}{a - 1}$$

Assuming $A(n)$ is true, we add $a^{n+1}$ to both sides of $A(n)$, obtaining

$$\begin{aligned} 1 + a + a^2 + a^3 + \cdots + a^n + a^{n+1} &= \frac{a^{n+1} - 1}{a - 1} + a^{n+1} \\ &= \frac{a^{n+1} - 1 + (a-1)a^{n+1}}{a - 1} \\ &= \frac{a^{n+2} - 1}{a - 1} \end{aligned}$$

which is $A(n+1)$. Thus $A(n+1)$ is true whenever $A(n)$ is true. By the principle of induction, $A$ is true for all $n \in \mathbf{P}$.

**2.30.** Prove: If $n \in \mathbf{Z}$ and $n$ is a positive integer, then $n \geq 1$. (This is not true for the rational numbers $\mathbf{Q}$.) In other words, if $A(n)$ is the statement that $n \geq 1$, then $A(n)$ is true for every $n \in \mathbf{P}$.

**Method 1:** (Mathematical Induction)
$A(n)$ holds for $n = 1$ since $1 \geq 1$. Assuming $A(n)$ is true, that is, $n \geq 1$, add 1 to both sides to obtain

$$n + 1 \geq 2 > 1$$

which is $A(n+1)$. That is, $A(n+1)$ is true whenever $A(n)$ is true. By the principle of mathematical induction, $A$ is true for every $n \in \mathbf{P}$.

**Method 2:** (Well-Ordering Principle)
Suppose there does exist a positive integer less than 1. By the well-ordering principle, there exists a least positive integer $a$ such that

$$0 < a < 1$$

Multiplying the inequality by the positive integer $a$ we obtain

$$0 < a^2 < a$$

Therefore, $a^2$ is a positive integer less than $a$ which is also less than 1. This contradicts $a$'s property of being the least positive integer less than 1. Thus there exists no positive integer less than 1.

**2.31.** Suppose $a$ and $b$ are positive integers. Prove: (*a*) If $b \neq 1$, then $a < ab$. (*b*) If $ab = 1$, then $a = 1$ and $b = 1$. (*c*) If $n$ is composite, then $n = ab$ where $1 < a, b < n$.

(*a*) By Problem 2.30, $b > 1$. Hence $b - 1 > 0$, that is, $b - 1$ is positive. By the property $[\mathbf{P}_1]$ of the positive integers $\mathbf{P}$, the following product is also positive:

$$a(b-1) = ab - a$$

Thus $a < ab$, as required.

(*b*) Suppose $b \neq 1$. By (*a*), $a < ab = 1$. This contradicts Problem 2.30; hence $b = 1$. It then follows that $a = 1$.

(*c*) If $n$ is not prime, then $n$ has a positive divisor $a$ such that $a \neq 1$ and $a \neq n$. Then $n = ab$ where $b \neq 1$ and $b \neq n$. Thus, by Problem 2.30 and by part (*a*), $1 < a, b < ab = n$.

**2.32.** Prove Theorem 2.6 (Well-Ordering Principle): Let $S$ be a nonempty set of positive integers. Then $S$ contains a least element.

Suppose $S$ has no least element. Let $M$ consist of those positive integers which are less than every element of $S$. Then $1 \in M$; otherwise, $1 \in S$ and 1 would be a least element of $S$. Suppose $k \in M$. Then $k$ is less than every element of $S$. Therefore $k + 1 \in M$; otherwise $k + 1$ would be a least element of $S$.

By the principle of mathematical induction, $M$ contains every positive integer. Thus $S$ is empty. This contradicts the hypothesis that $S$ is nonempty. Accordingly, the original assumption that $S$ has no least element cannot be true. Thus the theorem is true.

**2.33.** Prove the Principle of Mathematical Induction II: Let $A(n)$ be an assertion defined on the integers $n \geq 1$ such that:

(i) $A(1)$ is true.

(ii) $A(n)$ is true whenever $P(k)$ is true for all $1 \leq k < n$.

Then $A$ is true for all $n \geq 1$.

Let $S$ be the set of integers $n \geq 1$ for which $A$ is not true. Suppose $S$ is not empty. By the well-ordering principle, $S$ contains a least element $s_0$. By (i), $s_0 \neq 1$.

Since $s_0$ is the least element of $S$, $A$ is true for every integer $k$ where $1 \leq k < s_0$. By (ii), $A$ is true for $s_0$. This contradicts the fact that $s_0 \in S$. Hence $S$ is empty, and so $A$ is true for every integer $n \geq 1$.

## DIVISION ALGORITHM

**2.34.** For each pair of integers $a$ and $b$, find integers $q$ and $r$ such that $a = bq + r$ and $0 \leq r < |b|$:

(*a*) $a = 258$ and $b = 12$ (*c*) $a = -381$ and $b = 14$

(*b*) $a = 573$ and $b = -16$ (*d*) $a = -433$ and $b = -17$

(*a*) Here $a$ and $b$ are positive. Simply divide $a$ by $b$, that is, 258 by 12, to obtain the quotient $q = 21$ and the remainder $r = 6$.

(*b*) Here $a$ is positive, but $b$ is negative. Divide $a$ by $|b|$, that is, 573 by 16, to obtain a quotient $q' = 35$ and remainder $r' = 13$. Then

$$573 = (16)(35) + 13 = 573 = (-16)(-35) + 13$$

That is, $q = -q' = -35$ and $r = r' = 13$.

(*c*) Here $a$ is negative and $b$ is positive. Thus we have to make some adjustments to be sure that $0 \leq r < |b|$. Divide $|a| = 381$ by $b = 14$ to obtain the quotient $q' = 27$ and remainder $r' = 14$. Therefore,

$$381 = (14)(27) + 3 \quad \text{and so} \quad -381 = (14)(-27) - 3$$

We add and subtract $b = 14$ as follows:

$$-381 = (14)(-27) - 14 + 14 - 3 = (14)(-28) + 11$$

Thus $q = -28$ and $r = 11$. Alternatively, $q = -(q' + 1) = -28$ and $r = b - r' = 11$.

(*d*) Divide $|a| = 433$ by $|b| = 17$ to obtain a quotient $q' = 25$ and $r' = 8$. Then

$$433 = (17)(25) + 8 \quad \text{and so} \quad -433 = (-17)(25) - 8$$

We add and subtract $|b| = 17$ as follows:

$$-433 = (-17)(25) - 17 + 17 - 8 = (-17)(26) + 9$$

Thus $q = 26$ and $r = 9$. Thus $q = q' + 1$ and $r = b - r'$.

**2.35.** Prove $\sqrt{2}$ is not rational, that is, $\sqrt{2} \neq a/b$ where $a$ and $b$ are integers.

Suppose $\sqrt{2}$ is rational and $\sqrt{2} = a/b$ where $a$ and $b$ are integers reduced to lowest terms, i.e. gcd $(a, b) = 1$. Squaring both sides yields

$$2 = \frac{a^2}{b^2} \quad \text{or} \quad a^2 = 2b^2$$

Then 2 divides $a^2$. Since 2 is a prime, $2|a$. Say $a = 2c$. Then

$$2b^2 = a^2 = 4c^2 \quad \text{or} \quad b^2 = 2c^2$$

Then 2 divides $b^2$. Since 2 is a prime, $2|b$. Thus 2 divides both $a$ and $b$. This contradicts the assumption that gcd $(a, b) = 1$. Therefore, $\sqrt{2}$ is not rational.

**2.36.** Prove Theorem 2.8 (Division Algorithm) for the case of positive integers. That is, assuming $a$ and $b$ are positive integers, prove that there exist nonnegative integers $q$ and $r$ such that

$$a = bq + r \quad \text{and} \quad 0 \le r < b \qquad (*)$$

If $a < b$, choose $q = 0$ and $r = a$; and if $a = b$, choose $q = 1$ and $r = 0$. In either case, $q$ and $r$ satisfy $(*)$.

The proof is now by induction on $a$. If $a = 1$ then $a < b$ or $a = b$; hence the theorem holds when $a = 1$. Suppose $a > b$. Then $a - b$ is positive and $a - b < a$. By induction, the theorem holds for $a - b$. Thus there exist $q'$ and $r'$ such that

$$a - b = bq' + r' \quad \text{and} \quad 0 \le r' < b$$

Then

$$a = bq' + b + r' = b(q' + 1) + r'$$

Choose $q = q' + 1$ and $r = r'$. Then $q$ and $r$ are nonnegative integers and satisfy $(*)$.

**2.37.** Prove Theorem 2.8 (Division Algorithm): Let $a$ and $b$ be integers with $b \ne 0$. Then there exists integers $q$ and $r$ such that

$$a = bq + r \quad \text{and} \quad 0 \le r < |b|$$

Also, the integers $q$ and $r$ are unique.

Let $M$ be the set of nonnegative integers of the form $a - xb$ for some integer $x$. If $x = -|a|b$, then $a - xb$ is nonnegative (Problem 2.63); hence $M$ is nonempty. By the well-ordering principle, $M$ has a least element, say $r$. Since $r \in M$, we have

$$r \ge 0 \quad \text{and} \quad r = a - qb$$

for some integer $q$. We need only show that $r < |b|$.

Suppose $r \ge |b|$. Let $r' = r - |b|$. Then $r' \ge 0$ and also $r' < r$ because $b \ne 0$. Furthermore,

$$r' = r - |b| = a - qb - |b| = \begin{cases} a - (q + 1)b, & \text{if } b < 0 \\ a - (q - 1)b, & \text{if } b > 0 \end{cases}$$

In either case, $r'$ belongs to $M$. This contradicts the fact that $r$ is the least element of $M$. Accordingly, $r < |b|$. Thus the existence of $q$ and $r$ is proved.

We now show that $q$ and $r$ are unique. Suppose there exist integers $q$ and $r$ and $q'$ and $r'$ such that

$$a = bq + r \quad \text{and} \quad a = bq' + r' \quad \text{and} \quad 0 \le r,\ r' < |b|$$

Then $bq + r = bq' + r'$; hence

$$b(q - q') = r' - r$$

Thus $b$ divides $r' - r$. But $|r' - r| < |b|$ since $0 \le r, r' < |b|$. Accordingly, $r' - r = 0$. This implies $q - q' = 0$ since $b \ne 0$. Consequently, $r' = r$ and $q' = q$; that is, $q$ and $r$ are uniquely determined by $a$ and $b$.

## DIVISIBILITY, PRIMES, GREATEST COMMON DIVISOR

**2.38.** Find all positive divisors of: (*a*) 18, (*b*) $256 = 2^8$, (*c*) $392 = 2^3 \cdot 7^2$.

(*a*) Since 18 is relatively small, we simply write down all positive integers ($\le 18$) which divide 18. These are

$$1, 2, 3, 6, 9, 18$$

(*b*) Since 2 is a prime, the positive divisors of $256 = 2^8$ are simply the lower powers of 2, that is,

$$2^0, 2^1, 2^2, 2^3, 2^4, 2^5, 2^6, 2^7, 2^8$$

In other words, the positive divisors of 256 are

$$1, 2, 4, 8, 16, 32, 64, 128, 256$$

(*c*) Since 2 and 7 are prime, the positive divisors of $392 = 2^3 \cdot 7^2$ are the products of lower powers of 2 times lower powers of 7, that is:

$$2^0 \cdot 7^0, 2^1 \cdot 7^0, 2^2 \cdot 7^0, 2^3 \cdot 7^0, 2^0 \cdot 7^1, 2^1 \cdot 7^1, 2^2 \cdot 7^1, 2^3 \cdot 7^1, 2^0 \cdot 7^2, 2^1 \cdot 7^2, 2^2 \cdot 7^2, 2^3 \cdot 7^2$$

In other words, the positive powers of 392 are:

$$1, 2, 4, 8, 7, 14, 28, 56, 49, 98, 196, 392$$

(We have used the usual convention that $n^0 = 1$ for any nonzero number $n$.)

**2.39.** List all primes between 50 and 100.

Simply list all numbers $p$ between 50 and 100 which cannot be written as a product of two positive integers, excluding 1 and $p$. This yields:

$$51, 53, 57, 59, 61, 67, 71, 73, 79, 83, 87, 89, 91, 93, 97$$

**2.40.** Let $a = 8316$ and $b = 10\,920$.

(*a*) Find $d = \gcd(a, b)$, the greatest common divisor of $a$ and $b$.

(*b*) Find integers $m$ and $n$ such that $d = ma + nb$.

(*c*) Find $\operatorname{lcm}(a, b)$, the least common multiple of $a$ and $b$.

(*a*) Divide the smaller number $a = 8316$ into the larger number $b = 10\,920$, and then repeatedly divide each remainder into the divisor until obtaining a zero remainder. These steps are pictured in Fig. 2-10. The last nonzero remainder is 84. Thus

$$84 = \gcd(8316, 10\,920)$$

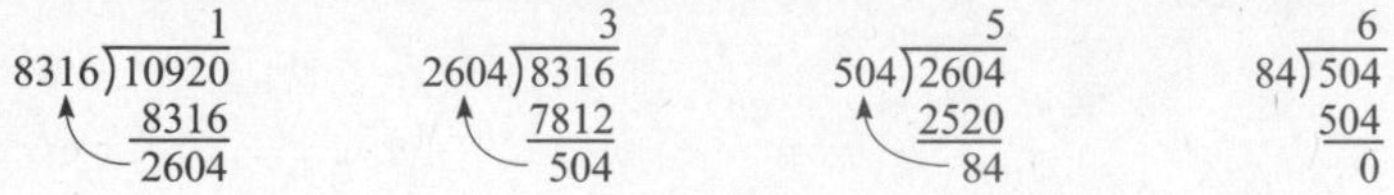

$$\begin{array}{r} 1 \\ 8316\overline{)10920} \\ \underline{8316} \\ 2604 \end{array} \qquad \begin{array}{r} 3 \\ 2604\overline{)8316} \\ \underline{7812} \\ 504 \end{array} \qquad \begin{array}{r} 5 \\ 504\overline{)2604} \\ \underline{2520} \\ 84 \end{array} \qquad \begin{array}{r} 6 \\ 84\overline{)504} \\ \underline{504} \\ 0 \end{array}$$

**Fig. 2-10**

(*b*) Now we find $m$ and $n$ such that

$$84 = 8316m + 10920n$$

The first three quotients in Fig. 2-10 yield the equations:

$$\begin{array}{llll} (1) & 10\,920 = 1(8316) + 2604; & \text{or} & 2604 = 10\,920 - 1(8316) \\ (2) & 8316 = 3(2604) + 504; & \text{or} & 504 = 8316 - 3(2604) \\ (3) & 2064 = 5(504) + 84; & \text{or} & 84 = 2604 - 5(504) \end{array}$$

Equation (*3*) tells us that 84 is a linear combination of 2604 and 504. We use (*2*) to replace 504 in (*3*) so we can write 84 as a linear combination of 2604 and 8316 as follows:

$$\begin{aligned} (4) \quad 84 &= 2604 - 5[8316 - 3(2604)] = 2604 - 5(8316) + 15(2604) \\ &= 16(2604) - 5(8316) \end{aligned}$$

We now use (*1*) to replace 2604 in (*4*) so we can write 84 as a linear combination of 8316 and 10 920 as follows:

$$\begin{aligned} 84 &= 16[10\,920 - 1(8316)] - 5(8316) \\ &= 16(10\,920) - 16(8316) - 5(8316) \\ &= -21(8316) + 16(10\,920) \end{aligned}$$

This is our desired linear combination. Thus $m = -21$ and $n = 16$.

(*c*) By Theorem 2.15,

$$\text{lcm}\ (a,b) = \frac{|ab|}{\text{gcd}\ (a,b)} = \frac{(8316)(10\,920)}{84} = 1\,081\,080$$

**2.41.** Suppose $a, b, c$ are integers. Prove:

(i) If $a|b$ and $b|c$, then $a|c$.

(ii) If $a|b$ then, for any integer $x, a|bx$.

(iii) If $a|b$ and $a|c$, then $a|(b+c)$ and $a|(b-c)$.

(iv) If $a|b$ and $b \neq 0$, then $a = \pm b$ or $|a| < |b|$.

(v) If $a|b$ and $b|a$, then $|a| = |b|$, i.e., $a = \pm b$.

(vi) If $a|1$, then $a = \pm 1$.

(i) If $a|b$ and $b|c$, then there exist integers $x$ and $y$ such that $ax = b$ and $by = c$. Replacing $b$ by $ax$, we obtain $axy = c$. Hence $a|c$.

(ii) If $a|b$, then there exists an integer $c$ such that $ac = b$. Multiplying the equation by $x$, we obtain $acx = bx$. Hence $a|bx$.

(iii) If $a|b$ and $a|c$, then there exist integers $x$ and $y$ such that $ax = b$ and $ay = c$. Adding the equalities, we obtain

$$ax + ay = b + c \qquad \text{and so} \qquad a(x+y) = b + c$$

Hence $a|(b+c)$. Subtracting the equalities, we obtain

$$ax - ay = b - c \qquad \text{and so} \qquad a(x-y) = b - c$$

Hence $a|(b-c)$.

(iv) If $a|b$, then there exists $c$ such that $ac = b$. Then

$$|b| = |ac| = |a||c|$$

By Problem 2.31(*b*), either $|c| = 1$ or $|a| < |a||c| = |b|$. If $|c| = 1$, then $c = \pm 1$; whence $a = \pm b$, as required.

(v) If $a|b$, then $a = \pm b$ or $|a| < |b|$. If $|a| < |b|$, $b \nmid a$. Hence $a = \pm b$.

(vi) If $a|1$, then $a = \pm 1$ or $|a| < |1| = 1$. By Problem 2.30, $|a| \geq 1$. Therefore, $a = \pm 1$.

**2.42.** A nonempty subset $J$ of $\mathbf{Z}$ is called an *ideal* if $J$ has the following two properties:

(1) If $a, b \in J$, then $a + b \in J$.

(2) If $a \in J$ and $n \in \mathbf{Z}$, then $na \in J$.

Let $d$ be the least positive integer in an ideal $J \neq \{0\}$. Prove that $d$ divides every element of $J$.

Since $J \neq \{0\}$, there exists $a \in J$ with $a \neq 0$. Then $-a = (-1)a \in J$. Thus $J$ contains positive elements. By the well-ordering principle, $J$ contains a least positive integer, so $d$ exists. Now let $b \in J$. Dividing $b$ by $d$, the division algorithm tells us there exist $q$ and $r$ such that

$$b = qd + r \qquad \text{and} \qquad 0 \leq r < d$$

Now $b$, $d \in J$, and $J$ is an ideal; hence $b + (-q)d = r$ also belongs to $J$. By the minimality of $d$, we must have $r = 0$. Hence $d|b$, as required.

**2.43.** Prove Theorem 2.12: Let $d$ be the smallest positive integer of the form $ax + by$. Then $d = \gcd(a, b)$.

Consider the set $J = \{ax + yb : x, y \in \mathbf{Z}\}$. Then

$$a = 1(a) + 0(b) \in J \quad \text{and} \quad b = 0(a) + 1(b) \in J$$

Also, suppose $s, t \in J$, say $s = x_1 a + y_1 b$ and $t = x_2 a + y_2 b$. Then, for any $n \in \mathbf{Z}$,

$$s + t = (x_1 + x_2)a + (y_1 + y_2)b \quad \text{and} \quad ns = (nx_1)a + (ny_1)b$$

also belong to $J$. Thus $J$ is an ideal. Let $d$ be the least positive element in $J$. We claim $d = \gcd(a, b)$.

By the preceding Problem 2.42, $d$ divides every element of $J$. Thus, in particular, $d$ divides $a$ and $b$. Now suppose $h$ divides both $a$ and $b$. Then $h$ divides $xa + yb$ for any $x$ and $y$; that is, $h$ divides every element of $J$. Thus $h$ divides $d$, and so $h < d$. Accordingly, $d = \gcd(a, b)$.

## FUNDAMENTAL THEOREM OF ARITHMETIC

**2.44.** Find the unique factorization of each number:

(*a*) 135, (*b*) 1330, (*c*) 3105, (*d*) 211.

(*a*) $135 = 5 \cdot 27 = 5 \cdot 3 \cdot 3 \cdot 3$ or $135 = 3^3 \cdot 5$.

(*b*) $1330 = 2 \cdot 665 = 2 \cdot 5 \cdot 133 = 2 \cdot 5 \cdot 7 \cdot 19$.

(*c*) $3105 = 5 \cdot 621 = 5 \cdot 3 \cdot 207 = 5 \cdot 3 \cdot 3 \cdot 69 = 5 \cdot 3 \cdot 3 \cdot 3 \cdot 23$ or $3105 = 3^3 \cdot 5 \cdot 23$.

(*d*) None of the primes 2, 3, 5, 7, 11 and 13 divides 211; hence 211 cannot be factored, that is, 211 is a prime.

**Remark**: We need only test those primes less than $\sqrt{211}$.

**2.45.** Let $a = 2^3 \cdot 3^5 \cdot 5^4 \cdot 11^6 \cdot 17^3$ and $b = 2^5 \cdot 5^3 \cdot 7^2 \cdot 114 \cdot 13^2$. Find $\gcd(a, b)$ and $\operatorname{lcm}(a, b)$.

Those primes $p_i$ which appear in both $a$ and $b$ will also appear in $\gcd(a, b)$. Furthermore, the exponent of $p_i$ in $\gcd(a, b)$ will be the smaller of its exponents in $a$ and $b$. Hence

$$\gcd(a, b) = 2^3 \cdot 5^3 \cdot 11^4$$

Those primes $p_i$ which appear in either $a$ or $b$ will also appear in $\operatorname{lcm}(a, b)$. Also, the exponent of $p_i$ in $\operatorname{lcm}(a, b)$ will be the larger of its exponents in $a$ and $b$. Hence

$$\operatorname{lcm}(a, b) = 2^5 \cdot 3^5 \cdot 5^4 \cdot 7^2 \cdot 11^6 \cdot 13^2 \cdot 17^3$$

**2.46.** Prove Theorem 2.16: Suppose $\gcd(a, b) = 1$, and $a$ and $b$ divide $c$. Then $ab$ divides $c$.

Since $\gcd(a, b) = 1$, there exist $x$ and $y$ such that $ax + by = 1$. Since $a|c$ and $b|c$, there exist $m$ and $n$ such that $c = ma$ and $c = nb$. Multiplying $ax + by = 1$ by $c$ yields

$$acx + bcy = c \quad \text{or} \quad a(nb)x + b(ma)y = c \quad \text{or} \quad ab(nx + my) = c$$

Thus $ab$ divides $c$.

**2.47.** Prove Corollary 2.18: Suppose a prime $p$ divides a product $ab$. Then $p|a$ or $p|b$.

Suppose $p$ does not divide $a$. Then $\gcd(p, a) = 1$ since the only divisors of $p$ are $\pm 1$ and $+p$. Thus there exist integers $m$ and $n$ such that $1 = mp + nq$. Multiplying by $b$ yields $b = mpb + nab$. By hypothesis, $p|ab$, say $ab = cp$. Then

$$b = mpb + nab = mpb + ncp = p(mb + nc)$$

Hence $p|b$, as required.

**2.48.** Prove: (*a*) Suppose $p|q$ where $p$ and $q$ are primes. Then $p = q$.
(*b*) Suppose $p|q_1 q_2 \cdots q_r$ where $p$ and the $q$'s are primes. Then $p$ is equal to one of the $q$'s.

(*a*) The only divisors of $q$ are $\pm 1$ and $\pm q$. Since $p > 1$, $p = q$.

(*b*) If $r = 1$, then $p = q_1$ by (*a*). Suppose $r > 1$. By Problem 2.47 (Corollary 2.18), $p|q_1$ or $p|(q_2 \cdots q_r)$. If $p|q_1$, then $p = q_1$ by (*a*). If not, then $p|(q_2 \cdots q_r)$. We repeat the argument. That is, we get $p = q_2$ or $p|(q_3 \cdots q_r)$. Finally (or by induction) $p$ must equal one of the $q$'s.

**2.49.** Prove the Fundamental Theorem of Arithmetic (Theorem 2.19): Every integer $n > 1$ can be expressed uniquely (except for order) as a product of primes.

We already proved Theorem 2.10 that such a product of primes exists. Hence we need only show that such a product is unique (except for order). Suppose

$$n = p_1 p_2 \cdots p_k = q_1 q_2 \cdots q_r$$

where the $p$'s and $q$'s are primes. Note that $p_1|(q_1 \cdots q_r)$. By the preceding Problem 2.48, $p_1$ equals one of the $q$'s. We arrange the $q$'s so that $p_1 = q_1$. Then

$$p_1 p_2 \cdots p_k = p_1 q_2 \cdots q_r \quad \text{and so} \quad p_2 \cdots p_k = q_2 \cdots q_r$$

By the same argument, we can rearrange the remaining $q$s so that $p_2 = q_2$. And so on. Thus $n$ can be expressed uniquely as a product of primes (except for order).

## Supplementary Problems

### REAL NUMBER SYSTEM R, SETS OF NUMBERS

**2.50.** Assuming $\mathbf{R}, \mathbf{Q}, \mathbf{Q}', \mathbf{Z}$, and $\mathbf{P}$ denote respectively, the real numbers, rational numbers, irrational numbers, integers, and positive integers, state whether each of the following is true or false.

(*a*) $\pi \in \mathbf{Q}$ (*c*) $-3 \in \mathbf{P}$ (*e*) $7 \in \mathbf{P}$ (*g*) $-6 \in \mathbf{Q}'$ (*i*) $7 \in \mathbf{Q}$

(*b*) $\sqrt{9} \in \mathbf{Q}$ (*d*) $\sqrt[3]{5} \in \mathbf{Q}$ (*f*) $\sqrt{-3} \in \mathbf{R}$ (*h*) $\sqrt{2} \in \mathbf{R}$ (*j*) $\frac{2}{3} \in \mathbf{Q}'$

**2.51.** State whether each is: (*a*) always true, (*b*) sometimes true, (*c*) never true. Here $a \neq 0$, $b \neq 0$.

(1) $a \in \mathbf{Z}$, $b \in \mathbf{Q}$, and $a - b \in \mathbf{P}$.
(2) $a \in \mathbf{Q}$, $b \in \mathbf{Q}'$, and $ab \in \mathbf{Q}'$.
(3) $a \in \mathbf{Q}'$, $b \in \mathbf{Q}'$, and $ab \in \mathbf{Q}'$.
(4) $a \in \mathbf{P}$, $b \in \mathbf{Q}$, and $ab \in \mathbf{P}$.
(5) $a \in \mathbf{Z}$, $b \in \mathbf{Q}$, and $a/b \in \mathbf{P}$.
(6) $a \in \mathbf{Z}$, $b \in \mathbf{Q}'$ and $a/b \in \mathbf{Q}$.
(7) $a \in \mathbf{P}$, $b \in \mathbf{R}$, and $a + b \in \mathbf{P}$.
(8) $a \in \mathbf{Z}$, $b \in \mathbf{Q}$, and $ab \in \mathbf{Q}'$.

**2.52.** Express each real number as an infinite decimal (without ending in zeros): (*a*) 5/6, (*b*) 3/11, (*c*) 3/5.

**2.53.** Consider the sets

$$A = \{2, 4, 8, \ldots, 2^n, \ldots\}, \qquad B = \{3, 6, 9, \ldots, 3n, \ldots\}, \qquad C = \{\ldots, -6, -3, 0, 3, 6, \ldots\}$$

Which of these sets are closed under the operations of:

(*a*) addition, (*b*) subtraction, (*c*) multiplication?

## ORDER AND INEQUALITIES, ABSOLUTE VALUE

**2.54.** Insert the correct symbol, <, >, or =, between each pair of integers:

(*a*) 2 ___ −6 (*c*) −7 ___ 3 (*e*) $2^3$ ___ 11 (*g*) −2 ___ −7
(*b*) −3 ___ −5 (*d*) −8 ___ −1 (*f*) $2^3$ ___ −9 (*h*) 4 ___ −9

**2.55.** Evaluate: (*a*) $|-6|$, $|5|$, $|0|$, (*b*) $|3-7|$, $|-3+7|$, $|-3-7|$.

**2.56.** Evaluate: (*a*) $|2-5|+|3+7|$, $|1-4|-|2-9|$;
(*b*) $|-4|+|2-3|$, $|-6-2|-|2-6|$.

**2.57.** Find the distance $d$ between each pair of real numbers:

(*a*) 2 and −5, (*b*) −6 and 3, (*c*) 2 and 8, (*d*) −7 and −1, (*e*) 3 and −3, (*f*) −7 and −9.

**2.58.** Find all integers $n$ such that: (*a*) $3 < 2n-4 < 10$, (*b*) $1 < 6-3n < 13$.

**2.59.** Prove Proposition 2.1: (i), $a \le a$, for any real number $a$.
(ii) If $a \le b$ and $b \le a$, then $a = b$

**2.60.** Prove Proposition 2.2: For any real numbers $a$ and $b$, exactly one of the following holds:

$$a < b, a = b, \text{ or } a > b.$$

**2.61.** Prove: (*a*) $2ab \le a^2 + b^2$, (*b*) $ab + ac + bc \le a^2 + b^2 + c^2$.

**2.62.** Prove Proposition 2.4: (i) $|a| \ge 0$, and $|a| = 0$ iff $a = 0$.
(ii) $-|a| \le a \le |a|$.
(v) $||a| - |b|| \le |a \pm b|$.

**2.63.** Show that $a - xb \ge 0$ if $b \ne 0$, and $x = -|a|b$.

## INTERVALS

**2.64.** Rewrite each interval in set-builder form:

(*a*) $A = [-1, 6)$, (*b*) $B = (2, 5)$, (*c*) $C = [-3, 0]$, (*d*) $D = (1, 4]$.

**2.65.** Which of the sets in Problem 2.64 is: (*a*) an open interval, (*b*) a closed interval?

**2.66.** Find the interval satisfying each inequality, i.e., rewrite the inequality in terms of $x$ alone:

(*a*) $1 \le x - 2 \le 4$, (*b*) $-3 \le x + 4 \le 7$, (*c*) $-6 \le 3x \le 12$, (*d*) $-4 \le -2x \le 6$.

**2.67.** Find the interval satisfying each inequality:

(*a*) $3 \le 2x - 5 \le 7$, (*b*) $-8 \le 4 - 3x \le 7$.

**2.68.** Rewrite without the absolute value sign:

(*a*) $|x| < 2$, (*b*) $|x - 3| < 5$, (*c*) $|2x - 5| < 9$.

**2.69.** Write each open interval in the form $|x - a| < r$:

(*a*) $3 < x < 9$, (*b*) $-5 < x < 1$.

**2.70.** Rewrite each set using an infinite interval notation:

(*a*) $x > -4$, (*b*) $x \leq 5$, (*c*) $x \geq 2$, (*d*) $x < -3$.

**2.71.** Let $A = [-4, 2)$, $B = (-1, 6)$, $C = (-\infty, 1]$. Find, and write in interval notation:

(*a*) $A \cup B$, (*c*) $A \backslash B$, (*e*) $A \cup C$, (*g*) $A \backslash C$, (*i*) $B \cup C$, (*k*) $B \backslash C$,

(*b*) $A \cap B$, (*d*) $B \backslash A$, (*f*) $A \cap C$, (*h*) $C \backslash A$, (*j*) $B \cap C$, (*l*) $C \backslash B$.

## BOUNDED AND UNBOUNDED SETS

**2.72.** State whether each set is bounded, bounded below, bounded above, unbounded:

(*a*) $A = \{x : x = \frac{1}{n}, n \in \mathbf{P}\}$, (*c*) $C = \{x : x = \frac{1}{2^n}, n \in \mathbf{Z}\}$, (*e*) $E = \{x : x = 2^n, n \in \mathbf{Z}\}$,

(*b*) $B = \{x : x = 3^n, n \in \mathbf{P}\}$, (*d*) $D = \{x : x \in \mathbf{P}, x < 2567\}$, (*f*) $F = \{x : |x| < 6\}$.

**2.73.** Are the following statements: (1) always true, (2) sometimes true, (3) never true?

(*a*) If $A$ is finite, then $A$ is bounded. (*c*) If $A$ is a subset of $[-23, 79]$, then $A$ is finite.

(*b*) If $A$ is infinite, then $A$ is bounded. (*d*) If $A$ is a subset of $[-23, 79]$, then $A$ is unbounded.

## INTEGERS Z, MATHEMATICAL INDUCTION, WELL-ORDERING PRINCIPLE

**2.74.** Prove the assertion $A$ that the sum of the first $n$ even positive integers is $n(n + 1)$; that is,

$$A(n)\colon\ 2 + 4 + 6 + \cdots + 2n = n(n + 1)$$

**2.75.** Prove: (*a*) $a^n a^m = a^{n+m}$, (*b*) $(a^n)^m = a^{nm}$, (*c*) $(ab)^n = a^n b^n$

**2.76.** Prove: $\dfrac{1}{1\cdot 2} + \dfrac{1}{2\cdot 3} + \dfrac{1}{3\cdot 4} + \cdots + \dfrac{1}{n(n+1)} = \dfrac{n}{n+1}$.

**2.77.** Prove: $|\mathscr{P}(A)| = 2^n$ where $|A| = n$. [Here $\mathscr{P}(A)$ is the power set of the finite set $A$ with $n$ elements.]

## DIVISION ALGORITHM

**2.78.** For each pair of integers $a$ and $b$, find integers $q$ and $r$ such that $a = bq + r$ and $0 \leq r < |b|$ :

(*a*) $a = 395$ and $b = 14$ (*c*) $a = -278$ and $b = 12$

(*b*) $a = 608$ and $b = -17$ (*d*) $a = -417$ and $b = -8$

**2.79.** Prove each of the following statements:

(*a*) The product of any three consecutive integers is divisible by 6.

(*b*) The product of any four consecutive integers is divisible by 24.

**2.80.** Show that each of the following numbers is not rational: (*a*) $\sqrt{3}$, (*b*) $\sqrt[3]{2}$.

**2.81.** Show that $\sqrt{p}$ is not rational, where $p$ is any prime number.

**DIVISIBILITY, GREATEST COMMON DIVISORS, PRIMES**

**2.82.** Find all possible divisors of: (*a*) 24, (*b*) $19\,683 = 3^9$, (*c*) $432 = 2^4 \cdot 3^3$.

**2.83.** List all prime numbers between 100 and 150.

**2.84.** For each pair of integers $a$ and $b$, find $d = \gcd(a, b)$ and express $d$ as a linear combination of $a$ and $b$:

(*a*) $a = 48$, $b = 356$ (*c*) $a = 2310$, $b = 168$
(*b*) $a = 165$, $b = 1287$ (*d*) $a = 195$, $b = 968$

**2.85.** Prove: (*a*) If $a|b$, then $a|-b$, $-a|b$, and $-a|-b$.
(*b*) If $ac|bc$, then $b|c$.

**2.86.** Prove: (*a*) If $am + bn = 1$, then $\gcd(a, b) = 1$
(*b*) If $a = bq + r$, then $\gcd(a, b) = \gcd(b, r)$.

**2.87.** Prove: (*a*) $\gcd(a, a + k)$ divides $k$.
(*b*) $\gcd(a, a + 2)$ equals 1 or 2.

**2.88.** Prove: If $n > 1$ is composite, then $n$ has a positive divisor $d$ such that $d \le \sqrt{n}$.

**FUNDAMENTAL THEOREM OF ARITHMETIC**

**2.89.** Express as a product of prime numbers:

(*a*) 2940, (*b*) 1485, (*c*) 8712, (*d*) 319410.

**2.90.** Suppose $a = 5880$ and $b = 8316$.

(*a*) Express $a$ and $b$ as products of primes.
(*b*) Find $\gcd(a, b)$ and $\operatorname{lcm}(a, b)$.
(*c*) Verify that $\operatorname{lcm}(a, b) = (|ab|)/\gcd(a, b)$.

**2.91.** Prove: If $a_1|n$ and $a_2|n, \ldots, a_k|n$, then $m|n$ where $m = \operatorname{lcm}(a_1, \ldots a_k)$.

**2.92.** Let $n$ be a positive integer. Prove:

(*a*) 3 divides $n$ if and only if 3 divides the sum of the digits of $n$.
(*b*) 9 divides $n$ if and only if 9 divides the sum of the digits of $n$.
(*c*) 8 divides $n$ if and only if 8 divides the integer formed by the last three digits of $n$.

## Answers to Supplementary Problems

**2.50.** Only (*b*), (*e*), (*h*), (*i*) are true.

**2.51.** (1) *b*; (2) *a*; (3) *b*; (4) *b*; (5) *b*; (6) *c*; (7) *b*; (8) *c*

**2.52.** (*a*) $5/6 = 0.8333\ldots$, (*b*) $3/11 = 0.2727\ldots$, (*c*) $3/5 = 0.5999\ldots$

**2.53.** (*a*) $B$ and $C$; (*b*) $C$; (*c*) $A, B, C$

**2.54.** (*a*) $2 > -6$; (*b*) $-3 > -5$; (*c*) $-7 < 3$; (*d*) $-8 < -1$; (*e*) $2^3 < 11$; (*f*) $2^3 > -9$; (*g*) $-2 > -7$; (*h*) $4 > -9$

**2.55.** (*a*) 6, 5, 0; (*b*) 4, 4, 10

**2.56.** (*a*) $3 + 10 = 13$, $3 - 7 = -4$; (*b*) $4 + 1 = 5$, $8 - 4 = 4$

**2.57.** (*a*) 7; (*b*) 9; (*c*) 6; (*d*) 6; (*e*) 6; (*f*) 3

**2.58.** (*a*) 4, 5, 6; (*b*) $-2, -1, 0, 1$

**2.64.** (*a*) $A = \{x : -1 \le x < 6\}$ (*c*) $C = \{x : -3 \le x \le 0\}$
(*b*) $B = \{x : 2 < x < 5\}$ (*d*) $D = \{x : 1 < x \le 4\}$

**2.65.** $B$ is open and $C$ is closed.

**2.66.** (*a*) $3 \le x \le 6$; (*b*) $-7 \le x \le 3$; (*c*) $-2 \le x \le 4$; (*d*) $-3 \le x \le 2$

**2.67.** (*a*) $4 \le x \le 6$; (*b*) $-1 \le x \le 4$

**2.68.** (*a*) $-2 < x < 2$; (*b*) $-2 < x < 8$; (*c*) $-2 < x < 7$

**2.69.** (*a*) $|x - 6| < 3$; (*b*) $|x + 2| < 3$

**2.70.** (*a*) $(-4, \infty)$; (*b*) $(-\infty, 5]$; (*c*) $[2, \infty)$; (*d*) $(-\infty, -3)$

**2.71.** (*a*) $[-4, 6)$, (*b*) $(-1, 2)$, (*c*) $[-4, -1]$, (*d*) $[2, 6)$, (*e*) $(-\infty, 2)$, (*f*) $[-4, -1]$, (*g*) $(1, 2)$, (*h*) $(-\infty, -4)$, (*i*) $(-\infty, 6)$, (*j*) $(-1, 1]$, (*k*) $(1, 6)$, (*l*) $(-\infty, -1]$.

**2.72.** (*a*) bounded; (*b*) only bounded below; (*c*) bounded; (*d*) bounded; (*e*) unbounded; (*f*) bounded

**2.73.** (*a*) 1; (*b*) 2; (*c*) 1; (*d*) 3

**2.74–2.77.** *Hint*: Use mathematical induction or well-ordering principle.

**2.78.** (*a*) $q = 28$, $r = 3$ (*b*) $q = -15$, $r = 13$ (*c*) $q = -24$, $r = 10$ (*d*) $q = 53$, $r = 7$

**2.79.** (*a*) One is divisible by 2 and one is divisible by 3.
(*b*) One is divisible by 4, another is divisible by 2, and one is divisible by 3.

**2.82.** (*a*) 1, 2, 3, 4, 6, 12, 24; (*b*) $3^n$ for $n = 0$ to 9; (*c*) $2^r 3^s$ for $r = 0$ to 4 and $s = 0$ to 3.

**2.83.** 101, 103, 107, 109, 113, 127, 131, 137, 139, 149

**2.84.** (*a*) $d = 4 = 5(356) - 37(48)$ (*c*) $d = 42 = 14(168) - 1(2310)$
(*b*) $d = 33 = 8(165) - 1(1287)$ (*d*) $d = 1 = 139(195) - 28(968)$

**2.89.** (*a*) $2940 = 2^2 \cdot 3 \cdot 5 \cdot 7^2$; (*b*) $1485 = 3^3 \cdot 5 \cdot 11$; (*c*) $8712 = 2^3 \cdot 3^2 \cdot 11^2$; (*d*) $319\,410 = 2 \cdot 3^3 \cdot 5 \cdot 7 \cdot 13^2$

**2.90.** (*a*) $a = 2^4 \cdot 3 \cdot 5 \cdot 7^2$, $b = 2^2 \cdot 3^3 \cdot 7 \cdot 11$; (*b*) $\gcd(a, b) = 2^2 \cdot 3 \cdot 7$, $\text{lcm}(a, b) = 2^4 \cdot 3^3 \cdot 5 \cdot 7^2 \cdot 11 = 1\,164\,240$

# Chapter 3

# Relations

## 3.1 INTRODUCTION

The reader is familiar with many relations which are used in mathematics and computer science, e.g., "less than", "is parallel to", "is a subset of", and so on. In a certain sense, these relations consider the existence or nonexistence of certain connections between pairs of objects taken in a definite order. Formally, we define a relation in terms of these "ordered pairs".

There are three kinds of relations which play a major role in our theory: (i) equivalence relations, (ii) order relations, (iii) functions. Equivalence relations are mainly covered in this chapter. Order relations are introduced here, but will also be discussed in Chapter 7. Functions are covered in the next chapter.

The connection between relations on finite sets and matrices are also included here for completeness. These sections, however, can be ignored at a first reading by those with no previous knowledge of matrix theory.

### Ordered Pairs

Relations, as noted above, will be defined in terms of ordered pairs $(a, b)$ of elements, where $a$ is designated as the first element and $b$ as the second element. Specifically:

$$(a, b) = (c, d) \text{ if and only if } a = c \text{ and } b = d$$

In particular, $(a, b) \neq (b, a)$ unless $a = b$. This contrasts with sets studied in Chapter 1 where the order of elements is irrelevant, for example, $\{3, 5\} = \{5, 3\}$.

## 3.2 PRODUCT SETS

Let $A$ and $B$ be two sets. The *product set* or *cartesian product* of $A$ and $B$, written $A \times B$ and read "$A$ cross $B$", is the set of all ordered pairs $(a, b)$ such that $a \in A$ and $b \in B$. Namely:

$$A \times B = \{(a, b) : a \in A, \ b \in B\}$$

One usually writes $A^2$ instead of $A \times A$.

**EXAMPLE 3.1** Recall that $\mathbf{R}$ denotes the set of real numbers, so $\mathbf{R}^2 = \mathbf{R} \times \mathbf{R}$ is the set of ordered pairs of real numbers. The reader may be familiar with the geometrical representation of $\mathbf{R}^2$ as points in the plane as in Fig. 3-1. Here each point $P$ represents an ordered pair $(a, b)$ of real numbers and vice versa; the vertical line through $P$ meets the (horizontal) $x$-axis at $a$, and the horizontal line through $P$ meets the (vertical) $y$-axis at $b$. $\mathbf{R}^2$ is frequently called the *cartesian* plane.

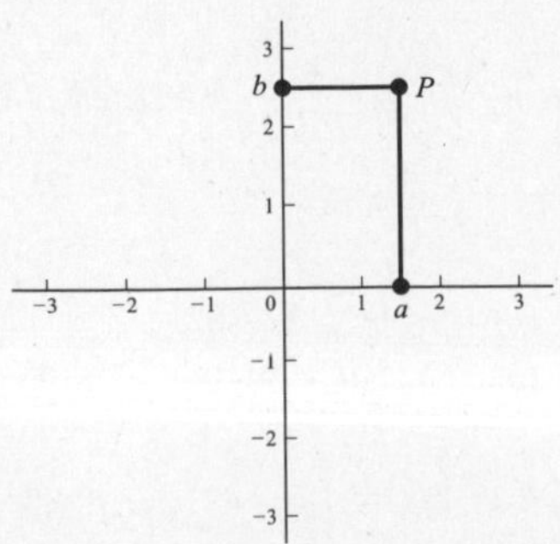


**Fig. 3-1**

**EXAMPLE 3.2** Let $A = \{1, 2\}$ and $B = \{a, b, c\}$. Then

$$A \times B = \{(1, a), (1, b), (1, c), (2, a), (2, b), (2, c)\}$$
$$B \times A = \{(a, 1), (a, 2), (b, 1), (b, 2), (c, 1), (c, 2)\}$$

Also,

$$A \times A = \{(1, 1)(1, 2), (2, 1), (2, 2)\}$$

There are two things worth noting in Example 3.2. First of all, $A \times B \neq B \times A$. The cartesian product deals with ordered pairs, so naturally the order in which the sets are considered is important. Secondly,

$$n(A \times B) = 6 = 2 \cdot 3 = n(A) \cdot n(B)$$

[where $n(A)$ = number of elements in $A$]. In fact:

$$\boxed{n(A \times B) = n(A) \cdot n(B)}$$

for any finite sets $A$ and $B$. This follows from the observation that, for any ordered pair $(a, b)$ in $A \times B$, there are $n(A)$ possibilities for $a$, and for each of these there are $n(B)$ possibilities for $b$.

### Product of Three or More Sets

The idea of a product of sets can be extended to any finite number of sets. Specifically, for any sets $A_1, A_2, \ldots, A_m$, the set of all $m$-element lists $(a_1, a_2, \ldots, a_m)$, where each $a_i \in A_i$, is called the (*cartesian*) *product* of the sets $A_1, A_2, \ldots, A_m$; it is denoted by

$$A_1 \times A_2 \times \cdots \times A_m \quad \text{or equivalently} \quad \prod_{i=1}^{m} A_i$$

Just as we write $A^2$ instead of $A \times A$, so we write $A^n$ for $A \times A \times \cdots \times A$ where there are $n$ factors. For example, $\mathbf{R}^3 = \mathbf{R} \times \mathbf{R} \times \mathbf{R}$ denotes the usual three-dimensional space.

## 3.3 RELATIONS

We begin with a definition.

**Definition:** Let $A$ and $B$ be sets. A *binary relation* or, simply, a *relation* from $A$ to $B$ is a subset of $A \times B$.

Suppose $R$ is a relation from $A$ to $B$. Then $R$ is a set of ordered pairs where each first element comes from $A$ and each second element comes from $B$. That is, for each pair $a \in A$ and $b \in B$, exactly one of the following is true:

(i) $(a, b) \in R$; we then say "$a$ is *R-related* to $b$", written $a\,R\,b$.

(ii) $(a, b) \notin R$; we then say "$a$ is not *R-related* to $b$", written $a \not R\, b$.

The *domain* of a relation $R$ from $A$ to $B$ is the set of all first elements of the ordered pairs which belong to $R$, and so it is a subset of $A$; and the *range* of $R$ is the set of all second elements, and so it is a subset of $B$.

Sometimes $R$ is a relation from a set $A$ to itself, that is, $R$ is a subset of $A^2 = A \times A$. In such a case, we say that $R$ is a relation *on* $A$.

Although $n$-ary relations, which involve ordered $n$-tuples, are introduced in Section 3.11, the term relation shall mean binary relation unless otherwise stated or implied.

**EXAMPLE 3.3**

(*a*) Let $A = \{1, 2, 3\}$ and $B = \{x, y, z\}$, and let $R = \{(1, y), (1, z), (3, y)\}$. Then $R$ is a relation from $A$ to $B$ since $R$ is a subset of $A \times B$. With respect to this relation,

$$1Ry, \ 1Rz, \ 3Ry, \qquad \text{but} \qquad 1\not{R}x, \ 2\not{R}x, \ 2\not{R}y, \ 2\not{R}z, \ 3\not{R}x, \ 3\not{R}z$$

The domain of $R$ is $\{1, 3\}$ and the range is $\{y, z\}$.

(*b*) Suppose we say that two countries are *adjacent* if they have some part of their boundaries in common. Then "is adjacent to" is a relation $R$ on the countries of the earth. Thus:

$$(\text{Italy, Switzerland}) \in R \qquad \text{but} \qquad (\text{Canada, Mexico}) \notin R$$

(*c*) Set inclusion $\subseteq$ is a relation on any collection of sets. For, given any pair of sets $A$ and $B$, either $A \subseteq B$ or $A \not\subseteq B$.

(*d*) A familiar relation on the set **Z** of integers is "$m$ divides $n$". A common notation for this relation is to write $m|n$ when $m$ divides $n$. Thus $6|30$ but $7 \nmid 25$.

(*e*) Consider the set $L$ of lines in the plane. Perpendicularity, written $\perp$, is a relation on $L$. That is, given any pair of lines $a$ and $b$, either $a \perp b$ or $a \not\perp b$. Similarly, "is parallel to", written $\|$, is a relation on $L$ since either $a \| b$ or $a \nparallel b$.

### Universal, Empty, Equality Relations

Let $A$ be any set. Then $A \times A$ and $\varnothing$ are subsets of $A \times A$ and hence are relations on $A$ called the *universal relation* and *empty relation*, respectively. Thus, for any relation $R$ on $A$, we have

$$\varnothing \subseteq R \subseteq A \times A$$

An important relation on the set $A$ is that of *equality*, that is, the relation

$$\{(a, a) : a \in A\}$$

which is usually denoted by "=". This relation is also called the *identity* or *diagonal relation* on $A$, and it may sometimes be denoted by $\Delta_A$ or simply $\Delta$.

### Inverse Relation

Let $R$ be any relation from a set $A$ to a set $B$. The *inverse* of $R$, denoted by $R^{-1}$, is the relation from $B$ to $A$ which consists of those ordered pairs which, when reversed, belong to $R$; that is,

$$R^{-1} = \{(b, a) : (a, b) \in R\}$$

For example:

$$\text{If} \quad R = \{(1, y), (1, z), (3, y)\}, \qquad \text{then} \qquad R^{-1} = \{(y, 1), (z, 1), (y, 3)\}.$$

[Here $R$ is the relation from $A = \{1, 2, 3\}$ to $B = \{x, y, z\}$ in Example 3.3(*a*).]

Clearly, if $R$ is any relation, then $(R^{-1})^{-1} = R$. Also, the domain of $R^{-1}$ is the range of $R$, and vice versa. Moreover, if $R$ is a relation on $A$, i.e., $R$ is a subset of $A \times A$, then $R^{-1}$ is also a relation on $A$.

## 3.4 PICTORIAL REPRESENTATIONS OF RELATIONS

This section discusses a number of ways of picturing and representing binary relations.

### Relations on R

Let $S$ be a relation on the set $\mathbf{R}$ of real numbers; that is, let $S$ be a subset of $\mathbf{R}^2 = \mathbf{R} \times \mathbf{R}$. Since $\mathbf{R}^2$ can be represented by the set of points in the plane, we can picture $S$ by emphasizing those points in the plane which belong to $S$. This pictorial representation of $S$ is sometimes called the *graph* of $S$.

Frequently, the relation $S$ consists of all ordered pairs of real numbers which satisfy some given equation

$$E(x, y) = 0$$

We usually identify the relation with the equation, i.e., we speak of the relation $E(x, y) = 0$.

**EXAMPLE 3.4** Consider the relation $S$ defined by the equation

$$x^2 + y^2 = 25 \qquad \text{or equivalently} \qquad x^2 + y^2 - 25 = 0$$

That is, $S$ consists of all ordered pairs $(x_0, y_0)$ which satisfy the given equation. The graph of the equation is a circle having its center at the origin and radius 5, as shown in Fig. 3-2.

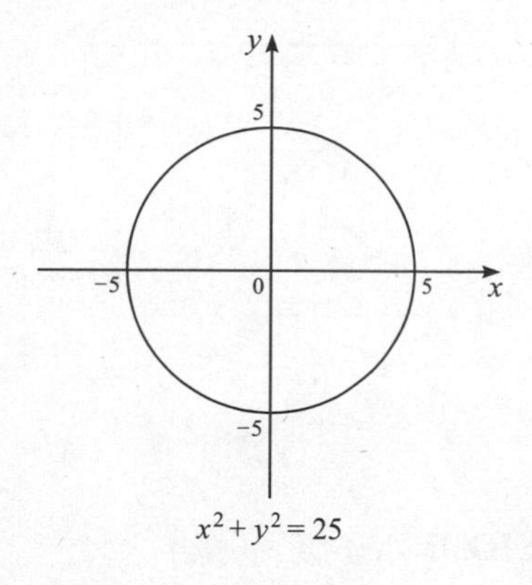


**Fig. 3-2**

### Representation of Relations on Finite Sets

Suppose $A$ and $B$ are finite sets. The following are two ways of picturing a relation $R$ from $A$ to $B$.

(i) Form a rectangular array whose rows are labeled by the elements of $A$ and whose columns are labeled by the elements of $B$. Put a 1 or 0 in each position of the array according as $a \in A$ is or is not related to $b \in B$. This array is called the *matrix* of the relation.

(ii) Write down the elements of $A$ and the elements of $B$ in two disjoint disks, and then draw an arrow from $a \in A$ to $b \in B$ whenever $a$ is related to $b$. This picture will be called the *arrow diagram* of the relation.

Consider, for example, the following relation $R$ from $A = \{1, 2, 3\}$ to $B = \{x, y, z\}$:

$$R = \{(1, y),\ (1, z), (3, y)\}$$

Figure 3-3 pictures this relation $R$ by the above two ways.

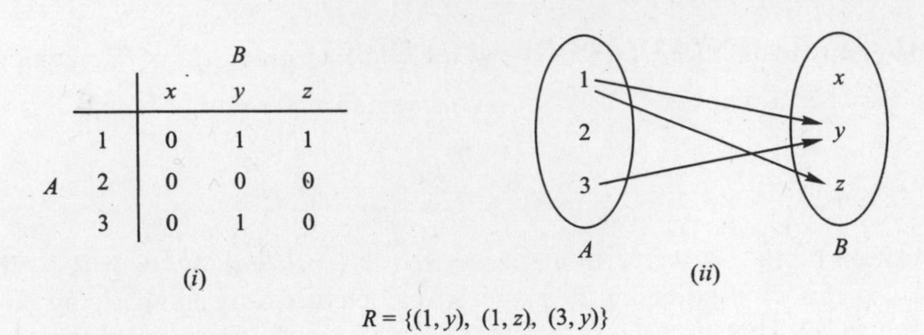


$R = \{(1, y),\ (1, z),\ (3, y)\}$

**Fig. 3-3**

### Directed Graphs of Relations on Sets

There is another way of picturing a relation $R$ when $R$ is a relation from a finite set $A$ to itself. First we write down the elements of the set $A$, and then we draw an arrow from each element $x$ to each element $y$ whenever $x$ is related to $y$. This diagram is called the *directed graph* of the relation $R$. Figure 3-4, for example, shows the directed graph of the following relation $R$ on the set $A = \{1, 2, 3, 4\}$:

$$R = \{(1,2),\ (2,2),\ (2,4),\ (3,2),\ (3,4),\ (4,1),\ (4,3)\}$$

Observe that there is an arrow from 2 to itself, since 2 is related to 2 under $R$.

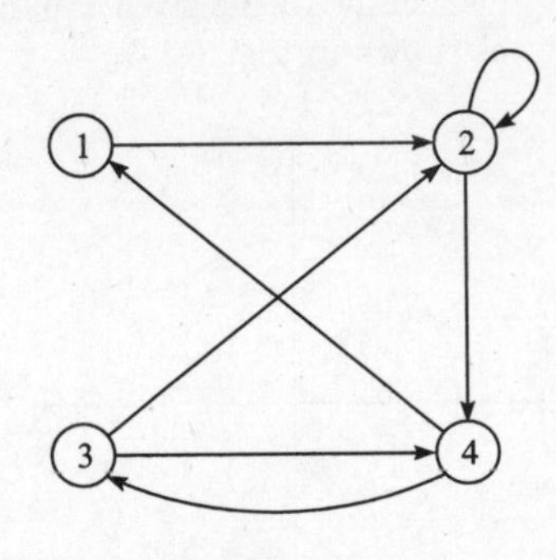


**Fig. 3-4**

## 3.5 COMPOSITION OF RELATIONS

Let $A$, $B$, $C$ be sets, and let $R$ be a relation from $A$ to $B$ and let $S$ be a relation from $B$ to $C$. Then $R$ and $S$ give rise to a relation from $A$ to $C$ denoted by $R \circ S$ and defined as follows:

$$R \circ S = \{(a, c) : \text{there exists } b \in B \text{ for which } (a, b) \in R \text{ and } (b, c) \in S\}$$

That is,

$$a(R \circ S)c \text{ whenever there exists } b \in B \text{ such that } a\,R\,b \text{ and } b\,S\,c$$

This relation $R \circ S$ is called the *composition* of $R$ and $S$; it is sometimes denoted by $RS$.

Our first theorem (proved in Problem 3.10) tells us that the composition of relations is associative. Namely:

**Theorem 3.1**: Let $A, B, C, D$ be sets. Suppose $R$ is a relation from $A$ to $B$, $S$ is a relation from $B$ to $C$, and $T$ is a relation from $C$ to $D$. Then

$$(R \circ S) \circ T = R \circ (S \circ T)$$

The arrow diagrams of relations give us a geometrical interpretation of the composition $R \circ S$ as seen in the following example.

**EXAMPLE 3.5** Let $A = \{1, 2, 3, 4\}$, $B = \{a, b, c, d\}$, $C = \{x, y, z\}$ and let

$$R = \{(1, a),\ (2, d),\ (3, a);\ (3, b),\ (3, d)\} \qquad \text{and} \qquad S = \{(b, x),\ (b, z),\ (c, y),\ (d, z)\}$$

Consider the arrow diagrams of $R$ and $S$ as in Fig. 3-5. Observe there is an arrow from 2 to $d$ which is followed by an arrow from $d$ to $x$. We can view these two arrows as a "path" which "connects" the element $2 \in A$ to the element $z \in C$. Thus

$$2(R \circ S)z \qquad \text{since} \qquad 2Rd \text{ and } dSz$$

Similarly there are paths from 3 to $x$ and from 3 to $z$. Hence

$$3(R \circ S)x \qquad \text{and} \qquad 3(R \circ S)z$$

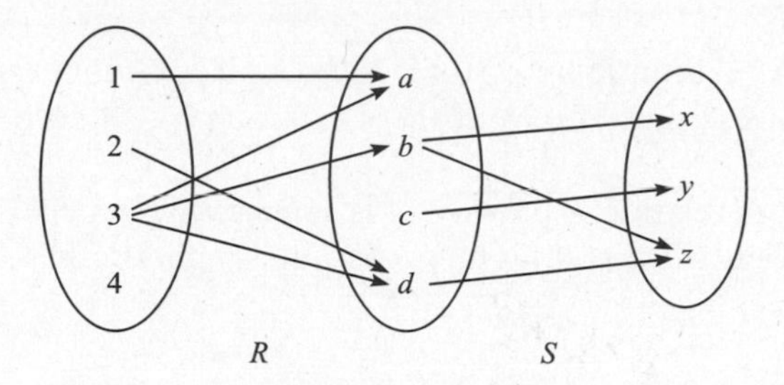


**Fig. 3-5**

No other element of $A$ is connected to an element of $C$. Accordingly,

$$R \circ S = \{(2, z),\ (3, x),\ (3, z)\}$$

Suppose $R$ is a relation on a set $A$, that is, $R$ is a relation from a set $A$ to itself. Then $R \circ R$, the composition of $R$ with itself, is always defined, and $R \circ R$ is sometimes denoted by $R^2$. Similarly, $R^3 = R^2 \circ R = R \circ R \circ R$, and so on. Thus $R^n$ is defined for all positive $n$.

**Warning**: Many texts denote the composition of relations $R$ and $S$ by $S \circ R$ rather than $R \circ S$. This is done in order to conform with the usual use of $g \circ f$ to denote the composition of $f$ and $g$ where $f$ and $g$ are functions. Thus the reader may have to adjust his notation when using this text as a supplement with another text. However, when a relation $R$ is composed with itself, then the meaning of $R \circ R$ is unambiguous.

### Composition of Relations and Matrices

There is a way of finding the composition $R \circ S$ of relations using matrices. Specifically, let $M_R$ and $M_S$ denote respectively the matrices of the relations $R$ and $S$ in Example 3.5. Then:

$$M_R = \begin{array}{c} \\ 1 \\ 2 \\ 3 \\ 4 \end{array} \begin{array}{c} \begin{array}{cccc} a & b & c & d \end{array} \\ \begin{pmatrix} 1 & 0 & 0 & 0 \\ 0 & 0 & 0 & 1 \\ 1 & 1 & 0 & 1 \\ 0 & 0 & 0 & 0 \end{pmatrix} \end{array} \qquad \text{and} \qquad M_S = \begin{array}{c} \\ a \\ b \\ c \\ d \end{array} \begin{array}{c} \begin{array}{ccc} x & y & z \end{array} \\ \begin{pmatrix} 0 & 0 & 0 \\ 1 & 0 & 1 \\ 0 & 1 & 0 \\ 0 & 0 & 1 \end{pmatrix} \end{array}$$

Multiplying $M_R$ and $M_S$ we obtain the matrix

$$M = M_R M_S = \begin{matrix} \\ 1 \\ 2 \\ 3 \\ 4 \end{matrix} \begin{matrix} x & y & z \\ \left( \begin{matrix} 0 & 0 & 0 \\ 0 & 0 & 1 \\ 1 & 0 & 2 \\ 0 & 0 & 0 \end{matrix} \right) \end{matrix}$$

The nonzero entries in this matrix tell us which elements are related by $R \circ S$. Thus $M = M_R M_S$ and $M_{R \circ S}$ have the same nonzero entries.

## 3.6 TYPES OF RELATIONS

Consider a given set $A$. This section discusses a number of important types of relations which are defined on $A$.

(1) **Reflexive Relations**: A relation $R$ on a set $A$ is *reflexive* if $a\,R\,a$ for every $a \in A$, that is, if $(a, a) \in R$ for every $a \in A$. Thus $R$ is not reflexive if there exists an $a \in A$ such that $(a, a) \notin R$.

(2) **Symmetric Relations**: A relation $R$ on a set $A$ is *symmetric* if whenever $a\,R\,b$ then $b\,R\,a$, that is, if whenever $(a, b) \in R$, then $(b, a) \in R$. Thus $R$ is not symmetric if there exists $a, b \in A$ such that $(a, b) \in R$ but $(b, a) \notin R$.

(3) **Antisymmetric Relations**: A relation $R$ on a set $A$ is *antisymmetric* if whenever $a\,R\,b$ and $b\,R\,a$ then $a = b$, that is, if whenever $(a, b)$ and $(b, a)$ belong to $R$ then $a = b$. Thus $R$ is not antisymmetric if there exist $a, b \in A$ such that $(a, b)$ and $(b, a)$ belong to $R$, but $a \neq b$.

(4) **Transitive Relations**: A relation $R$ on a set $A$ is *transitive* if whenever $a\,R\,b$ and $b\,R\,c$ then $a\,R\,c$, that is, if whenever $(a, b), (b, c) \in R$ then $(a, c) \in R$. Thus $R$ is not transitive if there exist $a, b, c \in A$ such that $(a, b), (b, c) \in R$, but $(a, c) \notin R$.

**EXAMPLE 3.6** Consider the following five relations on the set $A = \{1, 2, 3, 4\}$:

$$\begin{aligned}
R_1 &= \{(1,1),\ (1,2),\ (2,3),\ (1,3),\ (4,4)\} \\
R_2 &= \{(1,1),\ (1,2),\ (2,1)\,(2,2),\ (3,3), (4,4)\} \\
R_3 &= \{(1,3),\ (2,1)\} \\
R_4 &= \varnothing, \text{ the empty relation} \\
R_5 &= A \times A, \text{ the universal relation}
\end{aligned}$$

Determine which of the relations are: (*a*) reflexive, (*b*) symmetric, (*c*) antisymmetric, (*d*) transitive.

(*a*) Since $A$ contains the four elements 1, 2, 3, 4, a relation $R$ on $A$ is reflexive if it contains the four pairs $(1, 1)$, $(2, 2)$, $(3, 3)$, and $(4, 4)$. Thus only $R_2$ and the universal relation $R_5 = A \times A$ are reflexive. Note that $R_1$, $R_3$, and $R_4$ are not reflexive since, for example, $(2, 2)$ does not belong to any of them.

(*b*) $R_1$ is not symmetric since $(1, 2) \in R_1$ but $(2, 1) \notin R_1$. $R_3$ is not symmetric since $(1, 3) \in R_3$ but $(3, 1) \notin R_3$. The other relations are symmetric.

(*c*) $R_2$ is not antisymmetric since $(1, 2)$ and $(2, 1)$ belong to $R_2$, but $1 \neq 2$. Similarly, the universal relation $R_5$ is not antisymmetric. All the other relations are antisymmetric.

(*d*) The relation $R_3$ is not transitive since $(2, 1), (1, 3) \in R_3$ but $(2, 3) \notin R_3$. All the other relations are transitive.

**EXAMPLE 3.7** Consider the following five relations:

(1) Relation $\leq$ (less than or equal) on the set $\mathbf{Z}$ of integers.
(2) Set inclusion $\subseteq$ on a collection $\mathscr{C}$ of sets.
(3) Relation $\perp$ (perpendicular) on the set $L$ of lines in the plane.
(4) Relation $\parallel$ (parallel) on the set $L$ of lines in the plane.
(5) Relation $|$ of divisibility on the set $\mathbf{P}$ of positive integers. (Recall that $x|y$ if there exists $z$ such that $xz = y$.)

Determine which of the relations are: (*a*) reflexive, (*b*) symmetric, (*c*) antisymmetric, (*d*) transitive.

(*a*) The relation (3) is not reflexive since no line is perpendicular to itself. Also, (4) is not reflexive since no line is parallel to itself. The other relations are reflexive; that is, $x \leq x$ for every integer $x$ in $\mathbf{Z}$, $A \subseteq A$ for any set $A$ in $\mathscr{C}$, and $n|n$ for every positive integer $n$ in $\mathbf{P}$.

(*b*) The relation $\perp$ is symmetric since if line $a$ is perpendicular to line $b$ then $b$ is perpendicular to $a$. Also, $\parallel$ is symmetric since if line $a$ is parallel to line $b$ then $b$ is parallel to $a$. The other relations are not symmetric. For example, $3 \leq 4$ but $4 \leq 3$; $\{1,2\} \subseteq \{1,2,3\}$ but $\{1,2,3\} \not\subseteq \{1,2\}$; and $2|6$ but $6|2$.

(*c*) The relation $\leq$ is antisymmetric since whenever $a \leq b$ and $b \leq a$ then $a = b$. Set inclusion $\subseteq$ is antisymmetric since whenever $A \subseteq B$ and $B \subseteq A$ then $A = B$. Also, divisibility on $\mathbf{P}$ is antisymmetric since whenever $m|n$ and $n|m$ then $m = n$. (Note that divisibility on $\mathbf{Z}$ is not antisymmetric since $3|-3$ and $-3|3$ but $3 \neq -3$.) The relation $\perp$ is not antisymmetric since we can have distinct lines $a$ and $b$ such that $a \perp b$ and $b \perp a$. Similarly, $\parallel$ is not antisymmetric.

(*d*) The relations $\leq$, $\subseteq$ and $|$ are transitive. That is:

(i) If $a \leq b$ and $b \leq c$, then $a \leq c$.
(ii) If $A \subseteq B$ and $B \subseteq C$, then $A \subseteq C$.
(iii) If $a|b$ and $b|c$, then $a|c$.

On the other hand, the relation $\perp$ is not transitive. If $a \perp b$ and $b \perp c$, then it is not true that $a \perp c$. Since no line is parallel to itself, we can have $a \parallel b$ and $b \parallel a$, but $a \nparallel a$. Thus $\parallel$ is not transitive. (We note that the relation "is parallel or equal to" is a transitive relation on the set $L$ of lines in the plane.)

**Remark 1**: The properties of being symmetric and antisymmetric are not negatives of each other. For example, the relation $R = \{(1,3), (3,1), (2,3)\}$ is neither symmetric nor antisymmetric. On the other hand, the relation $R' = \{(1,1), (2,2)\}$ is both symmetric and antisymmetric.

**Remark 2**: The property of transitivity can also be expressed in terms of the composition of relations. Recall that, for a relation $R$ on a set $A$, we defined

$$R^2 = R \circ R \qquad \text{and, more generally,} \qquad R^n = R^{n-1} \circ R$$

Then one can show (Problem 3.66) that a relation $R$ is transitive if and only if $R^n \subseteq R$ for every $n \geq 1$.

## 3.7 CLOSURE PROPERTIES

Let $\mathscr{P}$ denote a property of relations on a set $A$ such as being symmetric or transitive. A relation on $A$ with property $\mathscr{P}$ will be called a *$\mathscr{P}$-relation*.

Now let $R$ be a given relation on $A$ with or without property $\mathscr{P}$. The *$\mathscr{P}$-closure* of $R$, written $\mathscr{P}(R)$, is a relation on $A$ containing $R$ such that

$$R \subseteq \mathscr{P}(R) \subseteq S$$

for any other $\mathscr{P}$-relation $S$ containing $R$. Clearly $R = \mathscr{P}(R)$ if $R$ itself has property $\mathscr{P}$.

The reflexive, symmetric, and transitive closures of a relation $R$ will be denoted respectively by:

$$\text{reflexive}(R), \qquad \text{symmetric}(R), \qquad \text{transitive}(R)$$

### Reflexive and Symmetric Closures

The next theorem tells us how to easily obtain the reflexive and symmetric closures of a relation. Here $\Delta_A = \{(a, a): a \in A\}$ is the *diagonal* or *equality* relation on $A$.

**Theorem 3.2**: Let $R$ be a relation on a set $A$. Then:

(i) $R \cup \Delta_A$ is the reflexive closure of $R$.

(ii) $R \cup R^{-1}$ is the symmetric closure of $R$.

In other words, reflexive($R$) is obtained by simply adding to $R$ those elements $(a, a)$ in the diagonal which do not already belong to $R$, and symmetric($R$) is obtained by adding to $R$ all pairs $(b, a)$ whenever $(a, b)$ belongs to $R$.

**EXAMPLE 3.8** Consider the following relation $R$ on the set $A = \{1, 2, 3, 4\}$:

$$R = \{(1,1),\ (1,3),\ (2,4),\ (3,1),\ (3,3),\ (4,3)\}$$

Then

$$\begin{aligned}\text{reflexive}(R) &= R \cup \{(2,2),\ (4,4)\} \\ &= \{(1,1),\ (1,3),\ (2,4),\ (3,1),\ (3,3),\ (4,3),\ (2,2),\ (4,4)\}\end{aligned}$$

and

$$\begin{aligned}\text{symmetric}(R) &= R \cup \{(4,2),\ (3,4)\} \\ &= \{(1,1),\ (1,3),\ (2,4),\ (3,1),\ (3,3),\ (4,3),\ (4,2),\ (3,4)\}\end{aligned}$$

### Transitive Closure

Let $R$ be a relation on a set $A$. Recall that $R^2 = R \circ R$ and $R^n = R^{n-1} \circ R$. We define

$$R^* = \bigcup_{i=1}^{\infty} R^i$$

The following theorem applies.

**Theorem 3.3**: $R^*$ is the transitive closure of a relation $R$.

Suppose $A$ is a finite set with $n$ elements. Using graph theory, one can easily show that

$$R^* = R \cup R^2 \cup \cdots \cup R^n$$

This gives us the following result.

**Theorem 3.4**: Let $R$ be a relation on a set $A$ with $n$ elements. Then

$$\text{transitive}(R) = R \cup R^2 \cup \cdots \cup R^n$$

Finding transitive($R$) can take a lot of time when $A$ has a large number of elements. Here we give a simple example where $A$ has only three elements.

**EXAMPLE 3.9** Consider the following relation $R$ on $A = \{1, 2, 3\}$:

$$R = \{(1,2),\ (2,3),\ (3,3)\}$$

Then

$$R^2 = R \circ R = \{(1,3),\ (2,3),\ (3,3)\} \qquad \text{and} \qquad R^3 = R^2 \circ R = \{(1,3),\ (2,3),\ (3,3)\}$$

Accordingly,

$$\text{transitive}(R) = R \cup R^2 \cup R^3 = \{(1,2),\ (2,3),\ (3,3),\ (1,3)\}$$

## 3.8 PARTITIONS

Let $S$ be a nonempty set. A *partition* of $S$ is a subdivision of $S$ into nonoverlapping, nonempty subsets. Precisely, a partition of $S$ is a collection $P = \{A_i\}$ of nonempty subsets of $S$ such that

(i) Each $a \in S$ belongs to one of the $A_i$.
(ii) The sets $\{A_i\}$ are mutually disjoint; that is,

$$\text{If } A_i \neq A_j, \text{ then } A_i \cap A_j = \varnothing$$

The subsets in a partition are called *cells*. Thus each $a \in S$ belongs to exactly one of the cells. Figure 3-6 is a Venn diagram of a partition of the rectangular set $S$ of points into five cells: $A_1$, $A_2$, $A_3$, $A_4$, $A_5$.

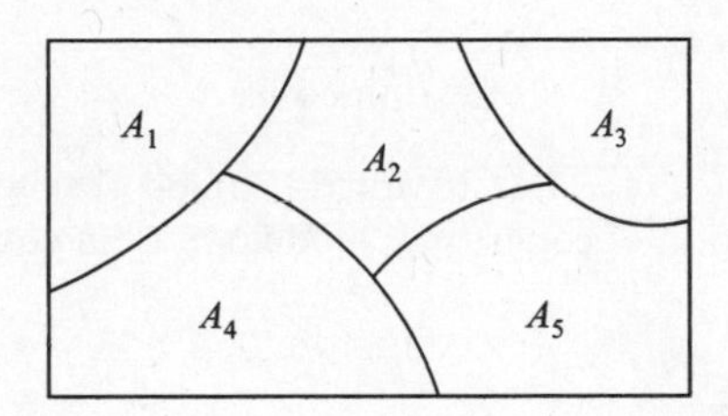


**Fig. 3-6**

**EXAMPLE 3.10** Consider the following collections of subsets of $S = \{1, 2, \ldots, 8, 9\}$:

(i) $P_1 = [\{1,3,5\}, \{2,6\}, \{4,8,9\}]$
(ii) $P_2 = [\{1,3,5\}, \{2,4,6,8\}, \{5,7,9\}]$
(iii) $P_3 = [\{1,3,5\}, \{2,4,6,8\}, \{7,9\}]$

Then $P_1$ is not a partition of $S$ since $7 \in S$ does not belong to any of the subsets. $P_2$ is not a partition of $S$ since $\{1,3,5\}$ and $\{5,7,9\}$ are not disjoint. On the other hand, $P_3$ is a partition of $S$.

**Remark**: Given a partition $P = \{A_i\}$ of a set $S$, any element $b \in A_i$ is called a *representative* of the cell $A_i$, and a subset $B$ of $S$ is called a *system* of representatives if $B$ contains exactly one element of each of the cells of $P$. Note $B = \{1, 2, 7\}$ is a system of representatives of the partition $P_3$ in Example 3.10.

## 3.9 EQUIVALENCE RELATIONS

Consider a nonempty set $S$. A relation $R$ on $S$ is an *equivalence relation* if $R$ is reflexive, symmetric, and transitive. That is, $R$ is an equivalence relation on $S$ if it has the following three properties:

(1) For every $a \in S$, $a\,R\,a$.
(2) If $a\,R\,b$, then $b\,R\,a$.
(3) If $a\,R\,b$ and $b\,R\,c$, then $a\,R\,c$.

The general idea behind an equivalence relation is that it is a classification of objects which are in some way "alike". In fact, the relation $=$ of equality on any set $S$ is an equivalence relation; that is,

(1) $a = a$ for every $a \in S$.
(2) If $a = b$, then $b = a$.
(3) If $a = b$ and $b = c$, then $a = c$.

For this reason, one frequently uses $\sim$ or $\equiv$ to denote an equivalence relation.

Examples of equivalence relations other than equality follow.

**EXAMPLE 3.11**

(*a*) Consider the set $L$ of lines and the set $T$ of triangles in the Euclidean plane. The relation "is parallel to or identical to" is an equivalence relation on $L$, and congruence and similarity are equivalence relations on $T$.

(*b*) The classification of animals by species, that is, the relation "is of the same species as," is an equivalence relation on the set of animals.

(*c*) The relation $\subseteq$ of set inclusion is not an equivalence relation. It is reflexive and transitive, but it is not symmetric since $A \subseteq B$ does not imply $B \subseteq A$.

(*d*) Let $m$ be a fixed positive integer. Two integers $a$ and $b$ are said to be *congruent modulo m*, written

$$a \equiv b \pmod{m}$$

if $m$ divides $a - b$. For example, for $m = 4$ we have $11 \equiv 3 \pmod{4}$ since 4 divides $11 - 3$, and $22 \equiv 6 \pmod{4}$ since 4 divides $22 - 6$. This relation of congruence modulo $m$ is an equivalence relation.

### Equivalence Relations and Partitions

Suppose $R$ is an equivalence relation on a set $S$. For each $a$ in $S$, let $[a]$ denote the set of elements of $S$ to which $a$ is related under $R$; that is,

$$[a] = \{x : (a, x) \in R\}$$

We call $[a]$ the *equivalence class* of $a$ in $S$ under $R$. The collection of all such equivalence classes is denoted by $S/R$, that is,

$$S/R = \{[a] : a \in S\}$$

It is called the *quotient* set of $S$ by $R$.

The fundamental property of an equivalence relation and its quotient set is contained in the following theorem (which is proved in Problem 3.28).

**Theorem 3.5**: Let $R$ be an equivalence relation on a set $S$. Then the quotient set $S/R$ is a partition of $S$. Specifically:

(i) For each $a$ in $S$, we have $a \in [a]$.

(ii) $[a] = [b]$ if and only if $(a, b) \in R$.

(iii) If $[a] \neq [b]$, then $[a]$ and $[b]$ are disjoint.

The converse of the above theorem (proved in Problem 3.29) is also true. That is,

**Theorem 3.6**: Suppose $P = \{A_i\}$ is a partition of a set $S$. Then there is an equivalence relation $\sim$ on $S$ such that the set $S/\sim$ of equivalence classes is the same as the partition $P = \{A_i\}$.

Specifically, for $a, b \in S$, the equivalence $\sim$ in Theorem 3.6 is defined by $a \sim b$ if $a$ and $b$ belong to the same cell in $P$.

Thus we see there is a one-to-one correspondence between the equivalence relations on a set $S$ and the partitions of $S$. Accordingly, for a given equivalence relation $R$ on a set $S$, we can talk about a system $B$ of representatives of the quotient set $S/R$ which would contain exactly one representative from each equivalence class.

**EXAMPLE 3.12**

(*a*) Consider the following relation $R$ on $S = \{1, 2, 3, 4\}$:

$$R = \{(1,1),\ (2,2),\ (1,3),\ (3,1),\ (3,3),\ (4,4)\}$$

One can show that $R$ is reflexive, symmetric and transitive, that is, that $R$ is an equivalence relation. Under the relation $R$,

$$[1] = \{1,3\}, \qquad [2] = \{2\}, \qquad [3] = \{1,3\}, \qquad [4] = \{4\}$$

Observe that $[1] = [3]$ and that $S/R = \{[1], [2], [4]\}$ is a partition of $S$. One can choose either $\{1, 2, 4\}$ or $\{2, 3, 4\}$ as a system of representatives of the equivalence classes.

(*b*) Let $R_5$ be the relation on the set $\mathbf{Z}$ of integers defined by

$$x \equiv y \ (\text{mod } 5)$$

which reads "$x$ is congruent to $y$ modulo 5" and which means that the difference $x - y$ is divisible by 5. Then $R_5$ is an equivalence relation on $\mathbf{Z}$. There are exactly five equivalence classes in the quotient set $\mathbf{Z}/R_5$ as follows:

$$\begin{aligned}
A_0 &= \{\ldots, -10, -5, 0, 5, 10, \ldots\}\\
A_1 &= \{\ldots, -9, -4, 1, 6, 11, \ldots\}\\
A_2 &= \{\ldots, -8, -3, 2, 7, 12, \ldots\}\\
A_3 &= \{\ldots, -7, -2, 3, 8, 13, \ldots\}\\
A_4 &= \{\ldots, -6, -1, 4, 9, 14, \ldots\}
\end{aligned}$$

Observe that any integer $x$, which can be uniquely expressed in the form $x = 5q + r$ where $0 \le r < 5$, is a member of the equivalence class $A_r$ where $r$ is the remainder. As expected, the equivalence classes are disjoint and

$$\mathbf{Z} = A_0 \cup A_1 \cup A_2 \cup A_3 \cup A_4$$

This quotient set $\mathbf{Z}/R_5$ is usually denoted by

$$\mathbf{Z}/5\mathbf{Z} \text{ or simply } \mathbf{Z}_5$$

Usually one chooses $\{0, 1, 2, 3, 4\}$ or $\{-2, -1, 0, 1, 2\}$ as a system of representatives of the equivalence classes.

## 3.10 PARTIAL ORDERING RELATIONS

This section defines another important class of relations. A relation $R$ on a set $S$ is called a *partial ordering* of $S$ or a *partial order* on $S$ if it has the following three properties:

(1) For every $a \in S$, we have $a \, R \, a$.
(2) If $a \, R \, b$ and $b \, R \, a$, then $a = b$.
(3) If $a \, R \, b$ and $b \, R \, c$, then $a \, R \, c$.

That is, $R$ is a partial ordering of $S$ if $R$ is reflexive, antisymmetric, and transitive.

A set $S$ together with a partial ordering $R$ is called a *partially ordered set* or *poset*. Partially ordered sets will be studied in more detail in Chapter 7, so here we simply give some examples.

**EXAMPLE 3.13**

(*a*) The relation $\subseteq$ of set inclusion is a partial ordering of any collection of sets since set inclusion has the three desired properties. That is,

(1) $A \subseteq A$ for any set $A$.
(2) If $A \subseteq B$ and $B \subseteq A$, then $A = B$.
(3) If $A \subseteq B$ and $B \subseteq C$, then $A \subseteq C$.

(*b*) The relation $\leq$ on the set $R$ of real numbers is reflexive, antisymmetric, and transitive. Thus $\leq$ is a partial ordering.

(*c*) The relation "$a$ divides $b$" is a partial ordering of the set $p$ of positive integers. However, "$a$ divides $b$" is not a partial ordering of the set $\mathbf{Z}$ of integers since $a|b$ and $b|a$ does not imply $a = b$. For example, $3|-3$ and $-3|3$ but $3 \neq -3$.

## 3.11 *n*-ARY RELATIONS

All the relations discussed above were binary relations. By an *n-ary relation*, we mean a set of ordered $n$-tuples. For any set $S$, a subset of the product set $S^n$ is called an $n$-ary relation on $S$. In particular, a subset of $S^3$ is called a *ternary relation* on $S$.

**EXAMPLE 3.14**

(*a*) Let $L$ be a line in the plane. Then "betweenness" is a ternary relation $R$ on the points of $L$; that is, $(a, b, c) \in R$ if $b$ lies between $a$ and $c$ on $L$.

(*b*) The equation $x^2 + y^2 + z^2 = 1$ determines a ternary relation $T$ on the set $\mathbf{R}$ of real numbers. That is, a triple $(x, y, z)$ belongs to $T$ if $(x, y, z)$ satisfies the equation which means that $(x, y, z)$ is the coordinates of a point in $\mathbf{R}^3$ on the sphere $S$ with radius 1 and center at the origin $0 = (0, 0, 0)$.

# Solved Problems

## ORDERED PAIRS AND PRODUCT SETS

**3.1.** Let $A = \{1, 2\}$, $B = \{x, y, z\}$, $C = \{3, 4\}$. Find $A \times B \times C$.

$A \times B \times C$ consists of all ordered triplets $(a, b, c)$ where $a \in A$, $b \in B$, $c \in C$. These elements of $A \times B \times C$ can be systematically obtained by a so-called tree diagram (Fig. 3-7). The elements of $A \times B \times C$ are precisely the 12 ordered triplets to the right of the tree diagram.

Observe that $n(A) = 2$, $n(B) = 3$, and $n(C) = 2$ and, as expected,

$$n(A \times B \times C) = 12 = n(A) \cdot n(B) \cdot n(C)$$

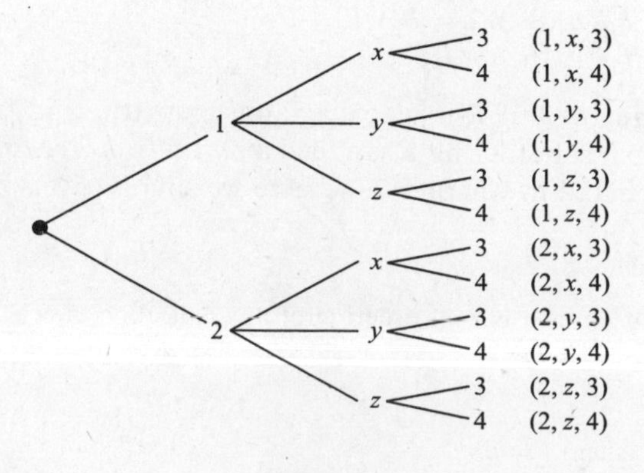


Fig. 3-7

**3.2.** Find $x$ and $y$ given $(2x,\ x+y) = (6, 2)$.

Two ordered pairs are equal if and only if the corresponding components are equal. Hence we obtain the equations

$$2x = 6 \quad \text{and} \quad x + y = 2$$

from which we derive the answer $x = 3$ and $y = -1$.

**3.3.** Let $A = \{1, 2\}$, $B = \{a, b, c\}$, $C = \{c, d\}$. Find $(A \times B) \cap (A \times C)$ and $A \times (B \cap C)$.

We have

$$A \times B = \{(1, a),\ (1, b),\ (1, c),\ (2, a),\ (2, b), (2, c)\}$$
$$A \times C = \{(1, c),\ (1, d),\ (2, c),\ (2, d)\}$$

Hence

$$(A \times B) \cap (A \times C) = \{(1, c),\ (2, c)\}$$

Since $B \cap C = \{c\}$,

$$A \times (B \cap C) = \{(1, c),\ (2, c)\}$$

Observe that $(A \times B) \cap (A \times C) = A \times (B \cap C)$. This is true for any sets $A$, $B$, and $C$ (see Problem 3.4).

**3.4.** Prove $(A \times B) \cap (A \times C) = A \times (B \cap C)$.

$$\begin{aligned}(A \times B) \cap (A \times C) &= \{(x, y) :\ (x, y) \in A \times B \text{ and } (x, y) \in A \times C\} \\ &= \{(x, y) :\ x \in A, y \in B \text{ and } x \in A, y \in C\} \\ &= \{(x, y) :\ x \in A, y \in B \cap C\} = A \times (B \cap C)\end{aligned}$$

## RELATIONS AND THEIR GRAPHS

**3.5.** Find the number of relations from $A = \{a, b, c\}$ to $B = \{1, 2\}$.

There are $3 \cdot 2 = 6$ elements in $A \times B$, and hence there are $m = 2^6 = 64$ subsets of $A \times B$. Thus there are $m = 64$ relations from $A$ to $B$.

**3.6.** Given $A = \{1, 2, 3, 4\}$ and $B = \{x, y, z\}$. Let $R$ be the following relation from $A$ to $B$:

$$R = \{(1, y),\ (1, z),\ (3, y),\ (4, x),\ (4, z)\}$$

(a) Determine the matrix of the relation.
(b) Draw the arrow diagram of $R$.
(c) Find the inverse relation $R^{-1}$ of $R$.
(d) Determine the domain and range of $R$.

(a) See Fig. 3-8(a). Observe that the rows of the matrix are labeled by the elements of $A$ and the columns by the elements of $B$. Also observe that the entry in the matrix corresponding to $a \in A$ and $b \in B$ is 1 if $a$ is related to $b$ and 0 otherwise.

(b) See Fig. 3-8(b). Observe that there is an arrow from $a \in A$ to $b \in B$ iff $a$ is related to $b$, i.e., iff $(a, b) \in R$.

(c) Reverse the ordered pairs of $R$ to obtain $R^{-1}$:

$$R^{-1} = \{(y, 1),\ (z, 1),\ (y, 3),\ (x, 4),\ (z, 4)\}$$

Observe that by reversing the arrows in Fig. 3-8(b) we obtain the arrow diagram of $R^{-1}$.

(*d*) The domain of $R$, $\text{Dom}(R)$, consists of the first elements of the ordered pairs of $R$, and the range of $R$, $\text{Ran}(R)$, consists of the second elements. Thus,

$$\text{Dom}(R) = \{1, 3, 4\} \qquad \text{and} \qquad \text{Ran}(R) = \{x, y, z\}$$

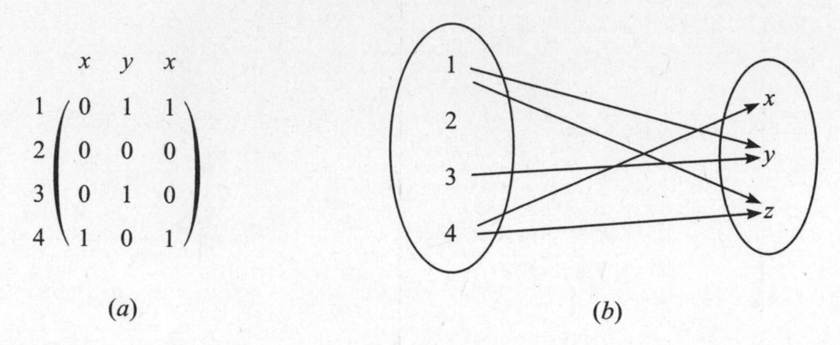


**Fig. 3-8**

**3.7.** Let $A = \{1, 2, 3, 4, 6\}$, and let $R$ be the relation on $A$ defined by "$x$ divides $y$", written $x|y$.

(*a*) Write $R$ as a set of ordered pairs.

(*b*) Draw its directed graph.

(*c*) Find the inverse relation $R^{-1}$ of $R$. Can $R^{-1}$ be described in words?

(*a*) Find those numbers in $A$ divisible by 1, 2, 3, 4, and then 6. These are:

$$1|1,\ 1|2,\ 1|3,\ 1|4,\ 1|6,\ 2|2,\ 2|4,\ 2|6,\ 3|3,\ 3|6,\ 4|4,\ 6|6$$

Hence

$$R = \{(1,1),\ (1,2),\ (1,3),\ (1,4),\ (1,6),\ (2,2),\ (2,4),\ (2,6),\ (3,3),\ (3,6),\ (4,4),\ (6,6)\}$$

(*b*) See Fig. 3-9.

(*c*) Reverse the ordered pairs of $R$ to obtain $R^{-1}$:

$$R^{-1} = \{(1,1),\ (2,1),\ (3,1),\ (4,1),\ (6,1),\ (2,2),\ (4,2),\ (6,2),\ (3,3),\ (6,3),\ (4,4),\ (6,6)\}$$

$R^{-1}$ can be described by the statement "$x$ is a multiple of $y$".

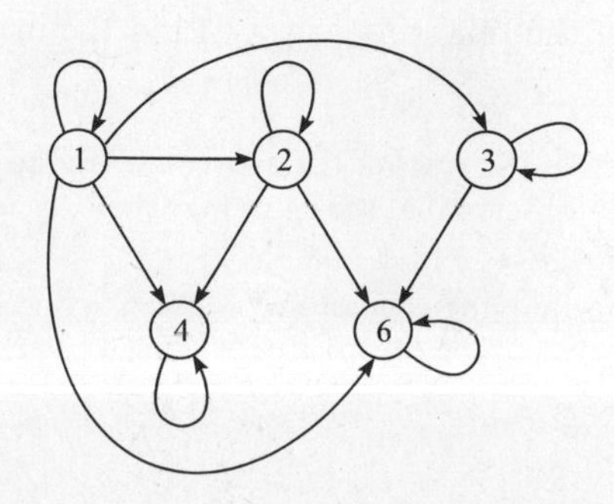


**Fig. 3-9**

**3.8.** Let $A = \{1, 2, 3\}$, $B = \{a, b, c\}$, $C = \{x, y, z\}$. Consider the following relation $R$ from $A$ to $B$ and relation $S$ from $B$ to $C$:

$$R = \{(1, b),\ (2, a),\ (2, c)\} \quad \text{and} \quad S = \{(a, y),\ (b, x),\ (c, y),\ (c, z)\}$$

(*a*) Find the composition relation $R \circ S$.

(*b*) Find the matrices $M_R$, $M_S$, and $M_{R \circ S}$ of the respective relations $R$, $S$, and $R \circ S$, and compare $M_{R \circ S}$ to the product $M_R M_S$.

(*a*) Draw the arrow diagram of the relations $R$ and $S$ as in Fig. 3-10. Observe that 1 in $A$ is "connected" to $x$ in $C$ by the path $1 \to b \to x$; hence $(1, x)$ belongs to $R \circ S$. Similarly, $(2, y)$ and $(2, z)$ belong to $R \circ S$. We have (as in Example 3.5)

$$R \circ S = \{(1, x),\ (2, y),\ (2, z)\}$$

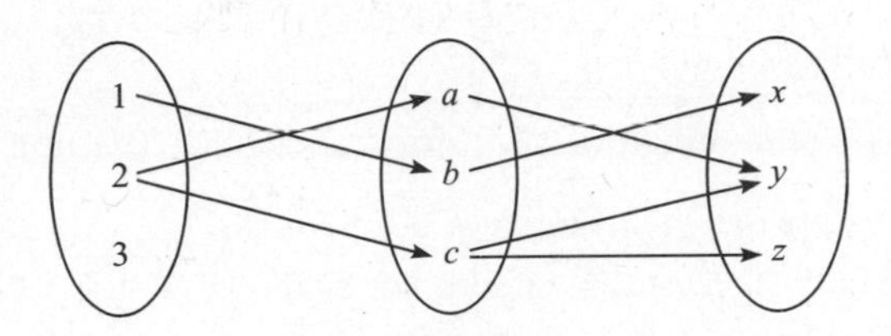


**Fig. 3-10**

(*b*) The matrices of $M_R$, $M_S$, and $M_{R \circ S}$ follow:

$$M_R = \begin{matrix} & a & b & c \\ 1 & 0 & 1 & 0 \\ 2 & 1 & 0 & 1 \\ 3 & 0 & 0 & 0 \end{matrix} \qquad M_S = \begin{matrix} & x & y & z \\ a & 0 & 1 & 0 \\ b & 1 & 0 & 0 \\ c & 0 & 1 & 1 \end{matrix} \qquad M_{R \circ S} = \begin{matrix} & x & y & z \\ 1 & 1 & 0 & 0 \\ 2 & 0 & 1 & 1 \\ 3 & 0 & 0 & 0 \end{matrix}$$

Multiplying $M_R$ and $M_S$ we obtain

$$M_R M_S = \begin{pmatrix} 1 & 0 & 0 \\ 0 & 2 & 1 \\ 0 & 0 & 0 \end{pmatrix}$$

Observe that $M_{R \circ S}$ and $M_R M_S$ have the same zero entries.

**3.9.** Let $R$ and $S$ be the following relations on $A = \{1, 2, 3\}$:

$$R = \{(1, 1),\ (1, 2),\ (2, 3),\ (3, 1),\ (3, )\}, \qquad S = \{(1, 2),\ (1, 3),\ (2, 1),\ (3, 3)\}$$

Find: (*a*) $R \cap S$, $R \cup S$, $R^c$; (*b*) $R \circ S$; (*c*) $S^2 = S \circ S$.

(*a*) Treat $R$ and $S$ simply as sets, and take the usual intersection and union. For $R^c$, use the fact that $A \times A$ is the universal relation on $A$.

$$R \cap S = \{(1, 2), (3, 3)\}, \quad R \cup S = \{(1, 1), (1, 2), (1, 3), (2, 1), (2, 3), (3, 1), (3, 3)\}$$
$$R^c = \{(1, 3), (2, 1), (2, 2), (3, 2)\}$$

(*b*) For each pair $(a, b) \in R$, find all pairs $(b, c) \in S$. Then $(a, c) \in R \circ S$. For example, $(1, 1) \in R$ and $(1, 2), (1, 3) \in S$; hence $(1, 2)$ and $(1, 3)$ belong to $R \circ S$. Thus,

$$R \circ S = \{(1, 2), (1, 3), (1, 1), (2, 3), (3, 2), (3, 3)\}$$

(*c*) Following the algorithm in (*b*), we get $S^2 = S \circ S = \{(1, 1), (1, 3), (2, 2), (2, 3), (3, 3)\}$.

**3.10.** Prove Theorem 3.1: Let $A, B, C, D$ be sets. Suppose $R$ is a relation from $A$ to $B$, $S$ is a relation from $B$ to $C$, and $T$ is a relation from $C$ to $D$. Then $(R \circ S) \circ T = R \circ (S \circ T)$.

We need to show that each ordered pair in $(R \circ S) \circ T$ belongs to $R \circ (S \circ T)$, i.e., that $(R \circ S) \circ T \subseteq R \circ (S \circ T)$, and vice versa.

Suppose $(a, d)$ belongs to $(R \circ S) \circ T$. Then there exists a $c$ in $C$ such that $(a, c) \in R \circ S$ and $(c, d) \in T$. Since $(a, c) \in R \circ S$, there exists a $b$ in $B$ such that $(a, b) \in R$ and $(b, c) \in S$. Since $(b, c) \in S$ and $(c, d) \in T$, we have $(b, d) \in S \circ T$; and since $(a, b) \in R$ and $(b, d) \in S \circ T$, we have $(a, d) \in R \circ (S \circ T)$. Thus

$$(R \circ S) \circ T \subseteq R \circ (S \circ T)$$

Similarly, $R \circ (S \circ T) \subseteq (R \circ S) \circ T$. Both inclusion relations prove $(R \circ S) \circ T = R \circ (S \circ T)$

## TYPES OF RELATIONS AND CLOSURE PROPERTIES

**3.11.** Determine when a relation $R$ on a set $A$ is:

(*a*) not reflexive, (*b*) not symmetric, (*c*) not transitive, (*d*) not antisymmetric.

(*a*) There exists $a \in A$ such that $(a, a)$ does not belong to $R$.

(*b*) There exists $(a, b)$ in $R$ such that $(b, a)$ does not belong to $R$.

(*c*) There exists $(a, b)$ and $(b, c)$ in $R$ such that $(a, c)$ does not belong to $R$.

(*d*) There exists distinct elements $a, b \in A$ such that $(a, b)$ and $(b, a)$ belong to $R$.

**3.12.** Let $A = \{1, 2, 3, 4\}$. Consider the following relation $R$ on $A$:

$$R = \{(1, 1), (2, 2), (2, 3), (3, 2), (4, 2), (4, 4)\}$$

(*a*) Draw its directed graph.

(*b*) Is $R$ (i) reflexive? (ii) symmetric? (iii) transitive? (iv) antisymmetric?

(*c*) Find $R^2 = R \circ R$.

(*a*) See Fig. 3-11.

(*b*) (i) $R$ is not reflexive because $3 \in A$ but $3 \not R 3$, i.e., $(3, 3) \notin R$.

(ii) $R$ is not symmetric because $4R2$ but $2 \not R 4$, i.e., $(4, 2) \in R$ but $(2, 4) \notin R$.

(iii) $R$ is not transitive because $4R2$ and $2R3$ but $4 \not R 3$, i.e., $(4, 2) \in R$ and $(2, 3) \in R$ but $(4, 3) \notin R$.

(iv) $R$ is not antisymmetric because $2R3$ and $3R2$ but $2 \neq 3$.

(*c*) For each pair $(a, b) \in R$, find all $(b, c) \in R$. Since $(a, c) \in R^2$,

$$R^2 = \{(1, 1), (2, 2), (2, 3), (3, 2), (3, 3), (4, 2), (4, 3), (4, 4)\}$$

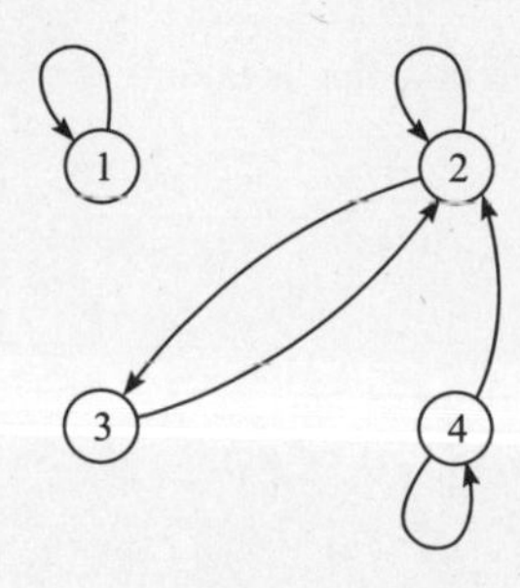


**Fig. 3-11**

**3.13.** Give examples of relations $R$ on $A = \{1, 2, 3\}$ having the stated property.

(*a*) $R$ is both symmetric and antisymmetric.

(*b*) $R$ is neither symmetric nor antisymmetric.

(*c*) $R$ is transitive but $R \cup R^{-1}$ is not transitive.

There are several possible examples for each answer. One possible set of examples follows:

(*a*) $R = \{(1, 1), (2, 2)\}$

(*b*) $R = \{(1, 2), (2, 1), (2, 3)\}$

(*c*) $R = \{(1, 2)\}$

**3.14.** Suppose $\mathscr{C}$ is a collection of relations $S$ on a set $A$ and let $T$ be the intersection of the relations $S$, that is, $T = \cap(S : S \in \mathscr{C})$. Prove:

(*a*) If every $S$ is symmetric, then $T$ is symmetric.

(*b*) If every $S$ is transitive, then $T$ is transitive.

(*a*) Suppose $(a, b) \in T$. Then $(a, b) \in S$ for every $S$. Since each $S$ is symmetric, $(b, a) \in S$ for every $S$. Hence $(b, a) \in T$ and $T$ is symmetric.

(*b*) Suppose $(a, b)$ and $(b, c)$ belong to $T$. Then $(a, b)$ and $(b, c)$ belong to $S$ for every $S$. Since each $S$ is transitive, $(a, c)$ belongs to $S$ for every $S$. Hence, $(a, c) \in T$ and $T$ is transitive.

**3.15.** Let $A = \{a, b, c\}$ and let $R$ be defined by

$$R = \{(a, a),\ (a, b),\ (b, c),\ (c, c)\}$$

Find: (*a*) reflexive($R$), (*b*) symmetric($R$), (*c*) transitive($R$).

(*a*) The reflexive closure of $R$ is obtained by adding all diagonal pairs of $A \times A$ to $R$ which are not currently in $R$. Hence

$$\text{reflexive}(R) = R \cup \{(b, b)\} = \{(a, a),\ (a, b),\ (b, c),\ (c, c),\ (b, b)\}$$

(*b*) The symmetric closure of $R$ is obtained by adding all pairs in $R^{-1}$ which are not currently in $R$. Hence

$$\begin{aligned}\text{symmetric}(R) &= R \cup \{(b, a),\ (c, b)\}\\ &= \{(a, a),\ (a, b),\ (b, a),\ (b, c),\ (c, b),\ (c, c)\}\end{aligned}$$

(*c*) Since $A$ has three elements, the transitive closure of $R$ is obtained by taking the union of $R$ with $R^2 = R \circ R$ and $R^3 = R \circ R \circ R$. We have:

$$\begin{aligned}R^2 &= R \circ R = \{(a, a),\ (a, b),\ (a, c),\ (b, c),\ (c, c)\}\\ R^3 &= R^2 \circ R = \{(a, a),\ (a, b),\ (a, c),\ (b, c),\ (c, c)\}\end{aligned}$$

Hence transitive$(R) = R \cup R^2 \cup R^3 = \{(a, a),\ (a, b),\ (a, c),\ (b, c),\ (c, c)\}$.

## PARTITIONS

**3.16.** Let $S = \{1, 2, 3, 4, 5, 6\}$. Determine which of the following are partitions of $S$:

(*a*) $P_1 = [\{1, 2, 3\},\ \{1, 4, 5, 6\}]$

(*b*) $P_2 = [\{1, 2\},\ \{3, 5, 6\}]$

(*c*) $P_3 = [\{1, 3, 5\},\ \{2, 4\},\ \{6\}]$

(*d*) $P_4 = [\{1, 3, 5\},\ \{2, 4, 6, 7\}]$

(*a*) No, since $1 \in S$ belongs to two cells.

(*b*) No, since $4 \in S$ does not belong to any cell.

(*c*) $P_3$ is a partition of $S$.

(*d*) No, since $\{2, 4, 6, 7\}$ is not a subset of $S$.

**3.17.** Find all partitions of $S = \{a, b, c, d\}$.

Note first that each partition of $S$ contains either 1, 2, 3, or 4 distinct cells. The partitions are as follows:

(1) $[S]$;

(2) $[\{a\}, \{b, c, d\}]$, $[\{b\}, \{a, c, d\}]$, $[\{c\}, \{a, b, d\}]$, $[\{d\}, \{a, b, c\}]$, $[\{a, b\}, \{c, d\}]$, $[\{a, c\}, \{b, d\}]$, $[\{a, d\}, \{b, c\}]$;

(3) $[\{a\}, \{b\}, \{c, d\}]$, $[\{a\}, \{c\}, \{b, d\}]$, $[\{a\}, \{d\}, \{b, c\}]$, $[\{b\}, \{c\}, \{a, d\}]$, $[\{b\}, \{d\}, \{a, c\}]$, $[\{c\}, \{d\}, \{a, b\}]$;

(4) $[\{a\}, \{b\}, \{c\}, \{d\}]$.

There are 15 different partitions of $S$.

**3.18.** Let $[A_1, A_2, \ldots, A_m]$ and $[B_1, B_2, \ldots, B_n]$ be partitions of $X$. Show that the collection of sets

$$P = [\{A_i \cap B_j\}] \backslash \varnothing$$

is also a partition (called the *cross partition*) of $X$. (Observe that we have deleted the empty set $\varnothing$.)

Let $x \in X$. Then $x$ belongs to $A_r$ for some $r$, and to $B_s$ for some $s$; hence $x$ belongs to $A_r \cap B_s$. Thus the union of the $A_i \cap B_j$ is equal to $X$. Now suppose $A_r \cap B_s$ and $A_{r'} \cap B_{s'}$ are not disjoint, say $y$ belongs to both sets. Then $y$ belongs to $A_r$ and $A_{r'}$; hence $A_r = A_{r'}$. Similarly $y$ belongs to $B_s$ and $B_{s'}$; hence $B_s = B_{s'}$. Accordingly, $A_r \cap B_s = A_{r'} \cap B_{s'}$. Thus the cells are mutually disjoint or equal. Accordingly, $P$ is a partition of $X$.

**3.19.** Let $X = \{1, 2, 3, \ldots, 8, 9\}$. Find the cross partition $P$ of the following partitions of $X$:

$$P_1 = [\{1, 3, 5, 7, 9\}, \{2, 4, 6, 8\}] \quad \text{and} \quad P_2 = [\{1, 2, 3, 4\}, \{5, 7\}, \{6, 8, 9\}]$$

Intersect each cell in $P_1$ with each cell in $P_2$ (omitting empty intersections) to obtain

$$P = [\{1, 3\}, \{5, 7\}, \{9\}, \{2, 4\}, \{8\}]$$

**3.20.** Let $f(n, k)$ represent the number of partitions of a set $S$ with $n$ elements into $k$ cells (for $k = 1, 2, \ldots, n$). Find a recursion formula for $f(n, k)$.

Note first that $f(n, 1) = 1$ and $f(n, n) = 1$ since there is only one way to partition $S$ with $n$ elements into either one cell or $n$ cells. Now suppose $n > 1$ and $1 < k < n$. Let $b$ be some distinguished element of $S$. If $\{b\}$ constitutes a cell, then $S \backslash \{b\}$ can be partitioned into $k - 1$ cells in $f(n-1, k-1)$ ways. On the other hand, each partition of $S \backslash \{b\}$ into $k$ cells allows $b$ to be admitted into a cell in $k$ ways. We have thus shown that

$$f(n, k) = f(n-1, k-1) + kf(n-1, k)$$

which is the desired recursion formula.

**3.21.** Consider the recursion formula in Problem 3.20. (*a*) Find the solution for $n = 1, 2, \ldots, 6$ in a form similar to Pascal's triangle. (*b*) Find the number $m$ of partitions of a set with $m = 6$ elements.

(*a*) Use the recursion formula to obtain the triangle in Fig. 3-12, for example:

$$f(6,4) = f(5,3) + 4f(5,4) = 25 + 4(10) = 65$$

(*b*) Use row 6 in Fig. 3-12 to obtain $m = 1 + 31 + 90 + 65 + 15 + 1 = 203$.

```
1
1   1
1   3   1
1   7   6   1
1  15  25  10   1
1  31  90  65  15   1
```

**Fig. 3-12**

## EQUIVALENCE RELATIONS AND PARTITIONS

**3.22.** Consider the set **Z** of integers and any integer $m > 1$. We say that $x$ is congruent to $y$ modulo $m$, written

$$x \equiv y \pmod{m}$$

if $x - y$ is divisible by $m$. Show that this defines an equivalence relation on **Z**.

We must show that the relation is reflexive, symmetric, and transitive.

(i) For any $x$ in **Z** we have $x \equiv x \pmod{m}$ because $x - x = 0$ is divisible by $m$. Hence the relation is reflexive.

(ii) Suppose $x \equiv y \pmod{m}$, so $x - y$ is divisible by $m$. Then $-(x - y) = y - x$ is also divisible by $m$, so $y \equiv x \pmod{m}$. Thus the relation is symmetric.

(iii) Now suppose $x \equiv y \pmod{m}$ and $y \equiv z \pmod{m}$, so $x - y$ and $y - z$ are each divisible by $m$. Then the sum

$$(x - y) + (y - z) = x - z$$

is also divisible by $m$; hence $x \equiv z \pmod{m}$. Thus the relation is transitive.

Accordingly, the relation of congruence modulo $m$ on **Z** is an equivalence relation.

**3.23.** Let $R$ be the following equivalence relation on the set $A = \{1, 2, 3, 4, 5, 6\}$:

$$R = \{(1,1), (1,5), (2,2), (2,3), (2,6), (3,2), (3,3), (3,6), (4,4), (5,1), (5,5), (6,2), (6,3), (6,6)\}$$

Find the partition of $A$ *induced* by $R$, i.e., find the equivalence classes of $R$.

Those elements related to 1 are 1 and 5, hence

$$[1] = \{1, 5\}$$

We pick an element which does not belong to [1], say 2. Those elements related to 2 are 2, 3, and 6, hence

$$[2] = \{2, 3, 6\}$$

The only element which does not belong to [1] or [2] is 4. The only element related to 4 is 4. Thus

$$[4] = \{4\}$$

Accordingly,

$$[\{1, 5\}, \{2, 3, 6\}, \{4\}]$$

is the partition of $A$ induced by $R$.

**3.24.** Let $A = \{1, 2, 3, \ldots, 14, 15\}$. Let $R$ be the equivalence relation on $A$ defined by congruence modulo 4.

(*a*) Find the equivalence classes determined by $R$.

(*b*) Find a system $B$ of equivalence class representatives which are multiples of 3.

(*a*) Recall (Problem 3.22) that $a \equiv b \pmod 4$ if 4 divides $a - b$ or, equivalently, if $a = b + 4k$ for some integer $k$. Accordingly:

(1) Add multiples of 4 to 1 to obtain $[1] = \{1, 5, 9, 13\}$.

(2) Add multiples of 4 to 2 to obtain $[2] = \{2, 6, 10, 14\}$.

(3) Add multiples of 4 to 3 to obtain $[3] = \{3, 7, 11, 15\}$.

(4) Add multiples of 4 to 4 to obtain $[4] = \{4, 8, 12\}$.

Then [1], [2], [3], [4] are all the equivalence classes since they include all the elements of $A$.

(*b*) Choose an element in each equivalence class which is a multiple of 3. Thus $B = \{9, 6, 3, 12\}$ or $B = \{9, 6, 15, 12\}$.

**3.25.** Consider the set of words $W$ = {sheet, last, sky, wash, wind, sit}. Find $W/R$ where $R$ is the equivalence relation defined by:

(*a*) "has the same number of letters", (*b*) "begins with the same letter".

(*a*) Those words with the same number of letters belong to the same cell; hence

$$W/R = [\{\text{sheet}\},\ \{\text{last, wash, wind}\},\ \{\text{sky, sit}\}]$$

(*b*) Those words beginning with the same letter belong to the same cell; hence

$$W/R = [\{\text{sheet, sky, sit}\},\ \{\text{last}\},\ \{\text{wash, wind}\}]$$

**3.26.** Let $A$ be a set of nonzero integers and let $\approx$ be the relation on $A \times A$ defined as follows:

$$(a, b) \approx (c, d) \qquad \text{whenever} \qquad ad = bc$$

Prove that $\approx$ is an equivalence relation.

We must show that $\approx$ is reflexive, symmetric, and transitive.

(i) *Reflexivity*: We have $(a, b) \approx (a, b)$ since $ab = ba$. Hence $\approx$ is reflexive.

(ii) *Symmetry*: Suppose $(a, b) \approx (c, d)$. Then $ad = bc$. Accordingly, $cb = da$ and hence $(c, d) = (a, b)$. Thus, $\approx$ is symmetric.

(iii) *Transitivity*: Suppose $(a, b) \approx (c, d)$ and $(c, d) \approx (e, f)$. Then $ad = bc$ and $cf = de$. Multiplying corresponding terms of the equations gives $(ad)(cf) = (bc)(de)$. Canceling $c \neq 0$ and $d \neq 0$ from both sides of the equation yields $af = be$, and hence $(a, b) \approx (e, f)$. Thus $\approx$ is transitive.

Accordingly, $\approx$ is an equivalence relation.

**3.27.** Let $A = \{1, 2, 3, \ldots, 14, 15\}$. Let $\approx$ be the equivalence relation on $A \times A$ defined by $(a, b) \approx (c, d)$ if $ad = bc$. (See Problem 3.26.) Find the equivalence class of $(3, 2)$.

We seek all $(m, n)$ such that $(3, 2) \approx (m, n)$, that is, such that $3n = 2m$ or $3/2 = m/n$. [In other words, if $(3, 2)$ is written as the fraction $3/2$, then we seek all fractions $m/n$ which are equal to $3/2$.] Thus:

$$[(3, 2)] = \{(3, 2),\ (6, 4),\ (9, 6),\ (12, 8),\ (15, 10)\}$$

**3.28.** Prove Theorem 3.5: Let $R$ be an equivalence relation on a set $S$. Then the quotient set $S/R$ is a partition of $S$. Specifically:

(i) For each $a \in S$, we have $a \in [a]$.

(ii) $[a] = [b]$ if and only if $(a, b) \in R$.

(iii) If $[a] \neq [b]$, then $[a]$ and $[b]$ are disjoint.

*Proof of* (i): Since $R$ is reflexive, $(a, a) \in R$ for every $a \in S$ and therefore $a \in [a]$.

*Proof of* (ii): Suppose $(a, b) \in R$. We want to show that $[a] = [b]$. Let $x \in [b]$; then $(b, x) \in R$. But by hypothesis $(a, b) \in R$ and so, by transitivity, $(a, x) \in R$. Accordingly $x \in [a]$. Thus $[b] \subseteq [a]$. To prove that $[a] \subseteq [b]$, we observe that $(a, b) \in R$ implies, by symmetry, that $(b, a) \in R$. Then, by a similar argument, we obtain $[a] \subseteq [b]$. Consequently, $[a] = [b]$.

On the other hand, if $[a] = [b]$, then, by (i), $b \in [b] = [a]$; hence $(a, b) \in R$.

*Proof of* (iii): We prove the equivalent contrapositive statement:

$$\text{If } [a] \cap [b] \neq \varnothing \quad \text{then} \quad [a] = [b]$$

If $[a] \cap [b] \neq \varnothing$, then there exists an element $x \in A$ with $x \in [a] \cap [b]$. Hence $(a, x) \in R$ and $(b, x) \in R$. By symmetry, $(x, b) \in R$ and by transitivity, $(a, b) \in R$. Consequently by (ii), $[a] = [b]$.

**3.29.** Prove Theorem 3.6: Suppose $P = \{A_i\}$ is a partition of a set $S$. Then there is an equivalence relation $\sim$ on $S$ such that $S/\sim$ is the same as the partition $P = \{A_i\}$.

For $a, b \in S$, define $a \sim b$ if $a$ and $b$ belong to the same cell $A_k$ in $P$. We need to show that $\sim$ is reflexive, symmetric, and transitive.

(i) Let $a \in S$. Since $P$ is a partition, there exists some $A_k$ in $P$ such that $a \in A_k$. Hence $a \sim a$. Thus $\sim$ is reflexive.

(ii) Symmetry follows from the fact that if $a, b \in A_k$, then $b, a \in A_k$.

(iii) Suppose $a \sim b$ and $b \sim c$. Then $a, b \in A_i$ and $b, c \in A_j$. Therefore $b \in A_i \cap B_j$. Since $P$ is a partition, $A_i = A_j$. Thus $a, c \in A_i$ and so $a \sim c$. Thus $\sim$ is transitive.

Accordingly, $\sim$ is an equivalence relation on $S$.

Furthermore,

$$[a] = \{x : a \sim x\} = \{x : x \text{ is in the same cell } A_k \text{ as } a\}$$

Thus the equivalence classes under $\sim$ are the same as the cells in the partition $P$.

## MISCELLANEOUS PROBLEMS

**3.30.** Consider the set $\mathbf{Z}$ of integers. Define $a \sim b$ if $b = a^r$ for some positive integer $r$. Show that $\sim$ is a partial ordering of $\mathbf{Z}$; that is, show that: (i) (Reflexive) $a \sim a$ for every $a \in \mathbf{Z}$. (ii) (Antisymmetric) If $a \sim b$ and $b \sim a$, then $a = b$. (iii) (Transitive) If $a \sim b$ and $b \sim c$, then $a \sim c$.

(i) Since $a = a^1$, we have $a \sim a$. Thus $\sim$ is reflexive.

(ii) Suppose $a \sim b$ and $b \sim a$, say $b = a^r$ and $a = b^s$. Then $a = (a^r)^s = a^{rs}$. There are four possibilities:

(1) $rs = 1$. Then $r = 1$ and $s = 1$ and so $a = b$.

(2) $a = 1$. Then $b = 1^r = 1 = a$.

(3) $b = 1$. Then $a = 1^s = 1 = b$.

(4) $a = -1$. Then $b = 1$ or $b = -1$. By (3), $b \neq 1$. Hence $b = -1 = a$.

In all cases $a = b$. Thus $\sim$ is antisymmetric.

(iii) Suppose $a \sim b$ and $b \sim c$, say $b = a^r$ and $c = b^s$. Then $c = (a^r)^s = a^{rs}$, and hence $a \sim c$. Thus $\sim$ is transitive.

Accordingly, $\sim$ is a partial ordering of $\mathbf{Z}$.

**3.31.** Let $A = \{1, 2, 3, \ldots, 14, 15\}$.

(*a*) Let $R$ be the ternary relation on $A$ defined by the equation $x^2 + 5y = z$. Write $R$ as a set of ordered triples.

(*b*) Let $S$ be the 4-ary relation on $A$ defined by

$$S = \{(x, y, z, t) : 4x + 3y + z^2 = t\}$$

Write $S$ as a set of 4-tuples.

(*a*) Since $x^2 > 15$ for $x > 3$, we need only find solutions for $y$ and $z$ when $x = 1, 2, 3$. This yields:

$$R = \{(1, 1, 6),\ (1, 2, 1),\ (2, 1, 9),\ (2, 2, 14),\ (3, 1, 14)\}$$

(*b*) Note we can only have $x = 1, 2, 3$. This yields:

$$S = \{(1, 1, 1, 8),\ (1, 1, 2, 11),\ (1, 2, 1, 11),\ (1, 2, 2, 14),\\ (1, 3, 1, 14),\ (2, 1, 1, 12),\ (2, 1, 2, 15),\ (2, 2, 1, 15)\}$$

**3.32.** Each of the following expressions defines a relation on $\mathbf{R}$:

(*a*) $y \le x^2$, (*b*) $y < 3 - x$, (*c*) $y > x^3$.

Sketch (by shading the appropriate area) each relation in the plane $\mathbf{R}^2$.

In order to sketch a relation on $\mathbf{R}$ defined by an expression of the form:

(1) $y > f(x)$, (2) $y \ge f(x)$, (3) $y < f(x)$, (4) $y \le f(x)$

first plot the equation $y = f(x)$ in the usual manner. Then the relation, i.e., the desired set, will consist, respectively, of the points:

(1) above, (2) above and on, (3) below, (4) below and on.

the equation $y = f(x)$.

Figure 3-13 shows the sketches of the three relations. The equations $y = f(x)$ in Fig. 3-13(*b*) and (*c*) are drawn with dashes to indicate that the points on the curve do not belong to the given relation.

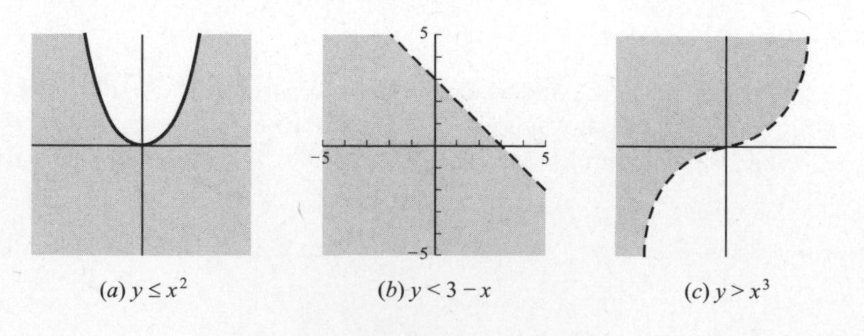


(*a*) $y \le x^2$ (*b*) $y < 3 - x$ (*c*) $y > x^3$

**Fig. 3-13**

**3.33.** Each of the following expressions defines a relation on $\mathbf{R}$:

(*a*) $x^2 + y^2 < 16$, (*b*) $x^2 - 4y^2 \ge 9$, (*c*) $x^2 + 4y^2 \le 16$.

Sketch (by shading the appropriate area) each relation in the plane $\mathbf{R}^2$.

In order to sketch a relation on **R** defined by an expression of the form $E(x, y) < k$ (respectively: $\leq, \equiv$, or $\geq$), first plot the equation $E(x, y) = k$. The curve $E(x, y) = k$ will, in simple situations, partition the plane into various regions. The relation will consist of all the points in one or more of the regions. Thus test at least one point in each region to determine whether or not all the points in that region belong to the relation. Also, use a dotted curve to indicate the points on the curve that do not belong to the relation.

Figure 3-14 shows each of the relations.

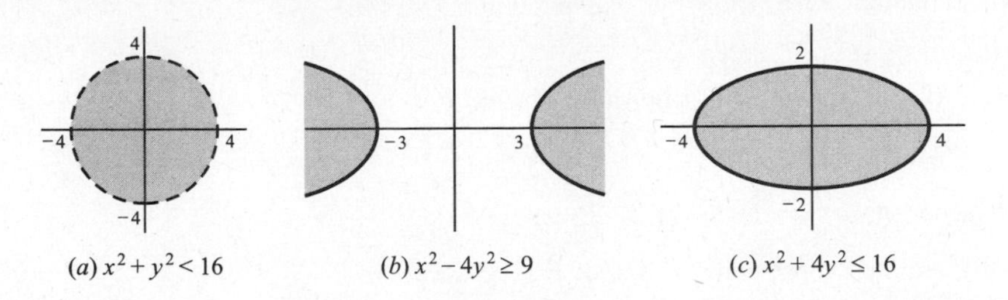


(a) $x^2 + y^2 < 16$ (b) $x^2 - 4y^2 \geq 9$ (c) $x^2 + 4y^2 \leq 16$

**Fig. 3-14**

## Supplementary Problems

### ORDERED PAIRS AND PRODUCT SETS

**3.34.** Let $S = \{a, b, c\}$, $T = \{b, c, d\}$, $W = \{a, d\}$. Find $S \times T \times W$ by constructing the tree diagram of $S \times T \times W$.

**3.35.** Let $C = \{\text{H}, \text{T}\}$, the set of possible outcomes if a coin is tossed. Find: (a) $C^2 = C \times C$; (b) $C^3$.

**3.36.** Find $x$ and $y$ if: (a) $(x + 2, 4) = (5, 2x + y)$; (b) $(y - 2,\ 2x + 1) = (x - 1,\ y + 2)$.

**3.37.** Suppose $n(A) = 3$ and $n(B) = 5$. Find the number of elements in:

(a) $A \times B$, $B \times A$, $A^2$, $B^2$; (b) $A \times B \times A$, $A^3$, $B^3$.

**3.38.** Sketch each of the following product sets in the plane $\mathbf{R}^2$ by shading the appropriate area:

(a) $[-3, 3] \times [-1, 2]$; (b) $[-3, 1) \times (-2, 2]$; (c) $(-2, 3] \times [-3, \infty)$.

[Here $[-3, \infty)$ is the infinite interval $\{x : x \geq -3\}$.]

**3.39.** Prove: $A \times (B \cup C) = (A \times B) \cup (A \times C)$.

**3.40.** Suppose $A = B \cap C$. Show that: (a) $A \times A = (B \times B) \cap (C \times C)$; (b) $A \times A = (B \times C) \cap (C \times B)$.

### RELATIONS

**3.41.** Consider the relation $R = \{(1, a), (1, b), (3, b), (3, d), (4, b)\}$ from $X = \{1, 2, 3, 4\}$ to $Y = \{a, b, c, d\}$.

(a) Find $E = \{x : x\,R\,b\}$ and $F = \{x : x\,R\,d\}$.
(b) Find $G = \{y : 1\,R\,y\}$ and $H = \{y : 2\,R\,y\}$.
(c) Find the domain and range of $R$.
(d) Find $R^{-1}$.

**3.42.** Let $R$ and $S$ be relations from $A = \{1,2,3\}$ to $B = \{a,b\}$ defined by

$$R = \{(1,a),\ (3,a),\ (2,b),\ (3,b)\} \quad \text{and} \quad S = \{(1,b),\ (2,b)\}$$

Find: (*a*) $R \cap S$; (*b*) $R \cup S$; (*c*) $R^c$, (*d*) composition $R \circ S$.

**3.43.** Find the number of relations from $A = \{a,b,c,d\}$ to $B = \{x,y\}$.

**3.44.** Let $R$ be the relation on $P$ defined by the equation $x + 3y = 12$.

(*a*) Write $R$ as a set of ordered pairs.

(*b*) Find: (i) domain of $R$, (ii) range of $R$, (iii) $R^{-1}$.

(*c*) Find the composition relation $R \circ R$.

**3.45.** Consider the relation $R = \{(1,3),\ (1,4),\ (3,2),\ (3,3),\ (3,4)\}$ on $A = \{1,2,3,4\}$.

(*a*) Find the matrix representation $M_R$ of $R$.

(*b*) Find the domain and range of $R$.

(*c*) Find $R^{-1}$.

(*d*) Draw the directed graph of $R$.

(*e*) Find the composition relation $R \circ R$.

**3.46.** Let $S$ be the following relation on $A = \{1,2,3,4,5\}$:

$$S = \{(1,2),\ (2,2),\ (2,4),\ (3,3),\ (3,5),\ (4,1),\ (5,2)\}$$

(*a*) Find the following subsets of $A$:

$$E = \{x : xS2\}, \quad F = \{x : xS3\}, \quad G = \{x : 2Sx\}, \quad H = \{x : 3Sx\}$$

(*b*) Find the matrix representation $M_S$ of $S$.

(*c*) Draw the directed graph of $S$.

(*d*) Find the composition relation $S \circ S$.

**3.47.** Let $R$ be the relation on $X = \{a,b,c,d,e,f\}$ defined by

$$R - \{(a,b),\ (b,b),\ (b,c),\ (c,f),\ (d,b),\ (e,a),\ (e,b),\ (e,f)\}$$

(*a*) Find each of the following subsets of $X$:

$$E = \{x : bRx\}, \quad F = \{x : xRb\}, \quad G = \{x : xRe\}, \quad H = \{x : eRx\}$$

(*b*) Find domain and range of $R$.

(*c*) Find the composition $R \circ R$.

## TYPES OF RELATIONS

**3.48.** Each of the following defines a relation on $\mathbf{P} = \{1,2,3,\ldots\}$:

$$(1)\ x > y, \quad (2)\ xy \text{ is a square}, \quad (3)\ x + y = 10, \quad (4)\ x + 4y = 10$$

Determine which relations are: (*a*) reflexive, (*b*) symmetric, (*c*) antisymmetric, (*d*) transitive.

**3.49.** Consider the relation $R = \{(1,1),\ (2,2),\ (2,3),\ (3,2),\ (4,2),\ (4,4)\}$ on $A = \{1,2,3,4\}$. Show that $R$ is not: (*a*) reflexive, (*b*) symmetric, (*c*) transitive, (*d*) antisymmetric.

**3.50.** Let $R, S, T$ be the relations on $A = \{1,2,3\}$ defined by:

$$R = \{(1,1),\ (2,2),\ (3,3)\} = \Delta_A, \quad S = \{(1,2),\ (2,1),\ (3,3)\} \quad T = \{(1,2),\ (2,3),\ (1,3)\}$$

Determine which of $R, S, T$ are: (*a*) reflexive, (*b*) symmetric, (*c*) antisymmetric, (*d*) transitive.

**3.51.** Let $R$ be a relation on a set $A$ where $n(A) \geq 3$. State whether each of the following is true or false. If it is false, give a counterexample on the set $A = \{1, 2, 3\}$:

(a) If $R$ is symmetric, then $R^c$ is symmetric.
(b) If $R$ is reflexive, then $R^c$ is reflexive.
(c) If $R$ is transitive, then $R^c$ is transitive.
(d) If $R$ is reflexive, then $R \cap R^{-1}$ is not empty.
(e) If $R$ is symmetric, then $R \cap R^{-1}$ is not empty.
(f) If $R$ is antisymmetric, then $R^{-1}$ is antisymmetric.

## CLOSURE PROPERTIES

**3.52.** Consider the relation $R = \{(1,1),\ (2,2),\ (2,3),\ (4,2)\}$ on $A = \{1, 2, 3, 4\}$. Find:

(a) reflexive closure of $R$; (b) symmetric closure of $R$; (c) transitive closure of $R$.

**3.53.** Find the transitive closure $R^*$ of the relation $R$ on $A = \{1, 2, 3, 4\}$ defined by the directed graph in:

(a) Fig. 3-15(a); (b) Fig. 3-15(b).

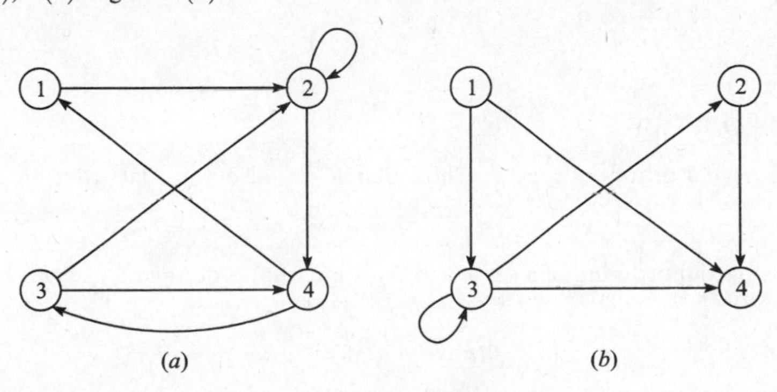


**Fig. 3-15**

**3.54.** Suppose $A$ has $n$ elements, say $A = \{1, 2, \ldots, n\}$.

(a) Suppose $R$ is a relation on $A$ with $r$ pairs. Find an upper bound for the number of pairs in: (i) reflexive closure of $R$; (ii) symmetric closure of $R$.
(b) Find a relation $R$ on $A$ with $n$ pairs such that the transitive closure $R^*$ of $R$ is the universal relation $A \times A$ (containing $n^2$ pairs).

## PARTITIONS

**3.55.** Let $S = \{1, 2, 3, 4, 5, 6\}$. Determine whether each of the following is a partition of $S$:

(a) $[\{1,3,5\},\ \{2,4\},\ \{3,6\}]$, (c) $[\{1\},\ \{3,6\},\ \{2,4,5\},\ \{3,6\}]$,
(b) $[\{1,5\},\ \{2\},\ \{3,6\}]$, (d) $[\{1\},\ \{2\},\ \{3\},\ \{4\},\ \{5\},\ \{6\}]$.

**3.56.** Find all partitions of $S = \{1, 2, 3\}$.

**3.57.** Let $P_1$ and $P_2$ be partitions of a set $S$, and let $P$ be the cross partition.

(a) Find bounds on the number $n$ of elements in $P$ if $P_1$ has $r$ elements and $P_2$ has $s$ elements.
(b) When will $P = P_1$?
(c) Find $P$ when $S = \{1, 2, 3, \ldots, 8, 9\}$ and

$$P_1 = [\{1,2,3,4,5\}, \{6,7,8,9\}] \quad \text{and} \quad P_2 = [\{1,3,5\}, \{2,6,7,9\}, \{4,8\}]$$

## EQUIVALENCE RELATIONS AND PARTITIONS

**3.58.** Let $S = \{1, 2, 3, \ldots, 19, 20\}$. Let $\equiv$ be the equivalence relation on $S$ defined by congruence modulo 7.

(*a*) Find the quotient set $S/\equiv$. (*b*) Find a system of equivalence class representatives consisting of even integers.

**3.59.** Let $A$ be a set of integers, and let $\sim$ be the relation on $A \times A$ defined by

$$(a, b) \sim (c, d) \qquad \text{if} \qquad a + d = b + c$$

(*a*) Prove that $\sim$ is an equivalence relation.

(*b*) Suppose $A = \{1, 2, 3, \ldots, 8, 9\}$. Find $[(2, 5)]$, the equivalence class of $(2, 5)$.

**3.60.** Let $\equiv$ be the relation on the set $R$ of real numbers defined by $a \equiv b$ if $b - a \in \mathbf{Z}$, that is, if $b - a$ is an integer.

(*a*) Show that $\equiv$ is an equivalence relation.

(*b*) Show that the half-open interval $A = [0, 1) = \{x : 0 \le x < 1\}$ is a system of equivalence class representatives.

## MISCELLANEOUS PROBLEMS

**3.61.** Suppose $R$ is a partial order on a set $A$. Show that $R^{-1}$ is also a partial order on $A$.

**3.62.** Suppose $R_1$ is a partial ordering of a set $A$ and $R_2$ is a partial ordering of a set $B$. Let $R$ be the relation on $A \times B$ defined by

$$(a, b)R(a', b') \qquad \text{if} \qquad aRa' \text{ and } bR_2b'$$

Show that $R$ is a partial ordering of $A \times B$.

**3.63.** Let $A = \{1, 2, 3, \ldots, 14, 15\}$.

(*a*) Let $R$ be the ternary relation on $A$ defined by the equation $x^3 + y = 5z$. Write $R$ as a set of ordered triples.

(*b*) Let $S$ be the 4-ary relation on $A$ defined by the equation $x_1^2 + 4x_2 + 5x_3 = x_4$. Write $S$ as a set of 4-tuples.

**3.64.** Sketch in the plane $R^2$ (by shading the appropriate area) each of the following relations on $R$:
(*a*) $y < x^2 - 4x + 2$; (*b*) $y \ge \dfrac{x}{2} + 2$.

**3.65.** For each of the following pairs of relations $S$ and $S'$ on $R$, sketch $S \cap S'$ in the plane $\mathbf{R}^2$ and find its domain and range:

(*a*) $S = \{(x, y) : x^2 + y^2 \le 25, \qquad S' = \{(x, y) : y \ge 4x^2/9\}$

(*b*) $S = \{(x, y) : x^2 + y^2 < 25, \qquad S' = \{(x, y) : y < 3x/4\}$

**3.66.** Show that a relation $R$ is transitive if and only if $R^n \subseteq R$ for every $n \ge 1$.

## Answers to Supplementary Problems

**3.34.** See Fig. 3-16. Using the notation: $aba = (a, b, a)$,

$$S \times T \times W = \{aba, abd, aca, acd, ada, add, bba, bbd, bca, bcd, bda, bdd, cba, cbd, cca, ccd, cda, cdd\}$$

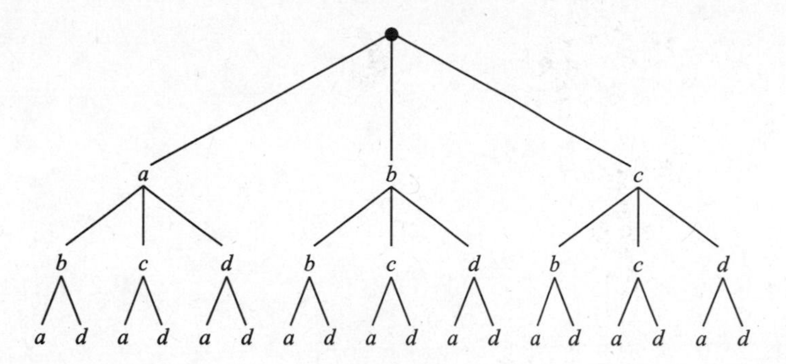


Fig. 3-16

**3.35.** (*a*) $C^2 = \{\text{HH}, \text{HT}, \text{TH}, \text{TT}\}$; (*b*) $C^3 = \{\text{HHH}, \text{HHT}, \text{HTH}, \text{HTT}, \text{THH}, \text{THT}, \text{TTH}, \text{TTT}\}$

**3.36.** (*a*) $x = 3$, $y = -2$; (*b*) $x = 2$, $y = 3$

**3.37.** (*a*) 15, 15, 9; (*b*) 45, 27, 125

**3.38.** See Fig. 3-17.

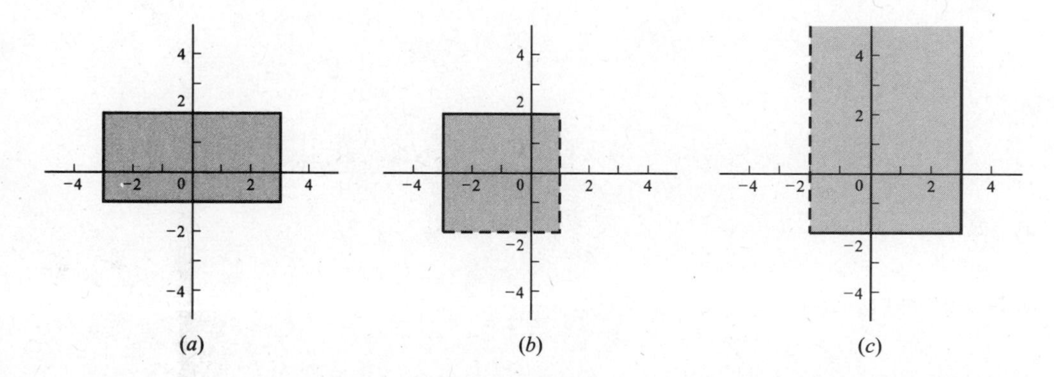


Fig. 3-17

**3.41.** (*a*) $E = \{1, 3, 4\}$, $F = \{3\}$; (*b*) $G = \{a, b\}$, $H = \varnothing$
(*c*) $\text{Dom}(R) = \{1, 3, 4\}$, $\text{Ran}(R) = \{a, b, d\}$
(*d*) $R^{-1} = \{(a, 1), (b, 1), (b, 3), (d, 3), (b, 4)\}$

**3.42.** (*a*) $\{(2, b)\}$; (*b*) $\{(1, a), (3, a), (2, b), (3, b), (1, b)\}$; (*c*) $\{(2, a), (1, b)\}$; (*d*) Not defined

**3.43.** $2^8 = 256$

**3.44.** (*a*) $R = \{(3, 3), (6, 2), (9, 1)\}$
(*b*) (i) $\{3, 6, 9\}$, (ii) $\{1, 2, 3\}$, (iii) $R^{-1} = \{(3, 3), (2, 6), (1, 9)\}$
(*c*) $\{3, 3\}$

**3.45.** (*a*) $M_R = \begin{bmatrix} 0 & 0 & 1 & 1 \\ 0 & 0 & 0 & 0 \\ 0 & 1 & 1 & 1 \\ 0 & 0 & 0 & 0 \end{bmatrix}$

(*b*) Domain $= \{1, 3\}$, range $= \{2, 3, 4\}$
(*c*) $R^{-1} = \{(3,1),\ (4,1),\ (2,3),\ (3,3),\ (4,3)\}$
(*d*) See Fig. 3-18(*a*)
(*e*) $R \circ R = \{(1,2),\ (1,3),\ (1,4),\ (3,2),\ (3,3),\ (3,4)\}$

**3.46.** (*a*) $E = \{1,2,5\},\ F = \{3\},\ G = \{2,4\},\ H = \{3,5\}$

(*b*) $M_S = \begin{bmatrix} 0 & 1 & 0 & 0 & 0 \\ 0 & 1 & 0 & 1 & 0 \\ 0 & 0 & 1 & 0 & 1 \\ 1 & 0 & 0 & 0 & 0 \\ 0 & 1 & 0 & 0 & 0 \end{bmatrix}$

(*c*) See Fig. 3-18(*b*)

(*d*) $S \circ S = \{(1,2),\ (1,4),\ (2,1),\ (2,2),\ (2,4),\ (3,2),\ (3,3),\ (3,5),\ (4,2),\ (5,2),\ (5,4)\}$

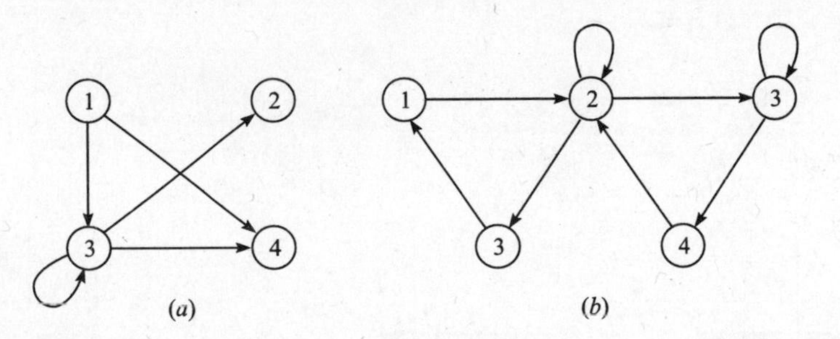


**Fig. 3-18**

**3.47.** (*a*) $E = \{b,c\},\ F = \{a,b,d,e\},\ G = \varnothing,\ H = \{a,b,f\}$

**3.48.** (*a*) None; (*b*) (2) and (3); (*c*) (1) and (4); (*d*) (1), (2), (4)

**3.49.** (*a*) $(3,3) \notin R$; (*b*) $(4,2) \in R$ but $(2,4) \notin R$; (*c*) $(2,3) \in R$, $(3,2) \in R$, but $2 \neq 3$;
(*d*) $(3,2) \in R$, $(2,3) \in R$, but $(3,3) \notin R$

**3.50.** (*a*) $R$; (*b*) $R$ and $S$; (*c*) $R$ and $T$; (*d*) $R$ and $T$

**3.51.** All true except: (*b*) $R = \{(1,1),\ (2,2),\ (3,3)\}$, so $(1,1) \notin R^c$; (*c*) and (*f*) $R = \{(2,2)\}$, so $(2,1),\ (1,2) \in R^c$, but $(2,2) \notin R^c$

**3.52.** (*a*) reflexive$(R) = \{(1,1),\ (2,2),\ (2,3),\ (4,2),\ (3,3),\ (4,4)\}$
(*b*) symmetric$(R) = \{(1,1),\ (2,2),\ (2,3),\ (4,2),\ (3,2),\ (2,4)\}$
(*c*) transitive$(R) = \{(1,1),\ (2,2),\ (2,3),\ (4,2),\ (4,3)\}$

**3.53.** (*a*) $A \times A$; (*b*) $\{(1,2),\ (1,3),\ (1,4),\ (3,3),\ (3,2),\ (3,4)\}$

**3.54.** (*a*) (i) $r+n$, (ii) $2r$
(*b*) $\{(1,2),\ (2,3), \ldots, (n-1,n),\ (n,1)\}$

**3.55.** (*a*) No; (*b*) no; (*c*) yes; (*d*) yes

**3.56.** There are five: $[S]$, $[\{1\},\ \{2,3\}]$, $[\{2\},\ \{1,3\}]$, $[\{3\},\ \{1,2\}]$, $[\{1\},\ \{2\},\ \{3\}]$.

**3.57.** (*a*) $\text{Max}(r,s) \le n \le rs$. (*b*) Every cell in $P_1$ is a subset of a cell in $P_2$.
(*c*) $[\{1,3,5\},\ \{2\},\ \{4\},\ \{6,7,9\},\ \{8\}]$

**3.58.** (*a*) $[\{1,8,15\},\ \{2,9,16\},\ \{3,10,17\},\ \{4,11,18\},\ \{5,12,19\},\ \{6,13,20\},\ \{7,14\}]$
(*b*) $\{8,2,10,4,12,6,14\}$

**3.59.** (*b*) $[(2,5)] = \{(1,4),\ (2,5),\ (3,6),\ (4,7),\ (5,8),\ (6,9)\}$

**3.60.** (*b*) If $a,b \in A$, then $b-a \notin A$. If $x \in R$, then $x = n+a$ where $n \in Z$ and $a \in A$.

**3.63.** (*a*) $\{(1,4,1),\ (1,9,2),\ (1,14,3),\ (2,2,2),\ (2,13,3)\}$;
(*b*) $\{(1,1,1,10),\ (1,1,2,15),\ (1,2,1,14),\ (2,1,1,13)\}$

**3.64.** See Fig. 3-19.

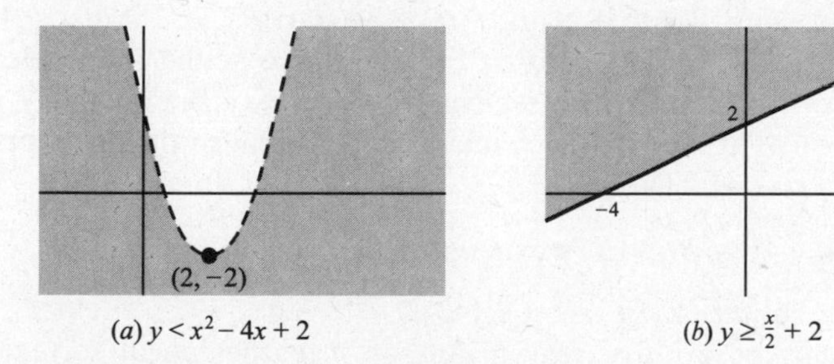


**Fig. 3-19**

**3.65.** (*a*) See Fig. 3-20(*a*); domain $= [-3,3]$, range $= [0,5]$
(*b*) See Fig. 3-20(*b*); domain $= (-4,5)$, range $= (-5,3)$.

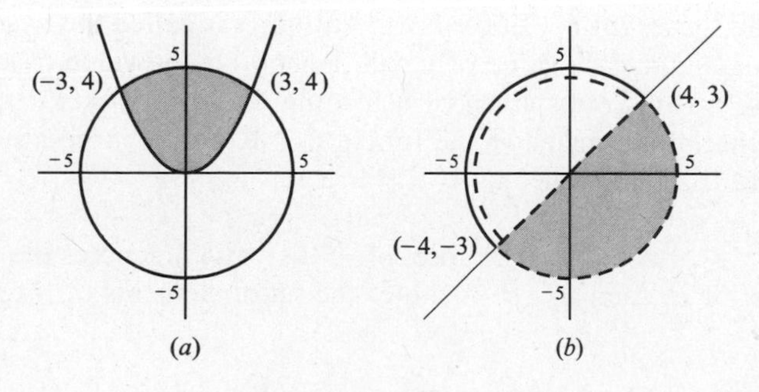


**Fig. 3-20**

# Chapter 4

# Functions

## 4.1 INTRODUCTION

One of the most important concepts in mathematics is that of a function. The terms "map", "mapping", "transformation", and many others mean the same thing; the choice of which word to use in a given situation is usually determined by tradition and the mathematical background of the person using the term.

## 4.2 FUNCTIONS

Suppose that to each element of a set $A$ we assign a unique element of a set $B$; the collection of such assignments is called a *function* from $A$ into $B$. The set $A$ is called the *domain* of the function, and the set $B$ is called the *target set*.

Functions are ordinarily denoted by symbols. For example, let $f$ denote a function from $A$ into $B$. Then we write

$$f: A \to B$$

which is read: "$f$ is a function from $A$ into $B$", or "$f$ takes $A$ into $B$", or "$f$ maps $A$ into $B$".

Suppose $f: A \to B$ and $a \in A$. Then $f(a)$ [read: "$f$ of $a$"] will denote the unique element of $B$ which $f$ assigns to $a$. This element $f(a)$ in $B$ is called the *image* of $a$ under $f$ or the *value* of $f$ at $a$. We also say that $f$ *sends* or *maps* $a$ into $f(a)$. The set of all such image values is called the *range* or *image* of $f$, and it is denoted by $\text{Ran}(f)$, $\text{Im}(f)$ or $f(A)$. That is,

$$\text{Im}(f) = \{b \in B : \text{there exists } a \in A \text{ for which } f(a) = b\}$$

We emphasize that $\text{Im}(f)$ is a subset of the target set $B$.

Frequently, a function can be expressed by means of a mathematical formula. For example, consider the function which sends each real number into its square. We may describe this function by writing

$$f(x) = x^2 \qquad \text{or} \qquad x \mapsto x^2 \qquad \text{or} \qquad y = x^2$$

In the first notation, $x$ is called a *variable* and the letter $f$ denotes the function. In the second notation, the barred arrow $\mapsto$ is read "goes into". In the last notation, $x$ is called the *independent variable* and $y$ is called the *dependent variable* since the value of $y$ will depend on the value of $x$.

Furthermore, suppose a function is given by a formula in terms of a variable $x$. Then we assume, unless otherwise stated, that the domain of the function is $\mathbf{R}$ or the largest subset of $\mathbf{R}$ for which the formula has meaning, and that the target set is $\mathbf{R}$.

**Remark**: Suppose $f: A \to B$. If $A'$ is a subset of $A$, then $f(A')$ denotes the set of images of elements in $A'$; and if $B'$ is a subset of $B$, then $f^{-1}(B')$ denotes the set of elements of $A$ each whose image belongs to $B'$. That is,

$$f(A') = \{f(a) : a \in A'\} \qquad \text{and} \qquad f^{-1}(B') = \{a \in A : f(a) \in B'\}$$

We call $f(A')$ the *image* of $A'$, and we call $f^{-1}(B')$ the *inverse image* or *preimage* of $B'$.

**EXAMPLE 4.1**

(*a*) Consider the function $f(x) = x^3$, i.e., $f$ assigns to each real number its cube. Then the image of 2 is 8, and so we may write $f(2) = 8$. Similarly, $f(-3) = -27$, and $f(0) = 0$.

(*b*) Let $g$ assign to each country in the world its capital city. Here the domain of $g$ is the set of all the countries in the world, and the target set is the list of cities in the world. The image of France under $g$ is Paris; that is $g(\text{France}) = \text{Paris}$. Similarly, $g(\text{Denmark}) = \text{Copenhagen}$ and $g(\text{England}) = \text{London}$.

(*c*) Figure 4-1 defines a function $f$ from $A = \{a, b, c, d\}$ into $B = \{r, s, t, u\}$ in the obvious way; that is,

$$f(a) = s, \qquad f(b) = u, \qquad f(c) = r, \qquad f(d) = s$$

The image of $f$ is the set $\{r, s, u\}$. Note that $t$ does not belong to the image of $f$ because $t$ is not the image of any element of $A$ under $f$.

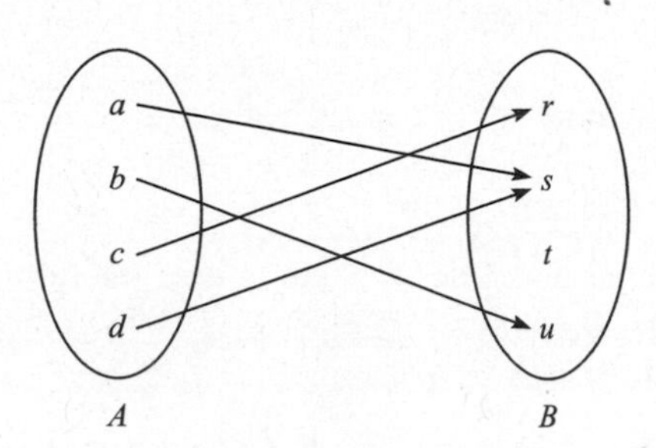


**Fig. 4-1**

### Identity Function

Consider any set $A$. Then there is a function from $A$ into $A$ which sends each element into itself. It is called the *identity function* on $A$ and it is usually denoted by $1_A$ or simply 1. In other words, the identity function $1_A : A \to A$ is defined by

$$1_A(a) = a$$

for every element $a \in A$.

### Functions as Relations

There is another point of view from which functions may be considered. First of all, every function $f: A \to B$ gvies rise to a relation from $A$ to $B$ called the *graph* of $f$ and defined by

$$\text{Graph of } f = \{(a, b) : a \in A, b = f(a)\}$$

Two functions $f: A \to B$ and $g: A \to B$ are defined to be equal, written $f = g$, if $f(a) = g(a)$ for every $a \in A$; that is, if they have the same graph. Accordingly, we do not distinguish between a function and its graph. Now, such a graph relation has the property that each $a$ in $A$ belongs to a unique ordered pair $(a, b)$ in the relation. On the other hand, any relation $f$ from $A$ to $B$ that has this property gives rise to a function $f: A \to B$, where $f(a) = b$ for each $(a, b)$ in $f$. Consequently, one may equivalently define a function as follows:

**Definition**: A function $f: A \to B$ is a relation from $A$ to $B$ (i.e., a subset of $A \times B$) such that each $a \in A$ belongs to a unique ordered pair $(a, b)$ in $f$.

Although we do not distinguish between a function and its graph, we will still use the terminology "graph of $f$" when referring to $f$ as a set of ordered pairs. Moreover, since the graph of $f$ is a relation, we can draw its picture as was done for relations in general, and this pictorial representation is itself sometimes called the graph of $f$. Also, the defining condition of a function, that each $a \in A$ belongs to a unique pair $(a, b)$ in $f$, is equivalent to the geometrical condition of each vertical line intersecting the graph in exactly one point.

**EXAMPLE 4.2**

(*a*) Let $f\colon A \to B$ be the function in Example 4.1(*c*). Then the graph of $f$ is the following set of ordered pairs:

$$f = \{(a, s),\ (b, u),\ (c, r),\ (d, s)\}$$

(*b*) Consider the following relations on the set $A = \{(1, 2, 3)\}$

$$f = \{(1, 3),\ (2, 3),\ (3, 1)\}, \qquad g = \{(1, 2),\ (3, 1)\}, \qquad h = \{(1, 3),\ (2, 1),\ (1, 2),\ (3, 1)\}$$

$f$ is a function from $A$ into $A$ since each member of $A$ appears as the first coordinate in exactly one ordered pair in $f$; here $f(1) = 3$, $f(2) = 3$ and $f(3) = 1$. $g$ is not a function from $A$ into $A$ since $2 \in A$ is not the first coordinate of any pair in $g$ and so $g$ does not assign any image to 2. Also $h$ is not a function from $A$ into $A$ since $1 \in A$ appears as the first coordinate of two distinct ordered pairs in $h$, $(1, 3)$ and $(1, 2)$. If $h$ is to be a function it cannot assign both 3 and 2 to the element $1 \in A$.

(*c*) By a *real polynomial function*, we mean a function $f : \mathbf{R} \to \mathbf{R}$ of the form

$$f(x) = a_n x^n + a_{n-1} x^{n-1} + \ldots + a_1 x + a_0$$

where the $a_i$ are real numbers. Since $\mathbf{R}$ is an infinite set, it would be impossible to plot each point of the graph. However, the graph of such a function can be approximated by first plotting some of its points and then drawing a smooth curve through these points. The points are usually obtained from a table where various values are assigned to $x$ and the corresponding values of $f(x)$ computed.

Figure 4-2 illustrates this technique using the function $f(x) = x^2 - 2x - 3$.

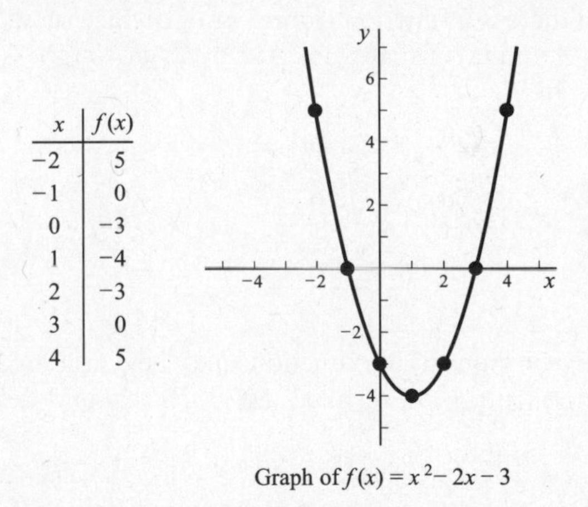

| $x$ | $f(x)$ |
|---|---|
| −2 | 5 |
| −1 | 0 |
| 0 | −3 |
| 1 | −4 |
| 2 | −3 |
| 3 | 0 |
| 4 | 5 |



Graph of $f(x) = x^2 - 2x - 3$

**Fig. 4-2**

## 4.3 COMPOSITION OF FUNCTIONS

Consider functions $f\colon A \to B$ and $g\colon B \to C$, that is, where the target set $B$ of $f$ is the domain of $g$. This relationship can be pictured by the following diagram:

$$A \xrightarrow{\ f\ } B \xrightarrow{\ g\ } C$$

Let $a \in A$; then its image $f(a)$ under $f$ is in $B$ which is the domain of $g$. Accordingly, we can find the image of $f(a)$ under the function $g$, that is, we can find $g(f(a))$. Thus we have a rule which assigns to each element $a$ in $A$ an element $g(f(a))$ in $C$ or, in other words, $f$ and $g$ give rise to a well defined function from $A$ to $C$. This new function is called the *composition* of $f$ and $g$, and it is denoted by

$$g \circ f$$

More briefly, if $f: A \to B$ and $g: B \to C$, then we define a new function $g \circ f: A \to C$ by

$$(g \circ f)(a) \equiv g(f(a))$$

Here $\equiv$ is used to mean equal by definition.

Note that we can now add the function $g \circ f$ to the above diagram of $f$ and $g$ as follows:

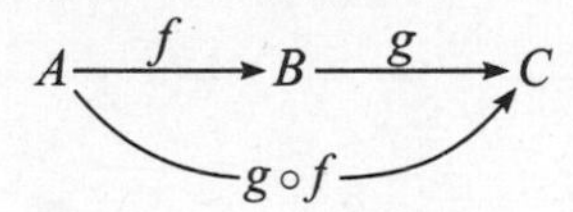


We emphasize that the composition of $f$ and $g$ is written $g \circ f$, and not $f \circ g$; that is, the composition of functions is read from right to left, and not from left to right.

**EXAMPLE 4.3**

(*a*) Let $f: A \to B$ and $g: B \to C$ be the functions defined by Fig. 4-3. We compute $g \circ f : A \to C$ by its definition:

$$(g \circ f)(a) \equiv g(f(a)) = g(y) = t, \quad (g \circ f)(b) \equiv g(f(b)) = g(z) = r, \quad (g \circ f)(c) \equiv g(f(c)) = g(y) = t$$

Observe that the composition $g \circ f$ is equivalent to "following the arrows" from $A$ to $C$ in the diagrams of the functions $f$ and $g$.

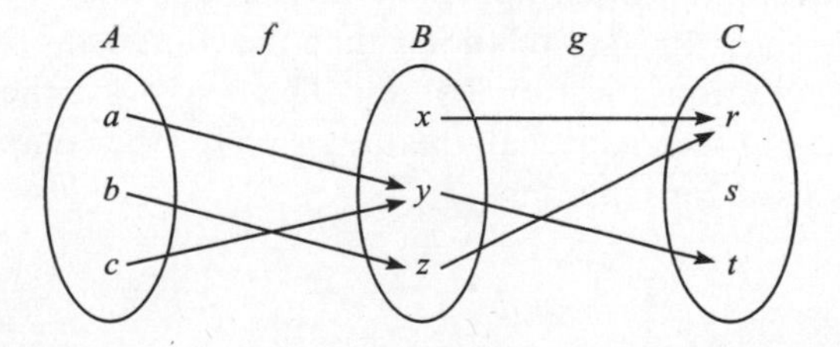


**Fig. 4-3**

(*b*) Let $f: \mathbf{R} \to \mathbf{R}$ and $g: \mathbf{R} \to \mathbf{R}$ be defined by $f(x) = x^2$ and $g(x) = x + 3$. Then

$$(g \circ f)(2) \equiv g(f(2)) = g(4) = 7; \quad (f \circ g)(2) \equiv f(g(2)) = f(5) = 25$$

Thus the composition functions $g \circ f$ and $f \circ g$ are not the same function. We compute a general formula for these functions:

$$(g \circ f)(x) \equiv g(f(x)) = g(x^2) = x^2 + 3$$
$$(f \circ g)(x) \equiv f(g(x)) = f(x + 3) = (x + 3)^2 = x^2 + 6x + 9$$

(*c*) Consider any function $f : A \to B$. Then one can easily show that

$$f \circ 1_A = f \quad \text{and} \quad 1_B \circ f = f$$

where $1_A$ and $1_B$ are the identity functions on $A$ and $B$, respectively. In other words, the composition of any function with the appropriate identity function is the function itself.

### Associativity of Composition of Functions

Consider functions $f: A \to B$, $g: B \to C$, and $h: C \to D$. Then, as pictured in Fig. 4-4(*a*), we can form the composition $g \circ f: A \to C$, and then the composition $h \circ (g \circ f): A \to D$. Similarly, as pictured in Fig. 4-4(*b*), we can form the composition $h \circ g: B \to D$, and then the composition

$(h \circ g) \circ f: A \to D$. Both $h \circ (g \circ f)$ and $(h \circ g) \circ f$ are functions with domain $A$ and target set $D$. The next theorem on functions (proved in Problem 4.15) states that these two functions are equal. That is:

**Theorem 4.1**: Let $f: A \to B$, $g: B \to C$, and $h: C \to D$. Then

$$h \circ (g \circ f) = (h \circ g) \circ f$$

Theorem 4.1 tells us that we can write $h \circ g \circ f: A \to D$ without any parentheses.

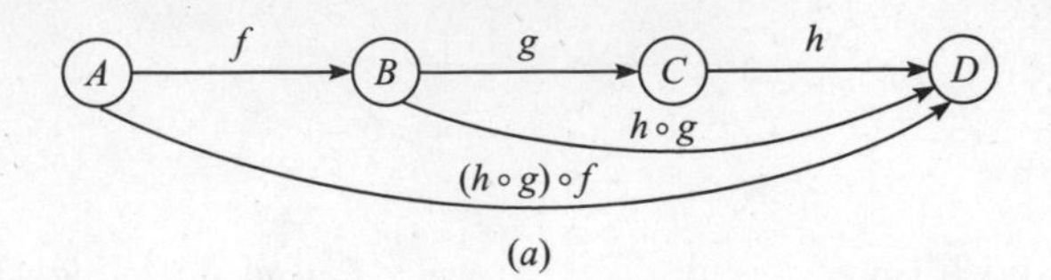


(a)

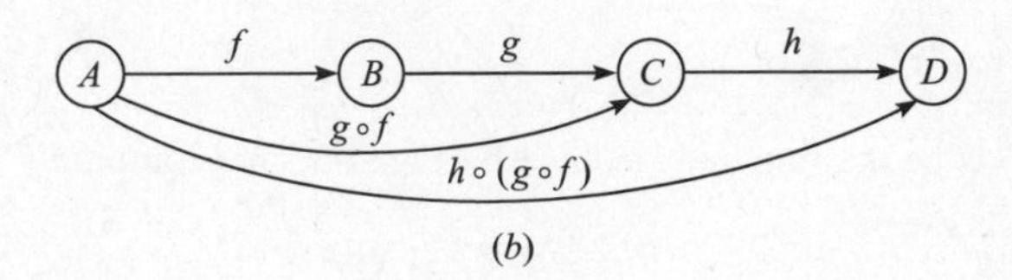


(b)

**Fig. 4-4**

**Remark**: The above definition of the composition of functions and Theorem 4.1 are not really new. Specifically, viewing the functions $f$ and $g$ as relations, then the composition function $g \circ f$ is the same as the composition of $f$ and $g$ as relations (Section 3.5) and Theorem 4.1 is the same as Theorem 3.1. One main difference is that here we use the functional notation $g \circ f$ for the composition of $f$ and $g$ instead of the notation $f \circ g$ which was used for relations.

## 4.4 ONE-TO-ONE, ONTO, AND INVERTIBLE FUNCTIONS

A function $f: A \to B$ is said to be *one-to-one* (written 1-1) if different elements in the domain $A$ have distinct images. Another way of saying the same thing follows:

$f$ is one-to-one if $f(a) = f(a')$ implies $a = a'$

A function $f: A \to B$ is said to be an *onto* function if every element of $B$ is the image of some element in $A$ or, in other words, if the image of $f$ is the entire target set $B$. In such a case we say that $f$ is a function of $A$ onto $B$ or that $f$ maps $A$ onto $B$. That is:

$f$ maps $A$ onto $B$ if $\forall b \in B$, $\exists a \in A$ such that $f(a) = b$

Here

$\forall$ means "for every", and $\exists$ means "there exist"

(These quantifiers are discussed in Chapter 10.)

A function $f: A \to B$ is said to be *invertible* if its inverse relation $f^{-1}$ is a function from $B$ to $A$. Equivalently, $f: A \to B$ is *invertible* if there exists a function $f^{-1}: B \to A$, called the *inverse* of $f$, such that

$$f^{-1} \circ f = 1_A \quad \text{and} \quad f \circ f^{-1} = 1_B$$

In general, an inverse function $f^{-1}$ need not exist or, equivalently, the inverse relation $f^{-1}$ may not be a function. The following theorem (proved in Problem 4.23) gives simple criteria which tell us when it is.

**Theorem 4.2**: A function $f: A \to B$ is invertible if and only if $f$ is both one-to-one and onto.

If $f: A \to B$ is both one-to-one and onto, then $f$ is called a *one-to-one correspondence* between $A$ and $B$. This terminology comes from the fact that each element of $A$ will correspond to a unique element of $B$ and vice versa.

Some texts use the term *injective* for a one-to-one function, *surjective* for an onto function, and *bijective* for a one-to-one correspondence.

**EXAMPLE 4.4** Consider functions $f_1: A \to B$, $f_2: B \to C$, $f_3: C \to D$, and $f_4: D \to E$ defined by Fig. 4-5. Now $f_1$ is one-to-one since no element of $B$ is the image of more than one element of $A$. Similarly, $f_2$ is one-to-one. However, neither $f_3$ nor $f_4$ is one-to-one since $f_3(r) = f_3(u)$ and $f_4(v) = f_4(w)$.

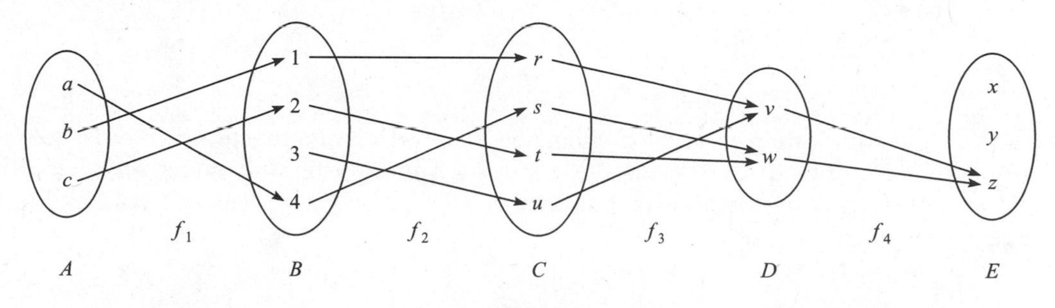


**Fig. 4-5**

As far as being onto is concerned, $f_2$ and $f_3$ are both onto functions since every element of $C$ is the image under $f_2$ of some element of $B$ and every element of $D$ is the image under $f_3$ of some element of $C$, i.e., $f_2(B) = C$ and $f_3(C) = D$. On the other hand, $f_1$ is not onto since $3 \in B$ but 3 is not the image under $f_1$ of any element of $A$, and $f_4$ is not onto since, for example, $x \in E$ but $x$ is not the image under $f_4$ of any element of $D$.

Thus $f_1$ is one-to-one but not onto, $f_3$ is onto but not one-to-one, and $f_4$ is neither one-to-one nor onto. However, $f_2$ is both one-to-one and onto, i.e., $f_2$ is a one-to-one correspondence between $A$ and $B$. Hence $f_2$ is invertible and $f_2^{-1}$ is a function from $C$ to $B$.

### Geometrical Characterization of One-to-One and Onto Functions

Consider now a real-valued function $f: \mathbf{R} \to \mathbf{R}$. Since $f$ may be identified with its graph and the graph may be plotted in the cartesian plane $\mathbf{R}^2$, we might wonder whether the concepts of being one-to-one and onto have some geometrical meaning. The answer is yes. Specifically:

(*a*) The function $f: \mathbf{R} \to \mathbf{R}$ is one-to-one means that there are no two distinct pairs $(a_1, b)$ and $(a_2, b)$ in the graph of $f$; hence each vertical line in $\mathbf{R}^2$ can intersect the graph of $f$ in at most one point.

(*b*) The function $f: \mathbf{R} \to \mathbf{R}$ is onto means that for every $b \in \mathbf{R}$ there is at least one point $a \in \mathbf{R}$ such that $(a, b)$ belongs to the graph of $f$; hence each vertical line in $\mathbf{R}^2$ must intersect the graph of $f$ at least once.

(*c*) Accordingly, the function $f: \mathbf{R} \to \mathbf{R}$ is one-to-one and onto, i.e., $f$ is invertible, if and only if each horizontal line in $\mathbf{R}^2$ will intersect the graph of $f$ in exactly one point.

We illustrate the above properties in the next example.

**EXAMPLE 4.5** Consider the following four functions from **R** into **R** whose graphs appear in Fig. 4-6:

$$f_1(x) = x^2, \qquad f_2(x) = 2^x, \qquad f_3(x) = x^3 - 2x^2 - 5x + 6, \qquad f_4(x) = x^3$$

Observe that there are horizontal lines which intersect the graph of $f_1$ twice and there are horizontal lines which do not intersect the graph of $f_1$ at all; hence $f_1$ is neither one-to-one nor onto. Similarly, $f_2$ is one-to-one but not onto, $f_3$ is onto but not one-to-one, and $f_4$ is both one-to-one and onto. The inverse of $f_4$ is the cube root function, that is,

$$f_4^{-1}(x) = \sqrt[3]{x}$$

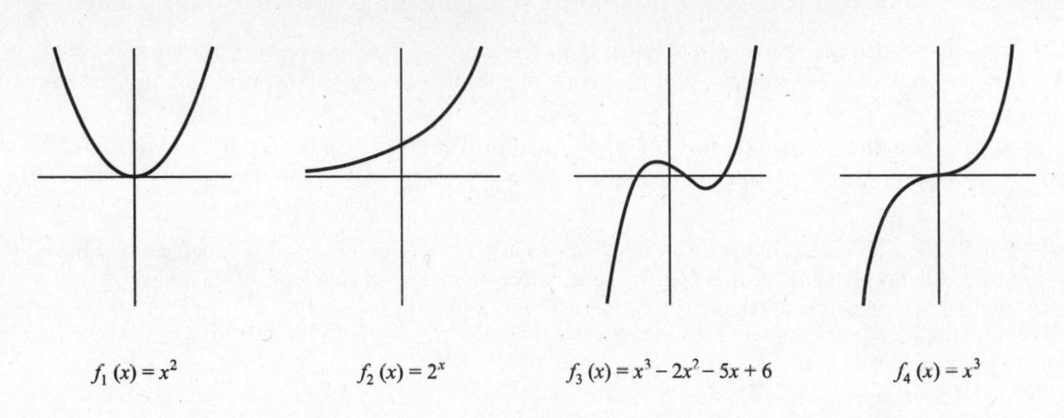


**Fig. 4-6**

**Remark**: Sometimes we restrict the domain and/or target set of a function $f$ in order to obtain an inverse function $f^{-1}$. For example, suppose we restrict the domain and target set of the function $f_1(x) = x^2$ to be the set $D$ of nonnegative real numbers. Then $f_1$ is one-to-one and onto and its inverse is the square root function, that is,

$$f_1^{-1}(x) = \sqrt{x}$$

Similarly, suppose we restrict the target set of the exponential function $f_2(x) = 2^x$ to be the set $\mathbf{R}^+$ of positive real numbers. Then $f_1$ is one-to-one and onto and its inverse is the logarithmic function (to the base 2), that is,

$$f_2^{-1}(x) = \log_2 x$$

(Exponential and logarithmic functions are investigated in Section 4.5.)

## 4.5 MATHEMATICAL FUNCTIONS, EXPONENTIAL AND LOGARITHMIC FUNCTIONS

This section presents various mathematical functions which appear often in mathematics and computer science, together with their notation. We also discuss the exponential and logarithmic functions, and their relationship.

### Integer and Absolute Value Functions

Let $x$ be any real number. The *integer value* of $x$, written $\mathrm{INT}(x)$, converts $x$ into an integer by deleting (truncating) the fractional part of the number. Thus

$$\mathrm{INT}(3.14) = 3, \qquad \mathrm{INT}(\sqrt{5}) = 2, \qquad \mathrm{INT}(-8.5) = -8, \qquad \mathrm{INT}(7) = 7$$

The *absolute value* of the real number $x$, written $\mathrm{ABS}(x)$ or $|x|$, is defined as the greater of $x$ or $-x$. Hence $\mathrm{ABS}(0) = 0$, and, for $x \neq 0$, $\mathrm{ABS}(x) = x$ or $\mathrm{ABS}(x) = -x$, depending on whether $x$ is positive or negative. Thus

$$|-15| = 15, \qquad |7| = 7, \qquad |-3.33| = 3.33, \qquad |4.44| = 4.44, \qquad |-0.975| = 0.075$$

We note that $|x| = |-x|$ and, for $x \neq 0$, $|x|$ is positive.

**Remainder Function; Modular Arithmetic**

Let $k$ be any integer and let $M$ be a positive integer. Then

$$k \text{ (mod } M)$$

(read $k$ *modulo* $M$) will denote the integer remainder when $k$ is divided by $M$. More exactly, $k$ (mod $M$) is the unique integer $r$ such that

$$k = Mq + r \qquad \text{where} \qquad 0 \le r < M$$

When $k$ is positive, simply divide $k$ by $M$ to obtain the remainder $r$. Thus

$$25 \text{ (mod } 7) = 4, \qquad 25 \text{ (mod } 5) = 0, \qquad 35 \text{ (mod } 11) = 2, \qquad 3 \text{ (mod } 8) = 3$$

Problem 4.25 shows a method to obtain $k$ (mod $M$) when $k$ is negative.

The term "mod" is also used for the mathematical congruence relation, which is denoted and defined as follows:

$$a \equiv b \text{ (mod } M) \qquad \text{if and only if} \qquad M \text{ divides } b - a$$

$M$ is called the *modulus*, and $a \equiv b$ (mod $M$) is read "$a$ is congruent to $b$ modulo $M$". The following aspects of the congruence relation are frequently useful:

$$0 \equiv M \text{ (mod } M) \qquad \text{and} \qquad a \pm M \equiv a \text{ (mod } M)$$

*Arithmetic modulo* $M$ refers to the arithmetic operations of addition, multiplication, and subtraction where the arithmetic value is replaced by its equivalent value in the set

$$\{0, 1, 2, \ldots, M - 1\}$$

or in the set

$$\{1, 2, 3, \ldots, M\}$$

For example, in arithmetic modulo 12, sometimes called "clock" arithmetic,

$$6 + 9 \equiv 3, \qquad 7 \times 5 \equiv 11, \qquad 1 - 5 \equiv 8, \qquad 2 + 10 \equiv 0 \equiv 12$$

(The use of 0 or $M$ depends on the application.)

**Exponential Functions**

Recall the following definitions for integer exponents (where $m$ is a positive integer):

$$a^m = a \cdot a \ldots a \,(m \text{ times}), \qquad a^0 = 1, \qquad a^{-m} = \frac{1}{a^m}$$

Exponents are extended to include all rational numbers by defining, for any rational number $m/n$,

$$a^{m/n} = \sqrt[n]{a^m} = \left(\sqrt[n]{a}\right)^m$$

For example,

$$2^4 = 16, \qquad 2^{-4} = \frac{1}{2^4} = \frac{1}{16}, \qquad 125^{2/3} = 5^2 = 25$$

In fact, exponents are extended to include all real numbers by defining, for any real number $x$,

$$a^x = \lim_{r \to x} a^r \qquad \text{where } r \text{ is a rational number}$$

Accordingly, the exponential function $f(x) = a^x$ is defined for all real numbers.

### Logarithmic Functions

Logarithms are related to exponents as follows. Let $b$ be a positive number. The logarithm of any positive number $x$ to the base $b$, written

$$\log_b x$$

represents the exponent to which $b$ must be raised to obtain $x$. That is,

$$y = \log_b x \quad \text{and} \quad b^y = x$$

are equivalent statements. Accordingly,

$$\log_2 8 = 3 \quad \text{since} \quad 2^3 = 8; \qquad \log_{10} 100 = 2 \quad \text{since} \quad 10^2 = 100$$

$$\log_2 64 = 6 \quad \text{since} \quad 2^6 = 64; \qquad \log_{10} 0.001 = -3 \quad \text{since} \quad 10^{-3} = 0.001$$

Furthermore, for any base $b$,

$$\log_b 1 = 0 \quad \text{since} \quad b^0 = 1$$

$$\log_b b = 1 \quad \text{since} \quad b^1 = b$$

The logarithm of a negative number and the logarithm of 0 are not defined.

Frequently, logarithms are expressed using approximate values. For example, using tables or calculators, one obtains

$$\log_{10} 300 = 2.4771 \quad \text{and} \quad \log_e 40 = 3.6889$$

as approximate answers. (Here $e = 2.718\,281 \cdots$.)

Three classes of logarithms are of special importance: logarithms to base 10, called *common logarithms*; logarithms to base $e$, called *natural logarithms*; and logarithms to base 2, called *binary logarithms*. Some texts write:

$$\ln x \text{ for } \log_e x \quad \text{and} \quad \lg x \text{ or } \log x \text{ for } \log_2 x$$

The term $\log x$, by itself, usually means $\log_{10} x$; but it is also used for $\log_e x$ in advanced mathematical texts and for $\log_2 x$ in computer science texts.

### Relationship between the Exponential and Logarithmic Functions

The basic relationship between the exponential and the logarithmic functions

$$f(x) = b^x \quad \text{and} \quad g(x) = \log_b x$$

is that they are inverses of each other; hence the graphs of these functions are related geometrically. This relationship is illustrated in Fig. 4-7 where the graphs of the exponential function $f(x) = 2^x$, the logarithmic function $g(x) = \log_2 x$, and the linear function $h(x) = x$ appear on the same coordinate axis. Since $f(x) = 2^x$ and $g(x) = \log_2 x$ are inverse functions, they are symmetric with respect to the linear function $h(x) = x$ or, in other words, the line $y = x$.

Figure 4-7 also indicates another important property of the exponential and logarithmic functions. Specifically, for any positive $c$, we have

$$g(c) < h(c) < f(c)$$

In fact, as $c$ increases in value, the vertical distances $h(c) - g(c)$ and $f(c) - g(c)$ increase in value. Moreover, the logarithmic function $g(x)$ grows very slowly compared with the linear function $h(x)$, and the exponential function $f(x)$ grows very quickly compared with $h(x)$.

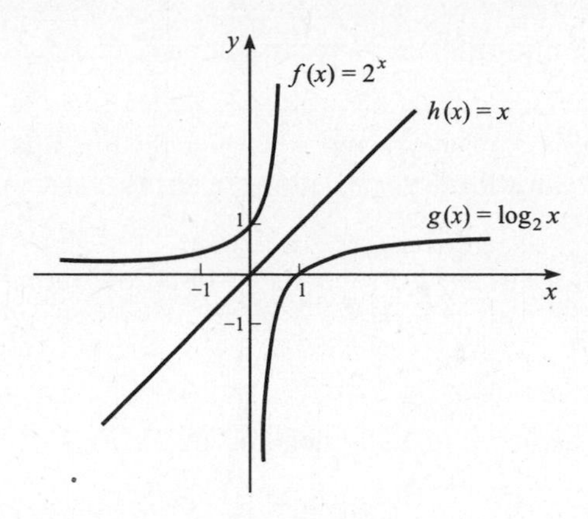


Fig. 4-7

## 4.6 RECURSIVELY DEFINED FUNCTIONS

A function is said to be *recursively defined* if the function definition refers to itself. In order for the definition not to be circular, the function definition must have the following two properties:

(1) There must be certain arguments, called *base values*, for which the function does not refer to itself.
(2) Each time the function does refer to itself, the argument of the function must be closer to a base value.

A recursive function with these two properties is said to be *well-defined*.

The following examples should help clarify these ideas.

### Factorial Function

The product of the positive integers from 1 to $n$, inclusive, is called "$n$ factorial" and is usually denoted by $n!$:

$$n! = 1 \cdot 2 \cdot 3 \ldots (n-2)(n-1)n$$

It is also convenient to define $0! = 1$, so that the function is defined for all nonnegative integers. Thus we have

$$0! = 1, \quad 1! = 1, \quad 2! = 1 \cdot 2 = 2, \quad 3! = 1 \cdot 2 \cdot 3 = 6, \quad 4! = 1 \cdot 2 \cdot 3 \cdot 4 = 24,$$
$$5! = 1 \cdot 2 \cdot 3 \cdot 4 \cdot 5 = 120, \quad 6! = 1 \cdot 2 \cdot 3 \cdot 4 \cdot 5 \cdot 6 = 720$$

and so on. Observe that

$$5! = 5 \cdot 4! = 5 \cdot 24 = 120 \quad \text{and} \quad 6! = 6 \cdot 5! = 6 \cdot 120 = 720$$

This is true for every positive integer $n$; that is,

$$n! = n \cdot (-1)!$$

Accordingly, the factorial function may also be defined as follows:

**Definition 4.1: (Factorial Function)**
(*a*) If $n = 0$, then $n! = 1$.
(*b*) If $n > 0$, then $n! = n \cdot (n-1)!$

Observe that the above definition of $n!$ is recursive, since it refers to itself when it uses $(n-1)!$ However:

(1) The value of $n!$ is explicitly given when $n = 0$ (thus 0 is a base value).
(2) The value of $n!$ for arbitrary $n$ is defined in terms of a smaller value of $n$ which is closer to the base value 0.

Accordingly, the definition is not circular, or, in other words, the function is well-defined.

**Fibonacci Sequence**

The celebrated Fibonacci sequence (usually denoted by $F_0, F_1, F_2, \ldots$) is as follows:

$$0, 1, 1, 2, 3, 5, 8, 13, 21, 34, 55, \ldots$$

That is, $F_0 = 0$ and $F_1 = 1$ and each succeeding term is the sum of the two preceding terms. For example, the next two terms of the sequence are

$$34 + 55 = 89 \qquad \text{and} \qquad 55 + 89 = 144$$

A formal definition of this function follows:

**Definition 4.2: (Fibonacci Sequence)**
(*a*) If $n = 0$ or $n = 1$, then $F_n = n$.
(*b*) If $n > 1$, then $F_n = F_{n-2} + F_{n-1}$.

This is another example of a recursive definition, since the definition refers to itself when it uses $F_{n-2}$ and $F_{n-1}$. However:

(1) The base values are 0 and 1.
(2) The value of $F_n$ is defined in terms of smaller values of $n$ which are closer to the base values.

Accordingly, this function is well-defined.

# Solved Problems

## FUNCTIONS

**4.1.** State whether or not each diagram in Fig. 4-8 defines a function from $A = \{a, b, c\}$ into $B = \{x, y, z\}$.

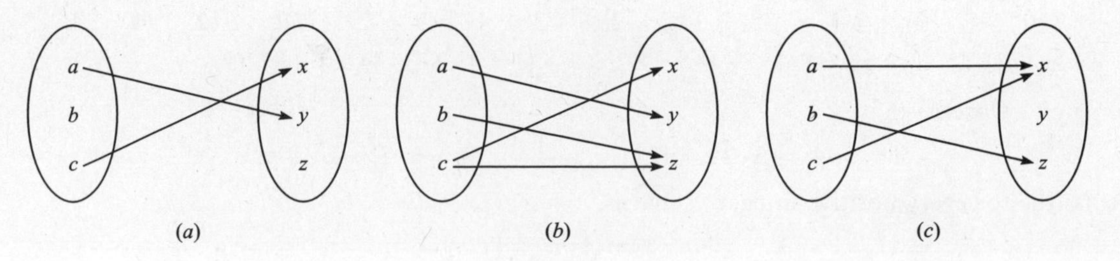


**Fig. 4-8**

(*a*) No. There is nothing assigned to the element $b \in A$.
(*b*) No. Two elements, $x$ and $z$, are assigned to $c \in A$.
(*c*) Yes. Every element in the domain $A = \{a, b, c\}$ is assigned a unique element in the target set $B$.

**4.2.** Let $X = \{1, 2, 3, 4\}$. Determine whether or not each relation below is a function from $X$ into $X$.

(a) $f = \{(2,3),\ (1,4),\ (2,1),\ (3,2),\ (4,4)\}$
(b) $g = \{(3,1),\ (4,2),\ (1,1)\}$
(c) $h = \{(2,1),\ (3,4),\ (1,4),\ (2,1),\ (4,4)\}$

Recall that a subset $f$ of $X \times X$ is a function $f: X \to X$ if and only if each $a \in X$ appears as the first coordinate in exactly one ordered pair in $f$.

(a) No. Two different ordered pairs $(2,3)$ and $(2,1)$ in $f$ have the same number 2 as their first coordinate.
(b) No. The element $2 \in X$ does not appear as the first coordinate in any ordered pair in $g$.
(c) Yes. Although $2 \in X$ appears as the first coordinate in two ordered pairs in $h$, these two ordered pairs are equal.

**4.3.** Let $A$ be the set of students in a school. Determine which of the following assignments defines a function on $A$:

(a) To each student assign his age.
(b) To each student assign his teacher.
(c) To each student assign his sex.
(d) To each student assign his spouse.

A collection of assignments is a function on $A$ if and only if each element $a$ in $A$ is assigned exactly one element. Thus:

(a) Yes, because each student has one and only one age.
(b) Yes, if each student has only one teacher; no, if any student has more than one teacher.
(c) Yes.
(d) No, unless every student is married.

**4.4.** Sketch the graph of: (a) $f(x) = x^2 + x - 6$; (b) $g(x) = x^3 - 3x^2 - x + 3$.

Set up a table of values for $x$ and then find the corresponding values of the function. Since the functions are polynomials, plot the points in a coordinate diagram and then draw a smooth continuous curve through the points. See Fig. 4-9.

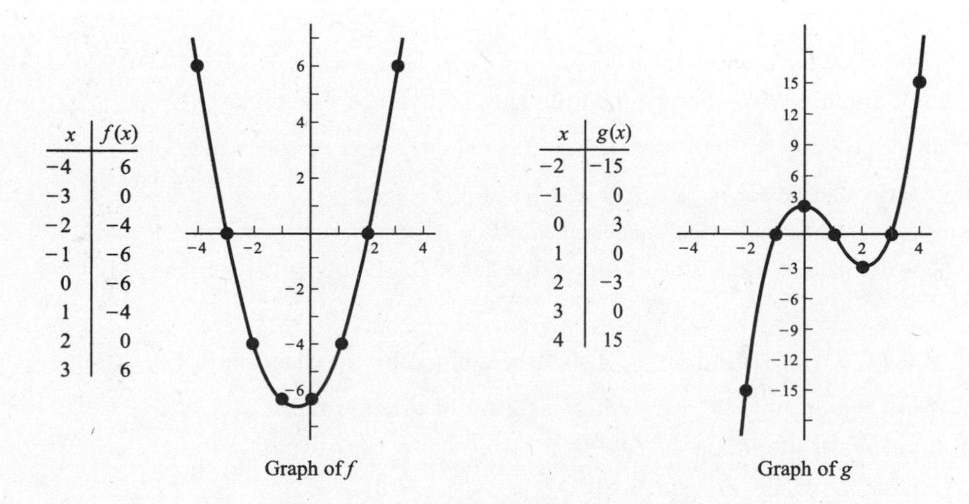

| $x$ | $f(x)$ |
|---|---|
| −4 | 6 |
| −3 | 0 |
| −2 | −4 |
| −1 | −6 |
| 0 | −6 |
| 1 | −4 |
| 2 | 0 |
| 3 | 6 |

| $x$ | $g(x)$ |
|---|---|
| −2 | −15 |
| −1 | 0 |
| 0 | 3 |
| 1 | 0 |
| 2 | −3 |
| 3 | 0 |
| 4 | 15 |



**Fig. 4-9**

**4.5.** Determine which of the graphs in Fig. 4-10 are functions from **R** into **R**.

Geometrically speaking, a set of points in the plane $\mathbf{R}^2$ is a function if and only if every vertical line contains exactly one point of the set. Thus: (*a*) Yes. (*b*) No. (*c*) No; however the graph does define a function from $D$ into **R** where $D = [-2, 2] = \{x : -2 \le x \le 2\}$.

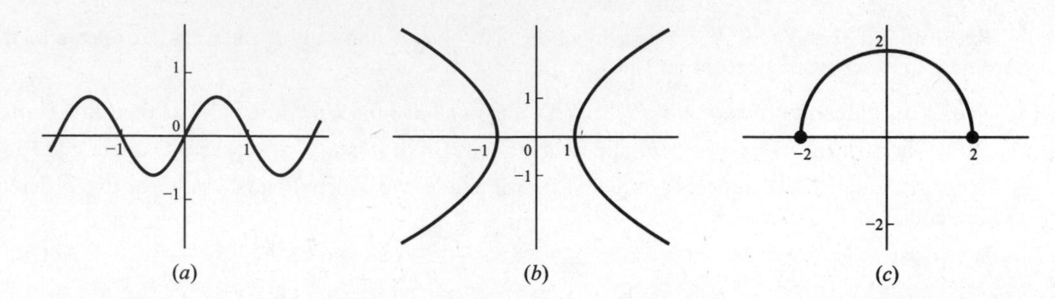


**Fig. 4-10**

**4.6.** Let the function $f : \mathbf{R} \to \mathbf{R}$ be defined as follows:

$$f(x) = \begin{cases} 3x - 1 & \text{if } x > 3 \\ x^2 - 2 & \text{if } -2 \le x \le 3 \\ 2x + 3 & \text{if } x < -2 \end{cases}$$

Find: (*a*) $f(2)$, (*b*) $f(4)$, (*c*) $f(-1)$, (*d*) $f(-3)$

Note that there are three formulas used to define the single function $f$. (The reader should not confuse formulas and functions.)

(*a*) Since 2 belongs to the closed interval $[-2, 3]$, we use the formula $f(x) = x^2 - 2$. Hence

$$f(2) = 2^2 - 2 = 4 - 2 = 2$$

(*b*) Since 4 belongs to $(3, \infty)$, we use the formula $f(x) = 3x - 1$. Thus $f(4) = 3(4) - 1 = 12 - 1 = 11$.

(*c*) Since $-1$ is in the interval $[-2, 3]$, we use the formula $f(x) = x^2 - 2$. Computing,

$$f(-1) = (-1)^2 - 2 = 1 - 2 = -1$$

(*d*) Since $-3$ is less than $-2$, i.e., $-3$ belongs to $(-\infty, -2)$, we use the formula $f(x) = 2x + 3$. Thus

$$f(-3) = 2(-3) + 3 = -6 + 3 = -3$$

**4.7.** Find the domain $D$ of each of the following real-valued functions:

(*a*) $f(x) = 1/(x - 2)$; (*b*) $g(x) = x^2 - 3x - 4$; (*c*) $h(x) = \sqrt{25 - x^2}$.

(*a*) $f$ is not defined for $x - 2 = 0$ or $x = 2$; hence $D = \mathbf{R}\backslash\{2\}$.

(*b*) $g$ is defined for every real number; hence $D = \mathbf{R}$.

(*c*) $h$ is not defined when $25 - x^2$ is negative; hence $D = [-5, 5] = \{x : -5 \le x \le 5\}$.

**4.8.** Let $A = \{1, 2, 3, 4, 5\}$ and let $f : A \to A$ be defined by the diagram in Fig. 4-11.

(*a*) Find the graph of $f$, i.e., write $f$ as a set of ordered pairs.

(*b*) Find $f(A)$, the image of $f$.

(*c*) Find $f(S)$ where $S = \{1, 3, 5\}$.

(*d*) Find $f^{-1}(T)$ where $T = \{2, 3\}$.

(*a*) The graph of $f$ consists of all pairs $(a, f(a))$ where $a \in A$. Hence

$$f = \{(1,3),\ (2,5),\ (3,5),\ (4,2),\ (5,3)\}$$

(*b*) $f(A)$ consists of all image points. Since only 2, 3, 5 appear as image points, $f(A) = \{2,3,5\}$.

(*c*) $f(S) = f(\{1,3,5\}) = \{f(1), f(3), f(5)\} = \{3,5,3\} = \{3,5\}$.

(*d*) The element 4 has image 2, and the elements 1 and 5 have image 3; hence $f^{-1}(T) = f^{-1}(\{2,3\}) = \{1,4,5\}$.

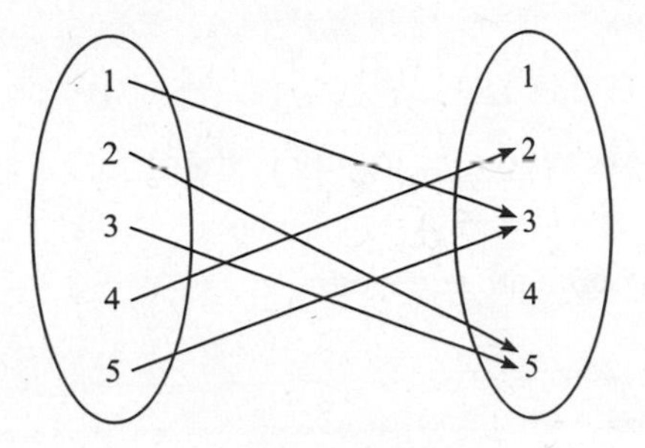


**Fig. 4-11**

**4.9.** Suppose $A = \{a, b\}$ and $B = \{1,2,3\}$. Find the number $m$ of functions:
(*a*) from $A$ into $B$, (*b*) from $B$ into $A$.

(*a*) There are three choices, 1, 2, or 3 for the image of $a$, and three choices for the image of $b$. Hence there are $m = 3 \cdot 3 = 9$ functions from $A$ into $B$.

(*b*) There are two choices, $a$ or $b$, for each of the three elements of $B$. Hence there are $m = 2 \cdot 2 \cdot 2 = 2^3 = 8$ functions from $B$ into $A$.

**4.10.** Suppose $A$ and $B$ are finite sets with $|A|$ elements and $|B|$ elements, respectively. Show there are $|B|^{|A|}$ functions from $A$ into $B$. (For this reason, one sometimes writes $B^A$ for the collection of all functions from $A$ into $B$.)

There are $|B|$ choices for each of the $|A|$ elements of $A$; hence there are $|B|^{|A|}$ possible functions from $A$ into $B$.

## COMPOSITION OF FUNCTIONS

**4.11.** Let the functions $f\colon A \to B$ and $g\colon B \to C$ be defined by Fig. 4-12. Find the composition function $g \circ f\colon A \to C$.

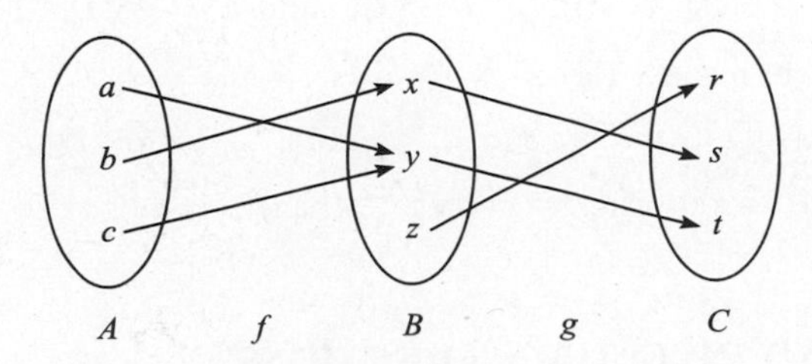


**Fig. 4-12**

Use the definition of the composition function to compute:

$$(g \circ f)(a) = g(f(a)) = g(y) = t, \quad (g \circ f)(b) = g(f(b)) = g(x) = s$$
$$(g \circ f)(c) = g(f(c)) = g(y) = t$$

Note that we arrive at the same answer if we "follow the arrows" in the diagram:

$$a \to y \to t, \qquad b \to x \to s, \qquad c \to y \to t$$

**4.12.** Let the functions $f$ and $g$ be defined by $f(x) = 2x + 1$ and $g(x) = x^2 - 2$. Find the formula defining the composition functions: (*a*) $g \circ f$, (*b*) $f \circ g$.

(*a*) Compute $g \circ f$ as follows:

$$(g \circ f)(x) = g(f(x)) = g(2x + 1) = (2x + 1)^2 - 2 = 4x^2 + 4x - 1.$$

Observe that the same answer can be found by writing

$$y = f(x) = 2x + 1 \qquad \text{and} \qquad z = g(y) = y^2 - 2$$

and then eliminating $y$ from both equations:

$$z = y^2 - 2 = (2x + 1)^2 - 2 = 4x^2 + 4x - 1$$

(*b*) Compute $f \circ g$ as follows:

$$(f \circ g)(x) = f(g(x)) = f(x^2 - 2) = 2(x^2 - 2) + 1 = 2x^2 - 3$$

**4.13.** Let $f\colon A \to B$. When is $f \circ f$ defined?

The composition $f \circ f$ is defined when the domain of $f$ is the same as the target set of $f$; that is, when $A = B$.

**4.14.** Let $f\colon \mathbf{R} \to \mathbf{R}$ be defined by $f(x) = x^2 + 2x$.

(*a*) Find $(f \circ f)$ (2) and $(f \circ f)$ (3). (*b*) Find a formula for $f \circ f$.

(*a*) $(f \circ f)\ (2) = f(f(2)) = f(8) = (8)^2 + 16 = 80$

$(f \circ f)\ (3) = f(f(3)) = f(15) = (15)^2 + 30 = 255$

(*b*) $$\begin{aligned}(f \circ f)\ (x) &= f(f(x)) = f(x^2 + 2x) = (x^2 + 2x)^2 + 2(x^2 + 2x)\\ &= x^4 + 4x^3 + 4x^2 + 2x^2 + 4x\\ &= x^4 + 4x^3 + 6x^2 + 4x\end{aligned}$$

**4.15.** Prove Theorem 4.1: Let $f\colon A \to B$, $g\colon B \to C$, and $h\colon C \to D$. Then $(f \circ g) \circ h = f \circ (g \circ h)$.

Consider any element $a \in A$. Then:

$$(h \circ (g \circ f))(a) = h((g \circ f)(a)) = h(g(f(a))) \qquad \text{and} \qquad ((h \circ g) \circ f)(a) = (h \circ g)(f(a)) = h(g(f(a)))$$

Thus $(h \circ (g \circ f))(a) = ((h \circ g) \circ f)(a)$ for every $a \in A$, and so $h \circ (g \circ f) = (h \circ g) \circ f$.

## ONE-TO-ONE, ONTO, AND INVERTIBLE FUNCTIONS

**4.16.** Suppose $f\colon A \to B$. Determine conditions under which:

(*a*) $f$ is not one-to-one (injective); (*b*) $f$ is not onto (surjective).

(*a*) $f$ is not one-to-one if there exist $a, a' \in A$ for which $f(a) = f(a')$ but $a \neq a'$.

(*b*) $f$ is not onto if there exists $b \in B$ such that $f(x) \neq b$ for every $x \in A$.

**4.17.** Determine if each function is one-to-one.

(*a*) To each person on the earth assign the number which corresponds to his age.
(*b*) To each country in the world assign the latitude and longitude of its capital.
(*c*) To each book written by only one author assign the author.
(*d*) To each country in the world which has a prime minister assign its prime minister.

(*a*) No. Many people in the world have the same age.
(*b*) Yes.
(*c*) No. There are different books with the same author.
(*d*) Yes. Different countries in the world have different prime ministers.

**4.18.** Let the functions $f: A \to B$, $g: B \to C$, and $h: C \to D$ be defined by Fig. 4-13.

(*a*) Determine if each function is one-to-one.
(*b*) Determine if each function is onto.
(*c*) Determine if each function is invertible.
(*d*) Find the composition $h \circ g \circ f$.

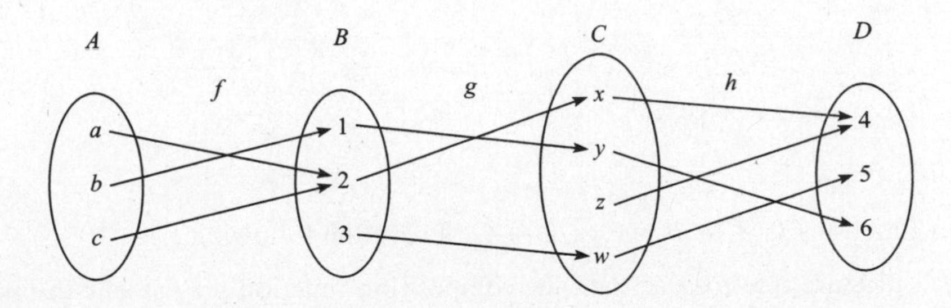


**Fig. 4-13**

(*a*) $f$ is not one-to-one since $f(a) = f(c)$ but $a \neq c$. $h$ is not one-to-one since $h(x) = h(z)$ but $x \neq z$. $g$ is one-to-one, the elements $1, 2, 3 \in B$ have distinct images.
(*b*) $f: A \to B$ is not onto since $3 \in B$ is not the image of any element in $A$.
$g: B \to C$ is not onto since $z \in C$ is not the image of any element in $B$.
$h: C \to D$ is onto since each element in $D$ is the image of some element of $C$.
(*c*) None of the functions are both one-to-one and onto; hence none of the functions are invertible.
(*d*) Now $a \to 2 \to x \to 4$, $b \to 1 \to y \to 6$, $c \to 2 \to x \to 4$. Hence $h \circ g \circ f = \{(a, 4),\ (b, 6),\ (c, 4)\}$.

**4.19.** Let $f: \mathbf{R} \to \mathbf{R}$ be defined by $f(x) = 2x - 3$. Now $f$ is one-to-one and onto; hence $f$ has an inverse function $f^{-1}$. Find a formula for $f^{-1}$.

Let $y$ be the image of $x$ under the function $f$; that is, set

$$y = f(x) = 2x - 3 \qquad (1)$$

Consequently, $x$ will be the image of $y$ under the inverse function $f^{-1}$.

**Method 1:** Solve for $x$ in terms of $y$ in equation (*1*) obtaining

$$x = (y + 3)/2$$

Then $f^{-1}(y) = (y + 3)/2$. Replace $y$ by $x$ to obtain

$$f^{-1}(x) = (x + 3)/2$$

which is the formula for $f^{-1}$ using the usual independent variable $x$.

**Method 2:** First interchange $x$ and $y$ in (*1*) obtaining

$$x = 2y - 3$$

Then solve for $y$ in terms of $x$ to obtain

$$y = (x + 3)/2 \qquad \text{and so} \qquad f^{-1}(x) = (x + 3)/2$$

**4.20.** Find a formula for the inverse of $g(x) = \dfrac{2x - 3}{5x - 7}$.

Set $y = g(x)$ and then interchange $x$ and $y$ as follows:

$$y = \frac{2x - 3}{5x - 7} \qquad \text{and then} \qquad x = \frac{2y - 3}{5y - 7}$$

Now solve for $y$ in terms of $x$:

$$5xy - 7x = 2y - 3 \quad \text{or} \quad 5xy - 2y = 7x - 3 \quad \text{or} \quad (5x - 2)y = 7x - 3$$

Thus

$$y = \frac{7x - 3}{5x - 2} \qquad \text{and so} \qquad g^{-1}(x) = \frac{7x - 3}{5x - 2}$$

(Here the domain of $g^{-1}$ excludes $x = 2/5$.)

**4.21.** Consider functions $f: A \to B$ and $g: B \to C$. Prove the following:

(*a*) If $f$ and $g$ are one-to-one, then the composition function $g \circ f$ is one-to-one.

(*b*) If $f$ and $g$ are onto functions, then $g \circ f$ is an onto function.

(*a*) Suppose $(g \circ f)(x) = (g \circ f)(y)$; then $g(f(x)) = g(f(y))$. Hence $f(x) = f(y)$ because $g$ is one-to-one. Furthermore, $x = y$ since $f$ is one-to-one. Accordingly, $g \circ f$ is one-to-one.

(*b*) Let $c$ be any arbitrary element of $C$. Since $g$ is onto, there exists a $b \in B$ such that $g(b) = c$. Since $f$ is onto, there exists an $a \in A$ such that $f(a) = b$. But then

$$(g \circ f)(a) = g(f(a)) = g(b) = c$$

Hence each $c \in C$ is the image of some element $a \in A$. Accordingly, $g \circ f$ is an onto function.

**4.22.** Consider functions $f: A \to B$ and $g: B \to C$. Prove the following:

(*a*) If $g \circ f$ is one-to-one, then $f$ is one-to-one.

(*b*) If $g \circ f$ is onto, then $g$ is onto.

(*a*) Suppose $f$ is not one-to-one. Then there exist distinct elements $x, y \in A$ for which $f(x) = f(y)$. Thus $(g \circ f)(x) = g(f(x)) = g(f(y)) = (g \circ f)(y)$; hence $g \circ f$ is not one-to-one. Therefore, if $g \circ f$ is one-to-one, then $f$ must be one-to-one.

(*b*) If $a \in A$, then $(g \circ f)(a) = g(f(a)) \in g(B)$; hence $(g \circ f)(A) \subseteq g(B)$. Suppose $g$ is not onto. Then $g(B)$ is properly contained in $C$ and so $(g \circ f)(A)$ is properly contained in $C$; thus $g \circ f$ is not onto. Accordingly, if $g \circ f$ is onto, then $g$ must be onto.

**4.23.** Prove Theorem 4.2: A function $f: A \to B$ is invertible if and only if $f$ is bijective (one-to-one and onto).

Suppose $f$ has an inverse, i.e., there exists a function $f^{-1}: B \to A$ for which $f^{-1} \circ f = 1_A$ and $f \circ f^{-1} = 1_B$. Since $1_A$ is one-to-one, $f$ is one-to-one by Problem 4.22; and since $1_B$ is onto, $f$ is onto by Problem 4.22. That is, $f$ is both one-to-one and onto.

Now suppose $f$ is both one-to-one and onto. Then each $b \in B$ is the image of a unique element in $A$, say $\hat{b}$. Thus if $f(a) = b$, then $a = \hat{b}$; hence $f(\hat{b}) = b$. Now let $g$ denote the mapping from $B$ to $A$ defined by $g(b) = \hat{b}$. We have:

(i) $(g \circ f)(a) = g(f(a)) = g(b) = \hat{b} = a$, for every $a \in A$; hence $g \circ f = 1_A$.

(ii) $(f \circ g)(b) = f(g(b)) = f(\hat{b}) = b$, for every $b \in B$; hence $f \circ g = 1_B$.

Accordingly, $f$ has an inverse. Its inverse is the mapping $g$.

## SPECIAL MATHEMATICAL FUNCTIONS, RECURSIVELY DEFINED FUNCTIONS

**4.24.** Find: (*a*) $\lfloor 7.5 \rfloor, \lfloor -7.5 \rfloor, \lfloor -18 \rfloor$, where $\lfloor x \rfloor$, called the *floor* of $x$, denotes the greatest integer that does not exceed $x$; (*b*) $\lceil 7.5 \rceil, \lceil -7.5 \rceil, \lceil -18 \rceil$, where $\lceil x \rceil$, called the *ceiling* of $x$, denotes the least integer that does not exceed $x$.

(*a*) $\lfloor 7.5 \rfloor = 7, \lfloor -7.5 \rfloor = -8, \lfloor -18 \rfloor = -18.$

(*b*) $\lceil 7.5 \rceil = 8, \lceil -7.5 \rceil = -7, \lceil -18 \rceil = -18.$

**4.25.** Find: (*a*) 26 (mod 7), 25 (mod 5), 35 (mod 11);
(*b*) −26 (mod 7), −371 (mod 8), −2345 (mod 6).

(*a*) When $k$ is positive, divide $k$ by the modulus $M$ to obtain the remainder $r$. Then $k \pmod M = r$. Thus:

$$26 \pmod 7 = 5, \qquad 25 \pmod 5 = 0, \qquad 35 \pmod{11} = 2$$

(*b*) When $k$ is negative, divide $|k|$ by the modulus $M$ to obtain the remainder $r'$. Then, when $r' \neq 0$, $k \pmod M = M - r'$. Thus:

$$-26 \pmod 7 = 7 - 5 = 2, \qquad -371 \pmod 8 = 8 - 3 = 5, \qquad -2345 \pmod 6 = 6 - 5 = 1$$

**4.26.** Using arithmetic modulo $M = 15$, evaluate: (*a*) $9 + 13$, (*b*) $7 + 11$, (*c*) $4 - 9$, (*d*) $2 - 10$.

Use $a \pm M \equiv a \pmod M$:

(*a*) $9 + 13 = 22 \equiv 22 - 15 = 7$ (*c*) $4 - 9 = -5 \equiv -5 + 15 = 10$

(*b*) $7 + 11 = 18 \equiv 18 - 15 = 3$ (*d*) $2 - 10 = -8 \equiv -8 + 15 = 7$

**4.27.** Evaluate: (*a*) $\log_2 8$; (*b*) $\log_2 64$; (*c*) $\log_{10} 100$; (*d*) $\log_{10} 0.001$.

(*a*) $\log_2 8 = 3$ since $2^3 = 8$ (*c*) $\log_{10} 100 = 2$ since $10^2 = 100$

(*b*) $\log_2 64 = 6$ since $2^6 = 64$ (*d*) $\log_{10} 0.001 = -3$ since $10^{-3} = 0.001$

**4.28.** Show that: (*a*) $\log_b AB = \log_b A + \log_b B$; (*b*) $\log_b A^n = n \log_b A$.

Let $\log_b A = x$ and $\log_b B = y$. Then $A = b^x$ and $B = b^y$.

(*a*) We have $AB = b^x b^y = b^{x+y}$. Hence

$$\log_b AB = x + y = \log_b A + \log_b B$$

(*b*) We have $A^n = (b^x)^n = b^{nx}$. Hence

$$\log_b A^n = nx = n \log_b A$$

**4.29.** Evaluate: (*a*) $2^5$, (*b*) $3^{-4}$, (*c*) $8^{2/3}$, (*d*) $25^{-3/2}$.

(*a*) $2^5 = 2 \cdot 2 \cdot 2 \cdot 2 \cdot 2 = 32$

(*b*) $3^{-4} = \dfrac{1}{3^4} = \dfrac{1}{81}$

(*c*) $8^{2/3} = (\sqrt[3]{8})^2 = 2^2 = 4$

(*d*) $25^{-3/2} = \dfrac{1}{25^{3/2}} = \dfrac{1}{(\sqrt{25})^3} = \dfrac{1}{5^3} = \dfrac{1}{125}$

**4.30.** Let $n$ denote a positive integer. Suppose a function $L$ is defined recursively as follows:

$$L(n) = \begin{cases} 0 & \text{if } n = 1 \\ L(\lfloor n/2 \rfloor) + 1 & \text{if } n > 1 \end{cases}$$

Find $L(25)$ and describe what this function does. (The *floor* function $\lfloor x \rfloor$ is defined in the above Problem 4.24.)

Find $L(25)$ recursively as follows:

$$\begin{aligned} L(25) &= L(12) + 1 \\ &= [L(6) + 1] + 1 = L(6) + 2 \\ &= [L(3) + 1] + 2 = L(3) + 3 \\ &= [L(1) + 1] + 3 = L(1) + 4 \\ &= 0 + 4 = 4 \end{aligned}$$

Each time $n$ is divided by 2, the value of $L$ is increased by 1. Hence $L$ is the greatest integer such that

$$2^L < n$$

Accordingly, $L(n) = \lfloor \log_2 n \rfloor$

# Supplementary Problems

## FUNCTIONS

**4.31.** Define each of the following functions from **R** into **R** by a formula:

(*a*) To each number let $f$ assign its square plus 3.

(*b*) To each number let $g$ assign its cube plus twice the number.

(*c*) To each number greater than or equal to 3 let $h$ assign the number squared; and to each number less than 3 let $h$ assign the number $-2$.

**4.32.** Let $f\colon \mathbf{R} \to \mathbf{R}$ be defined by

$$f(x) = \begin{cases} x^2 - 3x & \text{if } x \geq 2 \\ x + 2 & \text{if } x < 2 \end{cases}$$

Find $f(5)$, $f(0)$, and $f(-2)$.

**4.33.** Let $W = \{a, b, c, d\}$. Determine whether each set of ordered pairs is a function from $W$ into $W$:

(a) $\{(b,a),\ (c,d),\ (d,a),\ (c,d),\ (a,d)\}$, (c) $\{(a,b),\ (b,b),\ (c,b),\ (d,b)\}$,
(b) $\{(d,d),\ (c,a),\ (a,b), (d,b)\}$, (d) $\{(a,a),\ (b,a),\ (a,b),\ (c,d)\}$.

**4.34.** Let the function $g$ assign to each name in the following set $S$ the number of different letters needed to spell the name:

$$S = \{\text{Britt, Martin, Alan, Audrey, Julianna}\}$$

Find the graph of $g$, i.e., write $g$ as a set of ordered pairs.

**4.35.** Let $A = \{1, 2, 3, 4, 5\}$ and let $f\colon A \to A$ be defined by Fig. 4-14. (a) Write $f$ as a set of ordered pairs. (b) Find the image of $f$. (c) Find $f(S)$ where $S = \{1, 2, 4\}$. (d) Find $f^{-1}(T)$ where $T = \{1, 2, 3\}$.

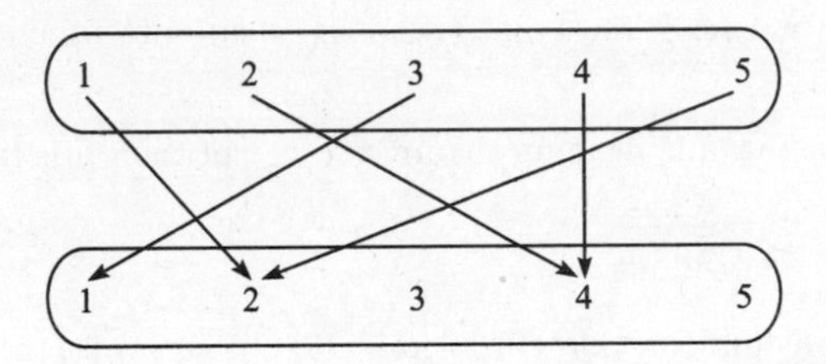


**Fig. 4-14**

**4.36.** Let $A = \{a, b, c\}$ and $B = \{1, 2, 3, 4\}$. Find the number of functions from: (a) $A$ into $B$; (b) $B$ into $A$.

**4.37.** Consider any function $f\colon A \to B$. Show $f^{-1}[f[A]] = A$.

**4.38.** A function with domain $A$ is called a *constant* function if every $a \in A$ is assigned the same element. Find the number of constant functions from $A$ into $B$.

## COMPOSITION FUNCTION

**4.39.** Figure 4-15 defines functions $f$, $g$, $h$ from $A = \{1, 2, 3, 4\}$ into itself.

(a) Find the images of $f$, $g$, $h$.
(b) Find the composition functions $f \circ g$, $h \circ f$, $g^2 = g \circ g$.
(c) Find the composition functions $h \circ g \circ f$ and $f \circ g \circ h$.

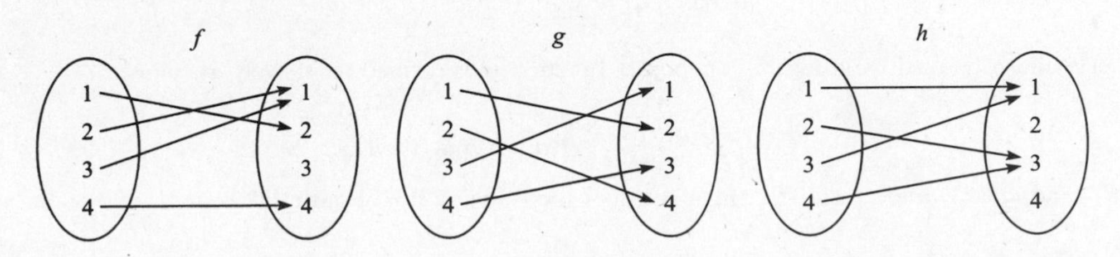


**Fig. 4-15**

**4.40.** Consider the functions $f(x) = x^2 + 3x + 1$ and $g(x) = 2x - 3$. Find a formula defining the composition function: (a) $f \circ g$; (b) $g \circ f$.

**4.41.** Let $V = \{1, 2, 3, 4\}$ and let

$$f = \{(1,3),\ (2,1),\ (3,4),\ (4,3)\} \quad \text{and} \quad g = \{(1,2),\ (2,3),\ (3,1),\ (4,1)\}$$

Find: (*a*) $f \circ g$; (*b*) $g \circ f$; (*c*) $f \circ f$.

**4.42.** Suppose $f\colon A \to B$ and $g\colon B \to C$. Show that $g \circ f$ is a constant function (Problem 4.38) if either $f$ or $g$ is a constant function.

## ONE-TO-ONE, ONTO AND INVERTIBLE FUNCTIONS

**4.43.** Which of the functions in Fig. 4-15 are: (*a*) one-to-one, (*b*) onto, (*c*) invertible?

**4.44.** Consider the formula $f(x) = x^2$.

(*a*) Find the largest interval $D$ such that $f\colon D \to \mathbf{R}$ is a one-to-one function.

(*b*) Find the smallest target set $T$ such that $f\colon \mathbf{R} \to T$ is an onto function.

**4.45.** Find the domain $D$ and a formula defining the inverse $f^{-1}$ of each function:

(*a*) $f(x) = x^3 + 5$; (*b*) $f(x) = \dfrac{x-2}{x-3}$.

**4.46.** Suppose $f\colon A \to B$ is a constant function (Problem 4.38). When will $f$ be: (*a*) one-to-one, (*b*) onto?

**4.47.** Suppose $f\colon A \to B$ and $g\colon B \to C$ are invertible functions. Show that $g \circ f\colon A \to C$ is invertible, and $(g \circ f)^{-1} = g^{-1} \circ f^{-1}$.

**4.48.** Let $W = [0, \infty) = \{x : x \geq 0\}$. Let $f\colon W \to W$, $g\colon W \to W$, $h\colon W \to W$ be defined as follows:

$$f(x) = x^4, \qquad g(x) = x^3 + 1, \qquad h(x) = x + 2$$

Which of the functions are (*a*) one-to-one, (*b*) onto, (*c*) invertible?

## SPECIAL MATHEMATICAL FUNCTIONS, RECURSIVELY DEFINED FUNCTIONS

**4.49.** Find: (*a*) $\lfloor 13.2 \rfloor$, $\lfloor -0.17 \rfloor$, $\lfloor 34 \rfloor$; (*b*) $\lceil 13.2 \rceil$, $\lceil -0.17 \rceil$, $\lceil 34 \rceil$. (See Problem 4.24.)

**4.50.** Find: (*a*) 10 (mod 3), 200 (mod 20), 29 (mod 6); (*b*) $-10$ (mod 3), $-29$ (mod 6), $-345$ (mod 11).

**4.51.** Find: (*a*) $3! + 4!$; (*b*) $3!(3! + 2!)$; (*c*) $6!/5!$; (*d*) $30!/28!$.

**4.52.** Evaluate: (*a*) $\log_2 16$; (*b*) $\log_3 27$; (*c*) $\log_{10} 0.01$.

**4.53.** Find: (*a*) $6^3$; (*b*) $7^{-2}$; (*c*) $4^{5/2}$; (*d*) $27^{-4/3}$.

**4.54.** Let $a$ and $b$ be positive integers. Suppose a function $Q$ is defined recursively as follows:

$$Q(a,b) = \begin{cases} 0 & \text{if } a < b \\ Q(a-b, b) + 1 & \text{if } a \geq b \end{cases}$$

(*a*) Find $Q(2,3)$ and $Q(14,3)$. (*b*) What does the function do? Find $Q(5861, 7)$.

## MISCELLANEOUS PROBLEMS

**4.55.** Find the domain $D$ of each of the following functions:

(*a*) $f(x) = 1/(x+3)$,

(*b*) $f(x) = 1/(x-3)$ where $x > 0$,

(*c*) $f(x) = \sqrt{16 - x^2}$,

(*d*) $f(x) = \log(x + 3)$.

**4.56.** Sketch the graph of each function:

(a) $f(x) = \frac{1}{2}x - 1$
(b) $g(x) = x^3 - 3x + 2$
(c) $h(x) = \begin{cases} 0 & \text{if } x = 0 \\ 1/x & \text{if } x \neq 0 \end{cases}$

## Answers to Supplementary Problems

**4.31.** (a) $f(x) = x^2 + 3$ (b) $g(x) = x^3 + 2x$;
(c) $h(x) = \begin{cases} x^2 & \text{if } x \geq 3 \\ -2 & \text{if } x < 3 \end{cases}$

**4.32.** $f(5) = 10$; $f(0) = 2$; $f(-2) = 0$

**4.33.** (a) Yes; (b) no; (c) yes; (d) no

**4.34.** $g = \{(\text{Britt}, 4), (\text{Martin}, 6), (\text{Alan}, 3), (\text{Audrey}, 6), (\text{Julianna}, 6)\}$

**4.35.** (a) $f = \{(1,2), (2,4), (3,1), (4,4), (5,2)\}$; (b) $\text{Im}(f) = \{1,2,4\}$; (c) $f(S) = \{2,4\}$;
(d) $f^{-1}(T) = \{1,3,5\}$

**4.36.** (a) $4^3 = 64$; (b) $3^4 = 81$

**4.37.** $f^{-1}(f(A)) = A$

**4.38.** Number of elements in $B$.

**4.39.** (a) $\text{Im}(f) = \{1,2,4\}$, $\text{Im}(g) = \{1,2,3,4\}$, $\text{Im}(h) = \{1,3\}$
(b) $f \circ g = \{(1,1), (2,4), (3,2), (4,1)\}$
$h \circ f = \{(1,3), (2,1), (3,1), (4,3)\}$
$g^2 = g \circ g = \{(1,4), (2,3), (3,2), (4,1)\}$
(c) $h \circ g \circ f = \{(1,3), (2,3), (3,3), (4,1)\}$
$f \circ g \circ h = \{(1,1), (2,2), (3,1), (4,2)\}$

**4.40.** (a) $(f \circ g)(x) = 4x^2 - 6x + 1$; (b) $(g \circ f)(x) = 2x^2 + 6x - 1$

**4.41.** (a) $f \circ g = \{(1,1), (2,4), (3,3), (4,3)\}$
(b) $g \circ f = \{(1,1), (2,2), (3,1), (4,1)\}$
(c) $f^2 = f \circ f = \{(1,4), (2,3), (3,3), (4,4)\}$

**4.43.** (a) Only $g$; (b) only $g$; (c) only $g$

**4.44.** (a) $D = [0, \infty)$ or $D = (-\infty, 0]$; (b) $T = [0, \infty)$

**4.45.** (a) $f^{-1}(x) = \sqrt[3]{x - 5}$, $D = \mathbf{R}$; (b) $f^{-1}(x) = (2 - 3x)/(1 - x)$, $D = \mathbf{R} \backslash \{1\}$

**4.46.** (a) $A$ has one element; (b) $B$ has one element

**4.48.** (*a*) $f, g, h$; (*b*) $f$; (*c*) $f$

**4.49.** (*a*) 13, $-1$, 34; (*b*) 14, 0, 34

**4.50.** (*a*) 1, 0, 5; (*b*) 2, 1, 7

**4.51.** (*a*) 30; (*b*) 48; (*c*) 6; (*d*) 870

**4.52.** (*a*) 4; (*b*) 3; (*c*) $-2$

**4.53.** (*a*) 216; (*b*) 1/49; (*c*) 32; (*d*) 1/81

**4.54.** (*a*) $Q(2,3) = 0$, $Q(14,3) = 2$; (*b*) $Q(a,b)$ is the remainder when $a$ is divided by $b$, so $Q(5861,7) = 2$.

**4.55.** (*a*) $R\backslash\{-3\}$; (*b*) $D = [0,\infty)\backslash\{-3\}$; (*c*) $D = [-4,4]$; (*d*) $D = (-3,\infty)$

**4.56.** See Fig. 4-16.

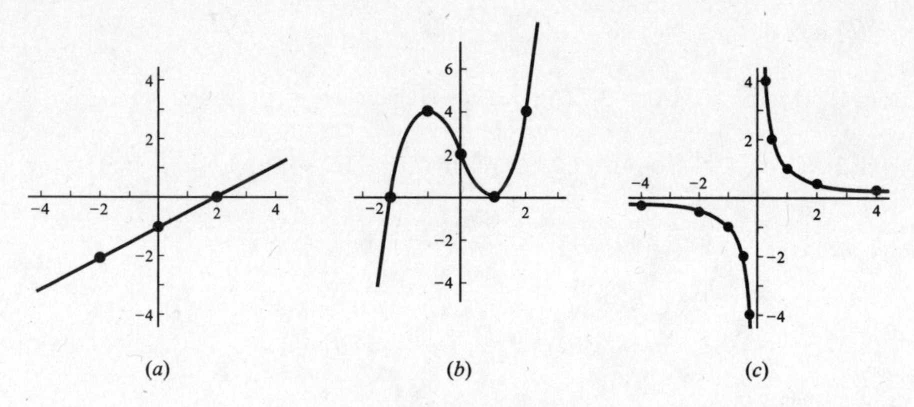


**Fig. 4-16**

# Chapter 5

# Further Theory of Sets and Functions

## 5.1 INTRODUCTION

This chapter investigates some additional properties of sets and functions including set operations on collections of sets and indexed sets. We also discuss the notion of a diagram of functions.

## 5.2 OPERATIONS ON COLLECTIONS OF SETS

Let $\mathscr{A}$ be a collection of sets. The *union* of $\mathscr{A}$, denoted by

$$\bigcup\{A : A \in \mathscr{A}\} \qquad \text{or} \qquad \bigcup_{A \in \mathscr{A}} A \qquad \text{or simply} \qquad \bigcup \mathscr{A}$$

consists of all elements $x$ such that $x$ belongs to at least one set in $\mathscr{A}$; that is,

$$\bigcup\{A : A \in \mathscr{A}\} = \{x : x \in A \text{ for some } A \text{ in } \mathscr{A}\}$$

Analogously, the *intersection* of $\mathscr{A}$, denoted by

$$\bigcap\{A : A \in \mathscr{A}\} \qquad \text{or} \qquad \bigcap_{A \in \mathscr{A}} A \qquad \text{or simply} \qquad \bigcap \mathscr{A}$$

consists of all elements $x$ such that $x$ belongs to all the sets in $\mathscr{A}$; that is,

$$\bigcap\{A : A \in \mathscr{A}\} = \{x : x \in A \text{ for every } A \text{ in } \mathscr{A}\}$$

If $\mathscr{A}$ is empty, then we do not define the intersection of $\mathscr{A}$. In case $\mathscr{A}$ is nonempty and finite, then the above are just the same as our previous definitions of union and intersection.

**EXAMPLE 5.1**

(*a*) Let $\mathscr{A} = [\{1,2,3\}, \{2,3,4\}, \{2,3,5\}]$. Then

$$\bigcup \mathscr{A} = \{1,2,3,4,5\} \qquad \text{and} \qquad \bigcap \mathscr{A} = \{2,3\}$$

(*b*) Let $A$ be any set and let $\mathscr{P} = \mathscr{P}(A)$ be the power set of $A$. Then:

$$\bigcup \mathscr{P} = A \qquad \text{and} \qquad \bigcap \mathscr{P} = \varnothing$$

(*c*) Let $\mathscr{A} = \{[-1,1], [-2,2], [-3,3], \ldots, [-n,n], \ldots\}$. Then

$$\bigcup \mathscr{A} = R \qquad \text{and} \qquad \bigcap \mathscr{A} = [-1,1]$$

## 5.3 INDEXED COLLECTIONS OF SETS

Algebraic properties of unions and intersections are usually presented in the context of one of the main ways of designating collections of sets, that is, as indexed collections of sets. Such collections of sets and the set operations on them are discussed in this section.

**Indexed Collections of Sets**

Let $I$ be any nonempty set, and let $\mathscr{L}$ be a collection of sets. An *indexing function* from $I$ to $\mathscr{L}$ is a function $f: I \to \mathscr{L}$. For any $i \in I$, we denote the image $f(i)$ by $A_i$. Thus the indexing function $f$ is usually denoted by

$$\{A_i: i \in I\} \qquad \text{or} \qquad \{A_i\}_{i \in I} \qquad \text{or simply} \qquad \{A_i\}$$

The set $I$ is called the *indexing set*, and the elements of $I$ are called *indices*. If $f$ is bijective, that is, one-to-one and onto, then we say that $\mathscr{L}$ is indexed by $I$.

**Remark**: Any nonempty collection $\mathscr{A}$ of distinct sets may be viewed as an indexed collection of sets by letting $\mathscr{A}$ be indexed by itself. Thus a collection of sets is usually given in the form $\{A_i: i \in I\}$, that is, as an indexed collection of sets.

**Operations on Indexed Collections of Sets**

Consider any indexed collection $\{A_i : i \in I\}$ of sets. The *union* of the collection $\{A_i : i \in I\}$, denoted by

$$\bigcup\{A_i : i \in I\} \qquad \text{or} \qquad \bigcup_{i \in I} A_i \qquad \text{or simply} \qquad \bigcup_i A_i$$

consists of those elements which belong to at least one of the $A_i$. Namely,

$$\bigcup\{A_i : i \in I\} = \{x : x \in A_i \text{ for some } i \in I\}$$

Analogously, the *intersection* of a collection set $\{A_i : i \in I\}$, denoted by

$$\bigcap\{A_i : i \in I\} \qquad \text{or} \qquad \bigcap_{i \in I} A_i \qquad \text{or, simply} \qquad \bigcap_i A_i$$

consists of those elements which belong to every $A_i$. Namely,

$$\bigcap\{A_i : i \in I\} = \{x : x \in A_i \text{ for every } i \in I\}$$

In the case that $I$ is a finite set, this is just the same as our previous definitions of union and intersection.

Suppose the indexing set $I$ is the set $\mathbf{P}$ of positive integers. Then $\{A_i\}$ is called a *sequence of sets*, usually denoted by $A_1, A_2, A_3, \ldots$, and the union and intersection of the sets may be denoted by

$$A_1 \cup A_2 \cup \cdots \qquad \text{and} \qquad A_1 \cap A_2 \cap \cdots$$

respectively.

Suppose $J \subseteq I$. Then the union and intersection of only those sets $A_i$ where $i \in J$ is denoted, respectively, by

$$\bigcup\{A_i : i \in J\} \quad \text{and} \quad \bigcap\{A_i : i \in J\} \qquad \text{or} \qquad \bigcup_{i \in J} A_i \quad \text{and} \quad \bigcap_{i \in I} A_i$$

We emphasize that $\bigcup_i A_i$ and $\bigcap_i A_i$ can only be used when the entire indexing set $I$ is used in the union and intersection.

**EXAMPLE 5.2**

(*a*) Let $I$ be the set $\mathbf{Z}$ of integers. To each integer $n$ we assign the following subset of $\mathbf{R}$:

$$A_n = \{x : x \leq n\}$$

In other words, $A_n$ is the infinite interval $(-\infty, n]$. For any real number $a$, there exist integers $n_1$ and $n_2$ such that $n_1 < a < n_2$. Hence

$$a \in \bigcup_n A_n \qquad \text{but} \qquad a \notin \bigcap_n A_n$$

Accordingly,

$$\bigcup_n A_n = \mathbf{R} \qquad \text{but} \qquad \bigcap_n A_n = \varnothing$$

(*b*) Let $I = \{1, 2, 3, 4, 5\}$ and $J = \{2, 3, 5\}$, and let

$$A_1 = \{1, 9\}, \quad A_2 = \{2, 4, 6, 9\}, \quad A_3 = \{3, 6, 7, 9\}, \quad A_4 = \{4, 8\}, \quad A_5 = \{5, 6, 9\}$$

Then

$$\bigcap_i A_i = \varnothing \quad \text{and} \quad \bigcup_i A_i = \{1, 2, \ldots, 9\}$$

However,

$$\bigcap_{i \in J} A_i = \{6, 9\} \quad \text{and} \quad \bigcup_{i \in J} A_i = \{2, 3, 4, 5, 6, 7, 9\}$$

The following theorem tells us, in particular, that the distributive laws and DeMorgan's law in Table 1-1 can be generalized to apply to indexed collections of sets.

**Theorem 5.1**: Let $B$ and $\{A_i\}$ with $i \in I$ be subsets of a universal set $\mathbf{U}$. Then:

(i) $B \cap (\cup\{A_i\}) = \cup\{B \cap A_i\}$ and $B \cup (\cap\{A_i\}) = \cap\{B \cup A_i\}$.

(ii) $(\cup\{A_i\})^c = \cap\{A_i^c\}$ and $(\cap\{A_i\})^c = \cup\{A_i^c\}$.

(iii) If $J$ is a subset of $I$, then

$$\bigcup_{i \in J} A_i \subseteq \bigcup_{i \in I} A_i \quad \text{and} \quad \bigcap_{i \in J} A_i \supseteq \bigcap_{i \in I} A_i$$

Since the empty set $\varnothing$ is a subset of any set, Theorem 5.1(iii) should imply that the empty intersection contains any set $A_i$. Accordingly, one sometimes defines

$$\bigcap_{\varnothing} A_i = \mathbf{U}$$

This may seem strange, but it is similar to defining $0! = 1$ and $a^0 = 1$ in order for general properties to be true.

We also note that Theorem 5.1(i) and (ii) apply to any collection $\mathcal{A}$ of sets.

## 5.4 SEQUENCES, SUMMATION SYMBOL

A *sequence* is a function from the set $\mathbf{P}$ of positive integers into a set $A$. The notation $a_n$ is used to denote the image of the integer $k$. Thus a sequence is usually denoted by

$$a_1, a_2, a_3, \ldots \quad \text{or} \quad \{a_n : n \in \mathbf{P}\} \quad \text{or simply} \quad \{a_n\}$$

Sometimes the domain of a sequence is the set $\mathbf{N} = \{0, 1, 2, \ldots\}$ of nonnegative integers rather than $\mathbf{P}$. In such a case we say that $n$ *begins* with 0 rather than 1.

A *finite sequence* over a set $A$ is a function from $\{1, 2, \ldots, m\}$ into $A$, and it is usually denoted by

$$a_1, a_2, \ldots, a_m$$

Such a finite sequence is sometimes called a *list* or an *m-tuple*.

**EXAMPLE 5.3**

(*a*) The familiar sequences

$$1, 1/2, 1/3, 1/4, \ldots \quad \text{and} \quad 1, 1/2, 1/4, 1/8, \ldots$$

may be formally defined, respectively, by

$$a_n = 1/n \quad \text{and} \quad b_n = 2^{-n}$$

where the first sequence begins with $n = 1$ and the second sequence begins with $n = 0$.

(*b*) The important sequence $1, -1, 1, -1, \ldots$ may be formally defined by

$$a_n = (-1)^{n+1} \quad \text{or, equivalently, by} \quad b_n = (-1)^n$$

where the first sequence begins with $n = 1$ and the second sequence begins with $n = 0$.

(c) (*Strings*): Suppose a set $A$ is finite and $A$ is viewed as a character set or an alphabet. Then a finite sequence over $A$ is called a *string* or *word*, and it is usually written in the form $a_1 a_2 \ldots a_m$, that is, without parentheses. The number $m$ of characters in the string is called its *length*. One also views the set with zero characters as a string; it is called the *empty string* or *null string*.

**Summation Symbol, Sums**

Consider a sequence $a_1, a_2, a_3, \ldots$. Frequently we want to form sums of elements from the sequence. Such sums may sometimes be conveniently represented using the summation symbol $\Sigma$ (the Greek letter sigma). Specifically, the sums

$$a_1 + a_2 + a_3 + \cdots + a_n \quad \text{and} \quad a_m + a_{m+1} + a_{m+2} + \cdots + a_n$$

will be denoted, respectively, by

$$\sum_{j=1}^{n} a_j \quad \text{and} \quad \sum_{j=m}^{n} a_j$$

The letter $j$ in the above expression is called a *dummy index* or *dummy variable*. Other letters frequently used as dummy variables are $i, k, s$, and $t$.

**EXAMPLE 5.4**

$$\sum_{i=1}^{n} a_i b_i = a_1 b_1 + a_2 b_2 + \cdots + a_n b_n$$

$$\sum_{j=2}^{5} j^2 = 2^2 + 3^2 + 4^2 + 5^2 = 4 + 9 + 16 + 25 = 54$$

$$\sum_{j=1}^{n} j = 1 + 2 + \cdots + n$$

The last sum in Example 5.4 appears often. It has the value $n(n+1)/2$. Namely,

$$1 + 2 + 3 + \cdots + n = \frac{n(n+1)}{2}$$

Thus, for example,

$$1 + 2 + 3 + \cdots + 50 = \frac{50(51)}{2} = 1275$$

The formula may be proved using mathematical induction.

## 5.5 FUNDAMENTAL PRODUCTS

Consider a list $A_1, A_2, \ldots, A_n$ of $n$ sets. A *fundamental product* of the sets is a set of the form

$$A_1^* \cap A_2^* \cap \cdots \cap A_m^*$$

where $A_i^*$ is either $A_i$ or $A_i^c$. We note that there are $2^n$ such fundamental products since there is a choice of two sets for each $A_i^*$. One can also show (Problem 5.54) that such fundamental products are disjoint and their union is the universal set $\mathbf{U}$.

There is a geometrical description of these fundamental products which is illustrated below.

**EXAMPLE 5.5** Consider three sets $A, B, C$. The following lists the eight fundamental products of the three sets:

$$P_1 = A \cap B \cap C \qquad P_3 = A \cap B^c \cap C \qquad P_5 = A^n \cap B \cap C \qquad P_7 = A^c \cap B^c \cap C$$
$$P_2 = A \cap B \cap C^c \qquad P_4 = A \cap B^c \cap C^c \qquad P_6 = A^c \cap B \cap C^c \qquad P_8 = A^c \cap B^c \cap C^c$$

These eight products correspond precisely to the eight disjoint regions in the Venn diagram of sets $A, B, C$ in Fig. 5-1 as indicated by the labeling of the regions.

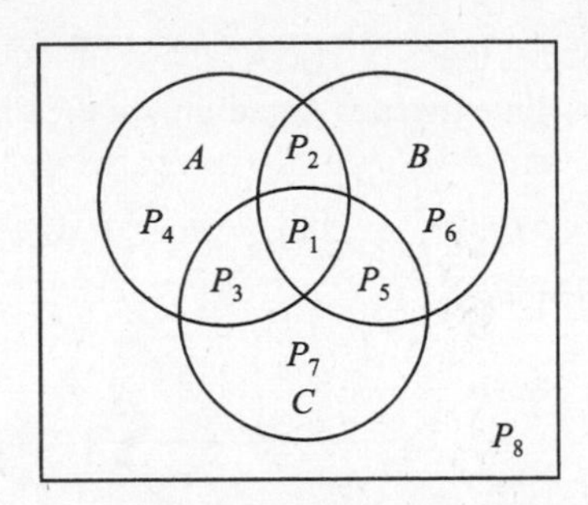


**Fig. 5-1**

A *Boolean expression* in the sets $A_1, A_2, \ldots, A_n$ is an expression $E = E(A_1, A_2, \ldots, A_n)$ which is built up from the sets using the operations of union, intersection, and complement. For example,

$$E_1 = (A \cup B^c)^c \cap (A^c \cap C)^c \cap (B^c \cup C) \qquad \text{and} \qquad E_2 = [(A \cap B^c) \cup (B^c \cap C)]^c$$

are Boolean expressions in the sets $A, B, C$.

The following theorem applies.

**Theorem 5.2**: Any Boolean expression $E = E(A_1, A_2, \ldots, A_n)$ is equal to the empty set $\varnothing$ or the unique union of a finite number of fundamental products.

This theorem is a special case of Theorem 11.8 on Boolean algebras. So its proof appears there. We indicate a geometrical interpretation here.

Consider sets $A, B, C$. Then any Boolean expression $E = E(A, B, C)$ will be uniquely represented by a finite number of regions in the Venn diagram in Fig. 5-1. Thus $E = E(A, B, C)$ is either the empty set or the union of one or more of the eight fundamental products in Fig. 5-1.

## 5.6 FUNCTIONS AND DIAGRAMS

Recall that we used the following diagram to represent functions $f\colon A \to B$ and $g\colon B \to C$:

$$A \xrightarrow{f} B \xrightarrow{g} C$$

Similarly, the following diagram represents functions $f\colon A \to B$, $g\colon B \to C$, and $h\colon A \to C$:

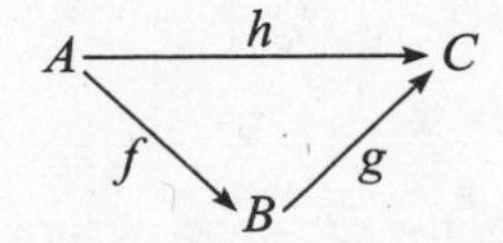


Note that the diagram defines two functions from $A$ to $C$, the function $h$ represented by a single arrow, and the composition function $g \circ f$ represented by a sequence of two connected arrows. Each arrow or sequence of arrows connecting $A$ to $C$ is called a *path* from $A$ to $C$.

**Definition:** A diagram of functions is said to be *commutative* if, for any pair of sets $X$ and $Y$ in the diagram, any two paths from $X$ to $Y$ are equal.

**EXAMPLE 5.6**

(*a*) Suppose the diagram of functions in Fig. 5-2(*a*) is commutative. Then:

$$i \circ h = f, \qquad g \circ i = j, \qquad g \circ f = j \circ h = g \circ i \circ h$$

(*b*) The functions $f: A \to B$ and $g: B \to A$ are inverses if and only if the diagrams in Fig. 5-2(*b*) are commutative, that is, if and only if

$$g \circ f = 1_A \qquad \text{and} \qquad f \circ g = 1_B$$

Here $1_A$ and $1_B$ are the identity functions.

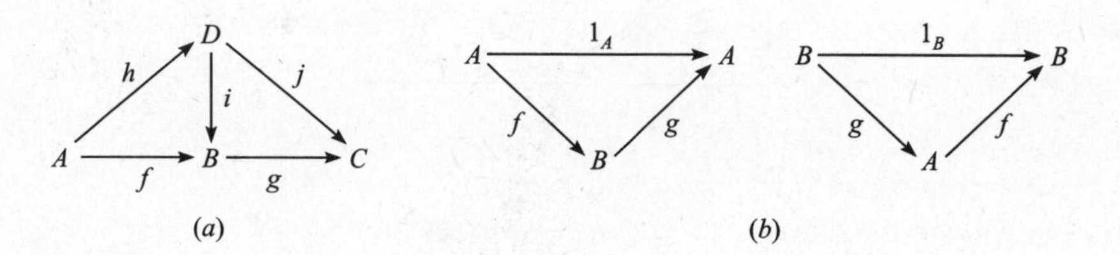

(*a*) (*b*)

**Fig. 5-2**

## 5.7 SPECIAL KINDS OF FUNCTIONS, FUNDAMENTAL FACTORIZATION

This section discusses a number of special kinds of functions which frequently occur in mathematics. We also define and discuss the fundamental factorization of a function.

### Restriction

Consider a function $f: A \to S$. Let $B$ be a subset of $A$. Then $f$ induces a function $f'$ on $B$ defined by

$$f'(b) = f(b)$$

for every $b \in B$. This function $f'$ is called the *restriction* of $f$ to $B$. It is sometimes denoted by

$$f|_B$$

**EXAMPLE 5.7**

(*a*) Let $f: \mathbf{R} \to \mathbf{R}$ be defined by $f(x) = x^2$. Recall that $f$ is not one-to-one, e.g., $f(2) = f(-2) = 4$. Consider the restriction of $f$ to the nonnegative real numbers $D = [0, \infty)$. Then $f|_D$ is one-to-one. [In fact, $f: D \to D$ is invertible and its inverse is the square root function $f^{-1}(x) = \sqrt{x}$.]

(*b*) Consider the functions

$$g = \{(1,3),\ (2,6),\ (3,11),\ (4,18),\ (5,27)\} \qquad \text{and} \qquad g' = \{(1,3),\ (3,11),\ (5,27)\}$$

Observe that $g'$ is a subset of $g$. Thus $g'$ is the restriction of $g$ to $B = \{1,3,5\}$, the set of first elements of $g'$. Note that $B$ is a subset of $A = \{1,2,3,4,5\}$, the set of first elements of $g$.

### Extension

Consider a function $f: A \to S$. Suppose $B$ to be a superset of $A$, that is, suppose $A \subseteq B$. Let $F: B \to S$ be a function on $B$ such that, for every $a \in A$,

$$F(a) = f(a)$$

This function $F$ is called an *extension* of $f$ to $B$. We note that such an extension is rarely unique.

**EXAMPLE 5.8**

(*a*) Let $f$ be the function on the nonnegative real numbers $D = [0, \infty)$ defined by $f(x) = x$. Then the absolute value function

$$|x| = \begin{cases} x & \text{if } x \geq 0 \\ -x & \text{if } x < 0 \end{cases}$$

is an extension of $f$ to the set **R** of all real numbers. Clearly, the identity function $1_R : \mathbf{R} \to \mathbf{R}$ is also an extension of $f$ to **R**.

(*b*) Consider the functions

$$f = \{(1,5),\ (3,11),\ (5,17)\} \qquad \text{and} \qquad F = \{(1,5),\ (2,8),\ (3,11),\ (4,14),\ (5,17)\}$$

Observe that $F$ is a superset of $f$. Thus the function $F$ is an extension of $f$ from $\text{dom}(f) = \{1,3,5\}$ to $\text{dom}(F) = \{1,2,3,4,5\}$.

### Inclusion Map

Let $A$ be a subset of a set $S$, that is, $A \subseteq S$. Let $\imath$ be the function from $A$ to $S$ defined by

$$\imath(a) = a$$

for every $a \in A$. Then $\imath$ is called the *inclusion map*. This map is frequently denoted by writing

$$\imath : A \hookrightarrow S$$

For example, the function $f$: $\mathbf{Z} \to \mathbf{R}$ defined by $f(n) = n$ is the inclusion map from the integers **Z** into the real numbers **R**.

### Characteristic Function

Consider a universal set **U**. For any subset $A$ of **U**, let $\chi_A$ be the function from **U** to $\{0, 1\}$ defined by

$$\chi_A(x) = \begin{cases} 1 & \text{if } x \in A \\ 0 & \text{if } x \notin A \end{cases}$$

Then $\chi_A$ is called the *characteristic function* of $A$.

**EXAMPLE 5.9** Let $\mathbf{U} = \{a, b, c, d, e\}$ and $A = \{a, d, e\}$. Then the function

$$\{(a,1),\ (b,0),\ (c,0),\ (d,1),\ (e,1)\}$$

is the characteristic function $\chi_A$.

On the other hand, any function $f$: $\mathbf{U} \to \{0, 1\}$ defines a subset $A_f$ of **U** as follows:

$$A_f = \{x : x \in \mathbf{U},\ f(x) = 1\}$$

Furthermore, the characteristic function $\chi_{A_f}$ of $A_f$ is the original function $f$. Thus there is a one-to-one correspondence between the power set $\mathscr{P}(\mathbf{U})$ of **U** and the set of all functions from **U** into $\{0, 1\}$.

### Equivalence Relation and Canonical Map

Let $\equiv$ be an equivalence relation on a set $S$. Recall that $\equiv$ induces a partition of $S$ into equivalence classes, called the quotient set of $S$ by $\equiv$, and denoted and defined by

$$s/\equiv\ = \{[a] : a \in S\}$$

Let $\eta : S \to S/\equiv$ be the function defined by

$$\eta(a) = [a]$$

that is, $\eta$ sends each element of $S$ into its equivalence class. Then $\eta$ is called the *canonical* or *natural map* from $S$ into $S/\equiv$.

**EXAMPLE 5.10** Consider the relation $\equiv$ of congruence modulo 5 on the set $\mathbf{Z}$ of integers; that is,

$$a \equiv b \pmod 5$$

if 5 divides $a - b$. Then $\equiv$ is an equivalence relation on $\mathbf{Z}$. There are five equivalence classes:

$$[0] = \{\ldots, -10, -5, 0, 5, 10, \ldots\} \qquad [3] = \{\ldots, -7, -2, 3, 8, 13, \ldots\}$$
$$[1] = \{\ldots, -9, -4, 1, 6, 11, \ldots\} \qquad [4] = \{\ldots, -6, -1, 4, 9, 14, \ldots\}$$
$$[2] = \{\ldots, -8, -3, 2, 7, 12, \ldots\}$$

Let $\eta : \mathbf{Z} \to \mathbf{Z}/\equiv$ be the canonical map. Then

$$\eta(7) = [7] = [2], \qquad \eta(19) = [19] = [4], \qquad \eta(-12) = [-12] = [3]$$

### Fundamental Factorization of a Function

Consider any function $f: A \to B$. Consider the relation $\sim$ on $A$ defined by

$$a \sim a' \quad \text{if} \quad f(a) = f(a')$$

We show (Problem 5.20) that $\sim$ is an equivalence relation on $A$. We will let $A/f$ denote the quotient set under this relation. Recall that $\text{Im}(f) = f(A)$ denotes the image of $f$ and it is a subset of the target set $B$.

The following lemma and theorem (proved in Problems 5.21 and 5.22) apply.

**Lemma 5.3**: The function $f^*: A/f \to f(A)$ defined by

$$f^*([a]) = f(a)$$

is well-defined and bijective.

**Theorem 5.4**: Let $f: A \to B$. Then the diagram in Fig. 5-3 is commutative; that is,

$$f = \imath \circ f^* \circ \eta$$

We note that, in Fig. 5-3, $\eta$ is the canonical mapping from $A$ into $A/f$, $f^*$ is the bijective function defined above, and $\imath$ is the inclusion map from $f(A)$ into $B$.

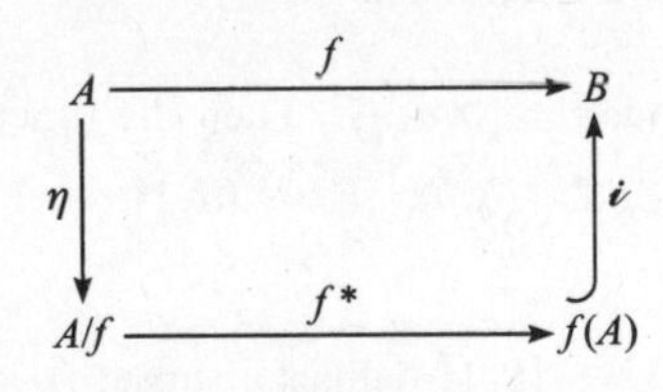


**Fig. 5-3**

## 5.8 ASSOCIATED SET FUNCTIONS

Consider a function $f: S \to T$. Recall that the *image* $f[A]$ of any subset $A$ of $S$ consists of the elements in $T$ which are images of elements in $A$, that is,

$$f[A] = \{b \in T : \text{there exists } a \in A \text{ such that } f(a) = b\}$$

Also recall that the *preimage* or *inverse image* $f^{-1}[B]$ of any subset $B$ of $T$ consists of all elements in $S$ whose images belong to $B$, that is,

$$f^{-1}[B] = \{a \in S : f(a) \in T\}$$

Thus $f[A]$ is a subset of $T$ and $f^{-1}[B]$ is a subset of $S$.

**EXAMPLE 5.11** Let $f\colon \mathbf{R} \to \mathbf{R}$ be defined by $f(x) = x^2$. Then

$$f[\{1,2,3,4\}] = \{1,4,9,16\} \quad \text{and} \quad f[(1,5)] = (1,25)$$

Also,

$$f^{-1}[\{4,9\}] = \{-3,-2,2,3\} \quad \text{and} \quad f^{-1}[(1,4)] = (1,2) \cup (-2,-1)$$

Accordingly, a function $f\colon S \to T$ induces a function, also denoted by $f$, from the power set $\mathscr{P}(S)$ of $S$ into the power set $\mathscr{P}(T)$ of $T$, and a function $f^{-1}$ from $\mathscr{P}(T)$ back to $\mathscr{P}(S)$. These functions $f$ and $f^{-1}$ are called *set functions* since they map sets into sets, i.e., their domains and target sets are collections of sets.

Observe that brackets [. .] rather than parentheses (. .) are used to distinguish between a function and its associated set functions, i.e., $f(a)$ denotes a value of the original function, whereas $f[A]$ and $f^{-1}[B]$ denote values of the associated set functions.

We note that the associated set function $f^{-1}$ is not in general the inverse of the associated set function $f$. For example, for the above function $f(x) = x^2$, we have

$$f^{-1} \circ f[(1,2)] = f^{-1}[(1,4)] = (1,2) \cup (-2,-1)$$

However, we do have the following theorem.

**Theorem 5.5**: Let $f\colon S \to T$, and let $A \subseteq S$ and $B \subseteq T$. Then:
(i) $A \subseteq f^{-1} \circ f[A]$.
(ii) $B = f \circ f^{-1}[B]$.

As noted above, the inclusion in (i) cannot in general be replaced by equality.

## 5.9 CHOICE FUNCTIONS

Consider a collection $\{A_i : i \in I\}$ of subsets of a set $B$. A function

$$f\colon \{A_i\} \to B$$

is called a *choice function* if, for every $i \in I$,

$$f(A_i) \in A_i$$

that is, if the image of each set is an element in the set.

**EXAMPLE 5.12** Consider the following subsets of $B = \{1,2,3,4,5\}$:

$$A_1 = \{1,2,3\}, \qquad A_2 = \{1,3,4\}, \qquad A_3 = \{2,5\}$$

Figure 5-4 shows functions $f$ and $g$ from $\{A_1, A_2, A_3\}$ into $B$. The function $f$ is not a choice function since $f(A_2) = 2$ does not belong to $A_2$, that is $f(A_2) \notin A_2$. On the other hand, $g$ is a choice function. Namely, $g(A_1) = 2$ belongs to $A_1$, $g(A_2) = 4$ belongs to $A_2$, and $g(A_3) = 2$ belongs to $A_3$, that is, $g(A_i) \in A_i$, for $i = 1,2,3$.

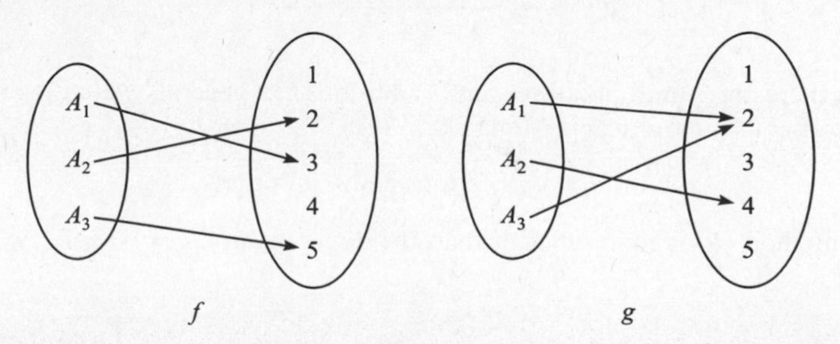


**Fig. 5-4**

**Remark**: Essentially, a choice function, for any collection of sets, "chooses" an element from each set in the collection. The question of whether or not a choice function exists for any collection of sets lies at the foundation of set theory. Chapter 9 will be devoted to this question.

## 5.10 ALGORITHMS AND FUNCTIONS

An algorithm $M$ is a finite step-by-step list of well-defined instructions for solving a particular problem, say, to find the output $f(X)$ for a given function $f$ with input $X$. (Here $X$ may be a list or set of values.) Frequently, there may be more than one way to obtain $f(X)$ as illustrated by the following examples. The particular choice of the algorithm $M$ to obtain $f(X)$ may depend on the "efficiency" or "complexity" of the algorithm; this question of the complexity of an algorithm $M$ is discussed in the next section.

**EXAMPLE 5.13** ***(Polynomial Evaluation)*** Suppose, for a given polynomial $f(x)$ and value $x = a$, we want to find $f(a)$, say,

$$f(x) = 2x^3 - 7x^2 + 4x - 15 \qquad \text{and} \qquad a = 5$$

This can be done in the following two ways.

(*a*) (***Direct Method***): Here we substitute $a = 5$ directly in the polynomial to obtain

$$f(5) = 2(125) - 7(25) + 4(5) - 7 = 250 - 175 + 20 - 15 = 80$$

Observe that there are $4 + 3 + 1 = 8$ multiplications and 3 additions. In general, evaluating a polynomial of degree $n$ directly would require approximately

$$n + (n-1) + \cdots + 1 = \frac{n(n-1)}{2} \text{ multiplications and } n \text{ additions.}$$

(*b*) (***Horner's Method or Synthetic Division***): Here we rewrite the polynomial by successively factoring out $x$ (on the right) as follows:

$$f(x) = (2x^2 - 7x + 4)x - 15 = [(2x - 7)x + 4]x - 15$$

Then

$$f(5) = [(3)5 + 4]5 - 15 = (19)5 - 15 = 95 - 15 = 80$$

For those familiar with synthetic division, the above arithmetic is equivalent to the following synthetic division:

$$\begin{array}{r|l} 5 & 2 - 7 + \ \ 4 - 15 \\ & \quad\ \ 10 + 15 + 95 \\ \hline & 2 + 3 + 19 + 80 \end{array}$$

Observe that here there are 3 multiplications and 3 additions. In general, evaluating a polynomial of degree $n$ by Horner's method would require approximately

$$n \text{ multiplications and } n \text{ additions}$$

Clearly Horner's method (*b*) is more efficient than the direct method (*a*).

**EXAMPLE 5.14** (***Greatest Common Divisor***) Let $a$ and $b$ be positive integers with, say, $b < a$; and suppose we want to find $d = \gcd(a, b)$, the greatest common divisor of $a$ and $b$. This can be done in the following two ways.

(*a*) (***Direct Method***): Here we find all the divisors of $a$ and all the divisors of $b$; say, by testing all the numbers from 2 to $a/2$ and from 2 to $b/2$. Then we pick the largest common divisor. For example, suppose $a = 258$ and $b = 60$. The divisors of $a$ and $b$ follow:

$$a = 258; \quad \text{divisors:} \quad 1, 2, 3, 6, 86, 129, 258$$
$$b = \ 60; \quad \text{divisors:} \quad 1, 2, 3, 4, 5, 6, 10, 12, 15, 20, 30, 60$$

Accordingly, $d = \gcd(258, 60) = 6$.

(*b*) (***Euclidean Algorithm***): Here we divide $a$ by $b$ to obtain a remainder $r_1$ (where $r_1 < b$). Then we divide $b$ by the remainder $r_1$ to obtain a second remainder $r_2$ (where $r_2 < r_1$). Next we divide $r_1$ by $r_2$ to obtain a third remainder $r_3$ (where $r_3 < r_2$). And so on. Since

$$a > b > r_1 > r_2 > r_3 > \cdots \qquad (*)$$

eventually we obtain a remainder $r_m = 0$. Then $r_{m-1} = \gcd(a, b)$. For example, suppose $a = 258$ and $b = 60$. Then:

(1) Dividing $a = 258$ by $b = 60$ yields the remainder $r_1 = 18$.
(2) Dividing $b = 60$ by $r_1 = 18$ yields the remainder $r_2 = 6$.
(3) Dividing $r_1 = 18$ by $r_2 = 6$ yields the remainder $r_3 = 0$.

Thus $r_2 = 6 = \gcd(258, 60)$.

**Remark**: The Euclidean algorithm is a very efficient way to find the greatest common divisor of two positive integers $a$ and $b$. The fact that the algorithm ends follows from (*). The fact that the algorithm yields $d = \gcd(a, b)$ follows from properties of the integers.

## 5.11 COMPLEXITY OF ALGORITHMS

The analysis of algorithms is a major task in mathematics and computer science. In order to compare algorithms, we must have some criteria to measure the efficiency of our algorithms. This section discusses this important topic.

Suppose $M$ is an algorithm, and suppose $n$ is the size of the input data. The time and space used by the algorithm are the two main measures for the efficiency of $M$. The time is measured by counting the number of "key operations"; for example:

(*a*) In sorting and searching, one counts the number of comparisons.
(*b*) In arithmetic, one counts multiplications and neglects additions.

Key operations are so defined when the time for the other operations is much less than or at most proportional to the time for the key operations. The space is measured by counting the maximum of memory needed by the algorithm.

The *complexity* of an algorithm $M$ is the function $f(n)$ which gives the running time and/or storage space requirement of the algorithm in terms of the size $n$ of the input data. Frequently, the storage space required by an algorithm is simply a multiple of the data size. Accordingly, unless otherwise stated or implied, the term "complexity" shall refer to the running time of the algorithm.

The complexity function $f(n)$, which we assume gives the running time of an algorithm, usually depends not only on the size $n$ of the input data but also on the particular data.

**EXAMPLE 5.15** Suppose we want to search through an English short story TEXT for the first occurrence of a given 3-letter word $W$. Clearly, if $W$ is the 3-letter word "the", then $W$ likely occurs near the beginning of TEXT, so $f(n)$ will be small. On the other hand, if $W$ is the 3-letter word "zoo", then $W$ may not appear in TEXT at all, so $f(n)$ will be large.

The above discussion leads us to the question of finding the complexity function $f(n)$ for certain cases. The two cases one usually investigates in complexity theory follow:

(1) *Worst case*: The maximum value of $f(n)$ for any possible input.
(2) *Average case*: The expected value of $f(n)$.

The analysis of the average case assumes a certain probabilistic distribution for the input data. The average case also uses the following concept in probability theory. Suppose the numbers $n_1, n_2, \ldots, n_k$ occur with respective probabilities $p_1, p_2, \ldots, p_k$. Then the *expectation* or *average value* $E$ is given by

$$E = n_1 p_1 + n_2 p_2 + \cdots + n_k p_k$$

**Remark**: The complexity of the average case of an algorithm is usually much more complicated to analyze than that of the worst case. Moreover, the probabilistic distribution that one assumes for the average case may not actually apply to real situations. Accordingly, unless otherwise stated or implied, the complexity of an algorithm shall mean the function which gives the running time of the worst case in terms of the input size. This is not too strong an assumption, since the complexity of the average case for many algorithms is proportional to the worst case.

## Rate of Growth; Big *O* Notation

Suppose $M$ is an algorithm, and suppose $n$ is the size of the input data. Clearly the complexity $f(n)$ of $M$ increases as $n$ increases. It is usually the rate of increase of $f(n)$ that we want to examine. This is usually done by comparing $f(n)$ with some standard function, such as

$$\log_2 n, \quad n, \quad n \log_2 n, \quad n^2, \quad n^3, \quad 2^n$$

The rates of growth for these standard functions are indicated in Fig. 5-5, which gives their approximate values for certain values of $n$. Observe that the functions are listed in the order of their rates of growth: the logarithmic function $\log_2 n$ grows most slowly, the exponential function $2^n$ grows most rapidly, and the polynomial functions $n^c$ grows according to the exponent $c$.

| $n$ \ $g(n)$ | $\log n$ | $n$ | $n \log n$ | $n^2$ | $n^3$ | $2^n$ |
|---|---|---|---|---|---|---|
| 5 | 3 | 5 | 15 | 25 | 125 | 32 |
| 10 | 4 | 10 | 40 | 100 | $10^3$ | $10^3$ |
| 100 | 7 | 100 | 700 | $10^4$ | $10^6$ | $10^{30}$ |
| 1000 | 10 | $10^3$ | $10^4$ | $10^6$ | $10^9$ | $10^{300}$ |

**Fig. 5-5** Rate of growth of standard functions.

The way we compare our complexity function $f(n)$ with one of the standard functions is to use the functional "big $O$" notation which we formally define below.

**Definition**: Let $f(x)$ and $g(x)$ be arbitrary functions defined on $R$ or a subset of $R$. We say "$f(x)$ is of order $g(x)$", written

$$f(x) = O(g(x))$$

if there exists a real number $k$ and a positive constant $C$ such that, for all $x > k$, we have

$$|f(x)| \le C|g(x)|$$

Assuming $f(n)$ and $g(n)$ are functions defined on the positive integers, then

$$f(n) = O(g(n))$$

means that $f(n)$ is bounded by a constant multiple of $g(n)$ for almost all $n$.

**Remark:** The above is called the "big $O$" notation since $f(x) = o(g(x))$ has an entirely different meaning. We also write

$$f(x) = h(x) + O(g(x)) \qquad \text{when} \qquad f(x) - h(x) = O(g(x)).$$

**EXAMPLE 5.16**

(*a*) Let $P(x)$ be a polynomial of degree $m$. We show (Problem 5.24) that $P(x) = O(x^m)$. Thus,

$$7x^2 - 9x + 4 = O(x^2) \quad \text{and} \quad 8x^3 - 576x^2 + 832x - 248 = O(x^3)$$

(*b*) The following gives the complexity of certain well-known searching and sorting algorithms in computer science:

(1) Linear search: $O(n)$ (3) Bubble sort: $O(n^2)$
(2) Binary search: $O(\log_n)$ (4) Merge-sort: $O(n \log n)$

# Solved Problems

## GENERALIZED OPERATIONS, INDEXED SETS

**5.1.** Let $\mathscr{A} = [\{1,2,3,4\},\ \{2,3,4,5\},\ \{3,4,5,6\},\ \{3,4,7,8,9\}]$.
Find: (*a*) $\bigcup \mathscr{A}$, (*b*) $\bigcap \mathscr{A}$.

(*a*) $\bigcup \mathscr{A}$ consists of all elements which belong to at least one of the sets in $\mathscr{A}$; hence

$$\bigcup \mathscr{A} = \{1, 2, 3, \ldots, 8, 9\}$$

(*b*) $\bigcap \mathscr{A}$ consists of those elements which belong to every set in $\mathscr{A}$; hence

$$\bigcap \mathscr{A} = \{3, 4\}$$

**5.2.** Let $A_m = \{m, 2m, 3m, \ldots\}$ where $m \in \mathbf{P}$; that is, $A_m$ consists of the positive multiples of $m$. Find: (*a*) $A_3 \cap A_5$; (*b*) $A_4 \cap A_6$; (*c*) $A_5 \cup A_{15}$; (*d*) $\bigcup(A_m : m \in S)$ where $S$ is the set of prime numbers.

(*a*) The numbers which are divisible by 3 and divisible by 5 are the multiples of 15. Thus $A_3 \cap A_5 = A_{15}$.

(*b*) The multiples of 12 and no other numbers are contained in $A_4$ and $A_6$; hence $A_4 \cap A_6 = A_{12}$.

(*c*) The multiples of 21 are contained in the multiples of 7, that is, $A_{21} \subseteq A_7$. Hence $A_7 \cup A_{21} = A_7$.

(*d*) Every positive integer except 1 is a multiple of a prime number. Thus

$$\bigcup(A_m : m \in S) = \{2, 3, 4, \ldots\} = \mathbf{P}\backslash\{1\}$$

**5.3.** Let $B_n = [n, n+1]$ where $n \in \mathbf{Z}$, the integers. Find:
(*a*) $B_1 \cup B_2$; (*b*) $B_3 \cap B_4$; (*c*) $\bigcup_{i=7}^{18} B_i = \bigcup(B_i : i \in \{7, 8, \ldots, 18\})$; (*d*) $\bigcup(B_i : i \in \mathbf{Z})$.

(*a*) $B_1 \cup B_2$ consists of all points in the intervals $[1, 2]$ and $[2, 3]$; hence $B_1 \cup B_2 = [1, 3]$.

(*b*) $B_3 \cap B_4$ consists of the points which lie in both $[3, 4]$ and $[4, 5]$; hence $B_3 \cap B_4 = \{4\}$.

(*c*) $\bigcup_{i=7}^{18} B_i$ means the union of the sets $[7, 8], [8, 9], \ldots, [18, 19]$. Hence

$$\bigcup_{i=7}^{18} B_i = [7, 19]$$

(*d*) Since every real number belongs to at least one interval $[i, i+1]$, we have $\bigcup(B_i : i \in \mathbf{Z}) = \mathbf{R}$.

**5.4.** Prove Theorem 5.1(i) (Distributive Law):

(*a*) $B \cap (\cup_i A_i) = \cup_i (B \cap A_i)$; (*b*) $B \cup (\cap_i A_i) = \cap_i (B \cup A_i)$.

(*a*)
$$\begin{aligned} B \cap (\cup_i A_i) &= \{x : x \in B,\ x \in \cup_i A_i\} \\ &= \{x : x \in B,\ \exists i_0 \text{ s.t. } x \in A_{i_0}\} \\ &= \{x : \exists i_0 \text{ s.t. } x \in B \cap A_{i_0}\} \\ &= \cup_i (B \cap A_i) \end{aligned}$$

(*b*)
$$\begin{aligned} B \cup (\cap_i A_i) &= \{x : x \in B \text{ or } \forall i,\ x \in A_i\} \\ &= \{x : \forall i,\ x \in B \text{ or } x \in A_i\} \\ &= \{x : \forall i,\ x \in (B \cup \in A_i)\} \\ &= \cap_i (B \cup A_i) \end{aligned}$$

Here $\exists$ means "there exists" and $\forall$ means "for every"; these quantifiers are discussed in Chapter 10.

**5.5.** Prove: Let $\{A_i : i \in I\}$ be an indexed collection of sets and let $i_0 \in I$. Then $\bigcap_i A_i \subseteq A_{i_0} \subseteq \bigcup_i A_i$.

Let $x \in \bigcap_i A_i$; then $x \in A_i$ for every $i \in I$. In particular, $x \in A_{i_0}$. Therefore, $\bigcap_i A_i \subseteq A_{i_0}$. Now let $y \in A_{i_0}$. Since $i_0 \in I$, $y \in \bigcup_i A_i$. Hence $A_{i_0} \subseteq \bigcup_i A_i$.

**5.6.** Prove Theorem 5.1(ii) (DeMorgan's law): $(\bigcup_i A_i)^c = \bigcap_i A_i^c$

$$\begin{aligned} (\textstyle\bigcup_i A_i)^c &= \{x : x \notin (\textstyle\bigcup_i A_i)\} \\ &= \{x : \forall i,\ x \notin A_i\} \\ &= \{x : \forall i,\ x \in A_i^c\} \\ &= \textstyle\bigcap_i A_i^c \end{aligned}$$

## SEQUENCES, SUMMATION SYMBOL

**5.7.** Write out the first six terms of each sequence:

(*a*) $a_n = (-1)^{n+1} n^2$ (*b*) $b_n = \frac{n}{n+1}$

(*c*) $c_n = \begin{cases} 3n & \text{if } n \text{ is odd} \\ 5 & \text{if } n \text{ is even} \end{cases}$

Assuming the sequence begins with $n = 1$, simply substitute $n = 1, 2, \ldots, 6$.

(*a*) 1, −4, 9, −16, 25, −36

(*b*) 1/2, 2/3, 3/4, 4/5, 5/6, 6/7

(*c*) 3, 5, 6, 5, 9, 5

**5.8.** Write out the first six terms of each sequence:

(*a*) $a_1 = 1,\ a_n = n + a_{n-1}$ for $n > 1$.

(*b*) $b_1 = 1,\ b_2 = 2,\ b_n = 3b_{n-2} + 2b_{n-1}$ for $n > 2$.

The sequences are defined recursively in terms of preceding terms of the sequence:

(*a*) 1, 2, 4, 7, 11, 16

(*b*) 1, 2, 7, 20, 61, 221

**5.9.** Find: (*a*) $\sum_{k=1}^{4} k^3$; (*b*) $\sum_{i=1}^{5} x_i$; (*c*) $\sum_{j=1}^{3} (j^4 - j^2)$.

(*a*) $\sum_{k=1}^{4} k^3 = 1^3 + 2^3 + 3^3 + 4^3 = 1 + 8 + 27 + 64 = 100$

(*b*) $\sum_{i=1}^{5} x_i = x_1 + x_2 + x_3 + x_4 + x_5$

(*c*) $\sum_{j=1}^{3} (j^4 - j^2) = (1 - 1) + (16 - 4) + (81 - 9) = 0 + 12 + 72 = 84$

**5.10.** Prove: $\sum_{k=1}^{n} [f(k) + g(k)] = \sum_{k=1}^{n} f(k) + \sum_{k=1}^{n} g(k)$.

The proof is by induction on $n$. For $n = 1$,

$$\sum_{k=1}^{1} [f(k) + g(k)] = f(1) + g(1) = \sum_{k=1}^{1} f(k) + \sum_{k=1}^{1} g(k)$$

Suppose $n > 1$, and the theorem holds for $n - 1$, that is, suppose

$$\sum_{k=1}^{n-1} [f(k) + g(k)] = \sum_{k=1}^{n-1} f(k) + \sum_{k=1}^{n-1} g(k)$$

Then

$$\begin{aligned}\sum_{k-1}^{n} [f(k) + g(k)] &= \sum_{k=1}^{n-1} [f(k) + g(k)] + [f(n) + g(n)] \\ &= \sum_{k=1}^{n-1} f(k) + \sum_{k=1}^{n-1} g(k) + [f(n)] + [g(n)] \\ &= \sum_{k=1}^{n-1} f(k) + f(n) + \sum_{k=1}^{n-1} g(k) + g(n) \\ &= \sum_{k=1}^{n} f(k) + \sum_{k=1}^{n} g(k)\end{aligned}$$

Thus the theorem is proved.

## DIAGRAMS AND FUNCTIONS

**5.11.** Consider the diagram in Fig. 5-6(*a*). (*a*) Find the number of paths from $A$ to $E$; what are they? (*b*) How many of the paths represent the same function?

(*a*) There are six paths from $A$ to $E$ as follows:

$A \to B \to E$ $\quad$ $A \to B \to C \to D \to E$ $\quad$ $A \to C \to D \to E$

$A \to B \to C \to E$ $\quad$ $A \to C \to E$ $\quad$ $A \to D \to E$

That is,

$$r \circ f, \quad s \circ i \circ f, \quad t \circ j \circ i \circ f, \quad s \circ h, \quad t \circ j \circ h, \quad t \circ g$$

As noted previously, the functions are written from right to left.

(*b*) If the diagram is commutative, then all six paths (functions) are equal. Otherwise, one cannot say anything about them.

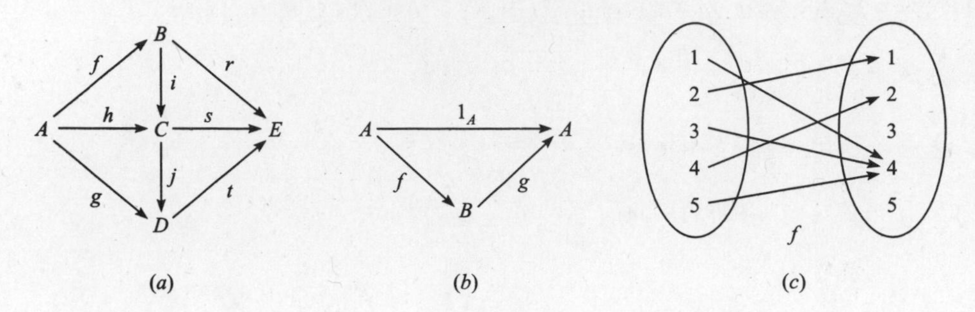


**Fig. 5-6**

**5.12.** Suppose the diagram in Fig. 5-6($b$) is commutative. (Recall that $1_A$ denotes the identity function on $A$.) State all information that is inferred by the diagram.

First, since the diagram is commutative, $g \circ f = 1_A$.

Furthermore, since $g \circ f$ is one-to-one, $f$ must be one-to-one; and since $g \circ f$ is onto, $g$ must also be onto. It need not be true that $g = f^{-1}$, since we do not know that $f \circ g = 1_B$.

## ASSOCIATED SET FUNCTIONS

**5.13.** Let $A = \{1, 2, 3, 4, 5\}$, and let $f\colon A \to A$ be defined by Fig. 5-6($c$). Find:

($a$) $f[\{1, 2, 5\}]$; ($b$) $f^{-1}[\{2, 3, 4\}]$; ($c$) $f^{-1}[\{3, 5\}]$.

($a$) $f[\{1, 3, 5\}] = \{f(1), f(2), f(5)\} = \{4, 1, 4\} = \{1, 4\}$.

($b$) $f^{-1}[\{2, 3, 4\}]$ consists of each element whose image is 2, 3 or 4. Hence $f^{-1}[\{2, 3, 4\} = \{4, 1, 3, 5\}]$.

($c$) $f^{-1}[\{3, 5\}] = \varnothing$ since no element has 3 or 5 as an image.

**5.14.** Consider the function $f\colon \mathbf{R} \to \mathbf{R}$ defined by $f(x) = x^2$. Find:

($a$) $f^{-1}[\{25\}]$; ($b$) $f^{-1}[\{-9\}]$; ($c$) $f^{-1}[\{x : x \leq 0\}]$; ($d$) $f^{-1}[[4, 25]] = f^{-1}[\{x : 4 \leq x \leq 25\}]$.

($a$) $f^{-1}[\{25\}] = \{5, -5\}$ since $f(5) = 25$ and $f(-5) = 25$ and since the square of no other number is 25.

($b$) $f^{-1}[\{-9\}] = \varnothing$ since the square of no real number is 9.

($c$) $f^{-1}[\{x : x \leq 0\}] = \{0\}$ since $f(0) = 0 \leq 0$ and since the square of every other real number is greater than 0.

($d$) $f^{-1}[\{x : 4 \leq x \leq 25\}]$ consists of those real numbers $x$ such that $4 \leq x^2 \leq 25$. Accordingly,

$$f^{-1}[\{x : 4 \leq x \leq 25\}] = [2, 5] \cup [-5, -2]$$

**5.15.** Suppose $f\colon S \to T$ is one-to-one. Prove that the associated set function $f\colon \mathscr{P}(S) \to \mathscr{P}(T)$ is also one-to-one.

Suppose $S = \varnothing$. Then $\mathscr{P}(S) = \{\varnothing\}$ has only one element. Hence $f\colon \mathscr{P}(S) \to \mathscr{P}(T)$ is one-to-one.

Suppose $S \neq \varnothing$. Then $\mathscr{P}(S)$ has at least two elements. Let $A, B \in \mathscr{P}(S)$, but $A \neq B$. Then there exists $p \in A$ such that $p \notin B$ (or $p \in B$ such that $p \notin A$). Then $f(p) \in f[A]$ and, since $f$ is one-to-one, $f(p) \notin f[B]$. Thus $f[A] \neq f[B]$, and so the associated set function is one-to-one.

**5.16.** Let $f: S \to T$ and let $A$ and $B$ be subsets of $S$. Prove $f[A \cup B] = f[A] \cup f[B]$.

We first show that $f[A \cup B] \subseteq f[A] \cup f[B]$. Let $y \in f[A \cup B]$. Then there exists $x \in S$ such that $f(x) = y$ and $x \in A \cup B$. Then $x \in A$ or $x \in B$. Hence $f(x) \in f[A]$ or $f(x) \in f[B]$. In either case, $y = f(x)$ belongs to $f[A] \cup f[B]$.

Next we prove the reverse inclusion, i.e., $f[A] \cup f[B] \subseteq f[A \cup B]$. Let $y \in f[A] \cup f[B]$. Then $y \in f[A]$ or $y \in f[B]$. If $y \in f[A]$, then there exists $x \in A$ such that $f(x) = y$, and if $y \in f[B]$, then there exists $x \in B$ such that $f(x) = y$. In either case, $y = f(x)$ with $x \in A \cup B$; hence $y \in f[A \cup B]$.

## SPECIAL FUNCTIONS: EXTENSION, CHOICE, CHARACTERISTIC

**5.17.** Consider the function $f(x) = x$ where $x \geq 0$, that is, where $D = [0, \infty)$ is the domain. State whether or not each of the following functions is an extension of $f$:

(*a*) $g_1(x) = x$ where $x \geq -2$; (*b*) $g_2(x) = (x + |x|)/2$; (*c*) $g_3(x) = x$ where $x \in [-1, 1]$.

A function $g$ is an extension of $f$ if the domain $D'$ of $g$ is an extension of the domain $D = [0, \infty)$ of $f$, and if $g(x) = x$ for every $x \in [0, \infty)$.

(*a*) Since $g_1$ satisfies both of the above conditions, $g_1$ is an extension of $f$.

(*b*) Note

$$g_2(x) = \begin{cases} (x + x)/2 = x & \text{if } x \geq 0 \\ (x - x)/2 = 0 & \text{if } x < 0 \end{cases}$$

Hence $g_2$ is an extension of $f$.

(*c*) The domain of $g_3$ is not a superset of the domain of $f$; hence $g_3$ is not an extension of $f$.

**5.18.** Consider the following subsets of $B = \{1, 2, 3, 4, 5\}$:

$$A_1 = \{1, 2, 3\}, \qquad A_2 = \{1, 5\}, \qquad A_3 = \{2, 4, 5\}, \qquad A_4 = \{3, 4\}$$

State whether or not each of the following functions from $\{A_1, A_2, A_3, A_4\}$ into $B$ is a choice function:

(*a*) $f_1 = \{(A_1, 1), (A_2, 2), (A_3, 3), (A_4, 4)\}$
(*b*) $f_2 = \{(A_1, 1), (A_2, 1), (A_3, 4), (A_4, 4)\}$
(*c*) $f_3 = \{(A_1, 2), (A_2, 1), (A_3, 4), (A_4, 3)\}$
(*d*) $f_4 = \{(A_1, 3), (A_2, 5), (A_3, 1), (A_4, 3)\}$

(*a*) Since $f_1(A_2) = 2$ is not an element in $A_2$, $f_1$ is not a choice function.
(*b*) Here $f_2(A_i)$ belongs to $A_i$, for each $i$; hence $f_2$ is a choice function.
(*c*) Also, $f_3(A_i)$ belongs to $A_i$, for each $i$, hence $f_3$ is a choice function.
(*d*) Note that $f_4(A_3) = 1$ does not belong to $A_3$, hence $f_4$ is not a choice function.

**5.19.** Let $A$ and $B$ be subsets of a universal set $\mathbf{U}$. Prove $\chi_{A \cap B} = \chi_A \chi_B$. [Here $\chi_A \chi_B$ is the product of the functions, not the composition, that is, $(\chi_A \chi_B)(x) = \chi_A(x)\chi_B(x)$.]

Let $x \in A \cap B$. Then $x \in A$ and $x \in B$. Hence

$$\chi_{A \cap B}(x) = 1 \qquad \text{and} \qquad (\chi_A \chi_B)(x) = \chi_A(x)\chi_B(x) = (1)(1) = 1$$

Suppose $y \notin A \cap B$. Then $\chi_{A \cap B}(y) = 0$. Also, $y \in (A \cap B)^c = A^c \cup B^c$, and so $y \in A^c$ or $y \in B^c$. This means $\chi_A(y) = 0$ or $\chi_B(y) = 0$, and therefore

$$(\chi_A \chi_B)(y) = \chi_A(y)\chi_B(y) = 0 = \chi_{A \cap B}(y)$$

Accordingly, $\chi_{A \cap B}$ and $\chi_A \chi_B$ assign the same number to each element in $\mathbf{U}$. Therefore, $\chi_{A \cap B} = \chi_A \chi_B$.

## FUNDAMENTAL FACTORIZATION

**5.20.** Let $f\colon A \to B$. Define $a \sim a'$ if $f(a) = f(a')$. Show that $\sim$ is an equivalence relation on $A$.

We must show that $\sim$ is: ($a$) reflexive, ($b$) symmetric, and ($c$) transitive.

($a$) For any $a \in A$, we have $f(a) = f(a)$. Hence $a \sim a$, and so $\sim$ is reflexive.

($b$) Suppose $a \sim a'$. Then $f(a) = f(a')$, and hence $f(a') = f(a)$. Thus $a' \sim a$, and so $\sim$ is symmetric.

($c$) Suppose $a \sim a'$ and $a' \sim a''$. Then $f(a) = f(a')$ and $f(a') = f(a'')$; hence $f(a) = f(a'')$. Thus $a \sim a''$, and so $\sim$ is transitive.

**5.21.** Prove Lemma 5.3: Let $f\colon A \to B$. Let $f^* : A/f \to f(A)$ be defined by $f^*([a]) = f(a)$. Then $f^*$ is well-defined and bijective.

Suppose $[a] = [a']$. Then $a \sim a'$. Hence $f(a) = f(a')$, and so $f^*$ is well-defined.

First we show that $f^*$ is one-to-one. Suppose $f^*([a]) = f^*([a'])$. Then $f(a) = f(a')$. Hence $a \sim a'$, and so $[a] = [a']$. Thus $f^*$ is one-to-one.

Next we show that $f^*$ is onto. Suppose $b \in f(A)$. Then there exists $a \in A$ such that $f(a) = b$. Then $f^*([a]) = f(a) = b$. Thus $f^*$ is onto. Therefore, $f^*$ is bijective (one-to-one and onto).

**5.22.** Prove Theorem 5.4: Let $f\colon A \to B$. Then diagram in Fig. 5-3 is commutative, that is, $f = \imath \circ f^* \circ \eta$.

Let $a \in A$. Then

$$\begin{aligned}(\imath \circ f^* \circ \eta)(a) &= (\imath \circ f^*)(\eta(a)) = (\imath \circ f^*)([a]) \\ &= \imath(f^*([a])) = \imath(f(a)) = f(a)\end{aligned}$$

Hence $f = \imath \circ f^* \circ \eta$.

**5.23.** Let $A = \{1, 2, 3, 4, 5\}$ and let $f$ be the function in Fig. 5-6($c$).

($a$) Find $A/f$ and $f(A)$. ($b$) Find the factorization $f = \imath \circ f^* \circ \eta$.

($a$) The elements with the same images are put in the same equivalence class. Hence $A/f = [\{1, 3, 5\}, \{2\}, \{4\}]$. Also, $f(A) = \{1, 2, 4\}$. [Note $|A/f| = |f(A)|$.]

($b$) We have:

$$\begin{array}{lllll}
1 \xrightarrow{\eta} & \{1, 3, 5\} & \xrightarrow{f^*} 4 & \xrightarrow{\imath} 4 \\
2 \xrightarrow{\eta} & \{2\} & \xrightarrow{f^*} 1 & \xrightarrow{\imath} 1 \\
3 \xrightarrow{\eta} & \{1, 3, 5\} & \xrightarrow{f^*} 4 & \xrightarrow{\imath} 4 \\
4 \xrightarrow{\eta} & \{4\} & \xrightarrow{f^*} 2 & \xrightarrow{\imath} 2 \\
5 \xrightarrow{\eta} & \{1, 3, 5\} & \xrightarrow{f^*} 4 & \xrightarrow{\imath} 4
\end{array}$$

On the other hand:

$$f(1) = 4, \quad f(2) = 1, \quad f(3) = 4, \quad f(4) = 2, \quad f(5) = 5$$

Thus $f = \imath \circ f^* \circ \eta$.

## ALGORITHMS AND COMPLEXITY

**5.24.** Suppose $P(x) = a_0 + a_1x + a_2x^2 + \cdots + a_mx^m$ has degree $m$. Prove $P(x) = O(x^m)$.

Let $b_0 = |a_0|, b_1 = |a_1|, \ldots, b_m = |a_m|$. Then, for $x \geq 1$, we have

$$\begin{aligned} |P(x)| &\leq b_0 + b_1x + b_2x^2 + \cdots + b_mx^m \\ &= \left[\frac{b_0}{x^m} + \frac{b_1}{x^{m-1}} + \cdots + b^m\right]x^m \\ &\leq (b_0 + b_1 + b_2 + \cdots + b_m)x^m = Mx^m \end{aligned}$$

where $M = |a_0| + |a_1| + \cdots + |a_m|$. Thus $P(x) = O(x^m)$.

**5.25.** Compare the factorial function $f(n) = n!$ to the functions in Fig. 5-5.

The factorial function $f(n) = n!$ grows faster than the exponential function $2^n$. Clearly, for $n \geq 4$,

$$2^n = 2 \cdot 2 \cdot \cdots \cdot 2 \geq 1 \cdot 2 \cdot 3 \cdot \cdots \cdot (n-1)n$$

Thus $f(n) = n!$ grows faster than every function in Fig. 5-5. In fact, $f(n) = n!$ grows faster than the exponential function $g(n) = c^n$ for any constant $c$.

**5.26.** Find $f(3)$ where $f(x) = 2x^4 - 5x^3 + 2x^2 - 6x - 7$.

Use synthetic division to obtain:

$$\begin{array}{r|rrrrr} 3 & 2 & -5 & 2 & -6 & -7 \\ & & 6 & 3 & 15 & 27 \\ \hline & 2 & 1 & 5 & 9 & 20 \end{array}$$

Thus $f(3) = 20$.

**5.27.** Suppose a list DATA contains $n$ elements, and suppose a specific NAME which appears in DATA is given. We want to find the location of NAME in the list using a linear search; that is, we compare NAME with DATA[1], DATA[2], and so on. Let $C(n)$ denote the number of comparisons. Find $C(n)$ for: (*a*) the worse case, (*b*) the average case.

(*a*) Clearly the worst case occurs when NAME is the last element in the list. Hence $C(n) = n$ is the worst-case complexity.

(*b*) Here we assume that it is equally likely for NAME to occur in any position in the list. Accordingly, the numbers of comparisons are $1, 2, \ldots, n$ and each number occurs with probability $p = 1/n$. (We do make the last comparison just to make sure that NAME is in the list.) Then:

$$\begin{aligned} C(n) &= 1 \cdot \frac{1}{n} + 2 \cdot \frac{1}{n} + \cdots + n \cdot \frac{1}{n} \\ &= (1 + 2 + \cdots + n) \cdot \frac{1}{n} \\ &= \frac{n(n+1)}{2} \cdot \frac{1}{n} = \frac{n+1}{2} \end{aligned}$$

This agrees with our intuitive feeling that the average number of comparisons needed to find NAME is approximately half the number of elements in the list.

## Supplementary Problems

### GENERALIZED OPERATIONS, INDEXED SETS

**5.28.** Let $\mathcal{A} = [\{1,2,3,4\}, \{1,3,5,7,9\}, \{1,2,3,6,8\}, \{1,3,7,8,9\}]$. Find: (*a*) $\bigcup\mathcal{A}$; (*b*) $\bigcap\mathcal{A}$.

**5.29.** For each $m \in \mathbf{P}$, let $A_m$ be the following subset of $\mathbf{P}$:

$$A_m = \{m, 2m, 3m, \ldots\} = \{\text{multiples of } m\}$$

(*a*) Find: (1) $A_2 \cap A_7$; (2) $A_6 \cap A_8$; (3) $A_3 \cup A_{12}$; (4) $A_3 \cap A_{12}$.

(*b*) Prove $\bigcap(A_i : i \in J) = \varnothing$, when $J$ is an infinite subset of $\mathbf{P}$.

**5.30.** For each $n \in \mathbf{Z}$, let $B_n = (n, n+1]$, a half-open interval. Find:

(*a*) $B_4 \cup B_5$; (*b*) $B_b \cap B_7$; (*c*) $\bigcup_{i=4}^{20} B_i$; (*d*) $B_s \cup B_{s+1} \cup B_{s+2}$; (*e*) $\bigcup_{i=0}^{15} B_{s+i}$; (*f*) $\bigcup(B_{s+i} : i \in \mathbf{Z})$.

**5.31.** For each $n \in \mathbf{P}$, let $D_n = [0, 1/n]$, $S_n = (0, 1/n]$, $T_n = [0, 1/n)$. Find: (*a*) $\bigcap_n D_n$; (*b*) $\bigcap_n S_n$; (*c*) $\bigcap_n T_n$.

**5.32.** Prove Theorem 5.1(iii): Suppose $J$ is a subset of $I$. Then

$$\bigcup_{i\in J} A_i \subseteq \bigcup_{i\in I} A_i \quad \text{and} \quad \bigcap_{i\in J} A_i \supseteq \bigcap_{i\in I} A_i$$

### SEQUENCES, SUMMATION SYMBOL

**5.33.** Write out the first six terms of each sequence:

(*a*) $a_n = (-1)^{n+1} n^3$

(*b*) $b_n = \dfrac{n^2}{2n+1}$

(*c*) $c_n = \begin{cases} n^2 & \text{if } n \text{ is odd} \\ n+4 & \text{if } n \text{ is even} \end{cases}$

**5.34.** Write out the first six terms of each sequence:

(*a*) $a_1 = 1$, $a_n = n^2 + 2a_{n-1}$ for $n > 1$.

(*b*) $b_1 = 1$, $b_2 = 2$, $b_n = 2b_{n-2} + 3b_{n-1}$ for $n > 2$.

**5.35.** Find: (*a*) $\displaystyle\sum_{k=3}^{5} k^4$; (*b*) $\displaystyle\sum_{k=0}^{4} a_k x^k$; (*c*) $\displaystyle\sum_{j=1}^{3} (j^3 + j^2 - j)$; (*d*) $\displaystyle\sum_{j=1}^{15} 1$.

**5.36.** Rewrite using the summation symbol:

(*a*) $\bar{x} = \dfrac{x_1 f_1 + x_2 f_2 + \cdots + x_n f_n}{f_1 + f_2 + \cdots + f_n}$

(*b*) $c_{ij} = a_{i1}b_{1j} + a_{i2}b_{2j} + \cdots + a_{in}b_{nj}$

## DIAGRAMS AND FUNCTIONS

**5.37.** Consider the diagram in Fig. 5-7($a$). Find the number of paths from $A$ to $D$ and state what they are.

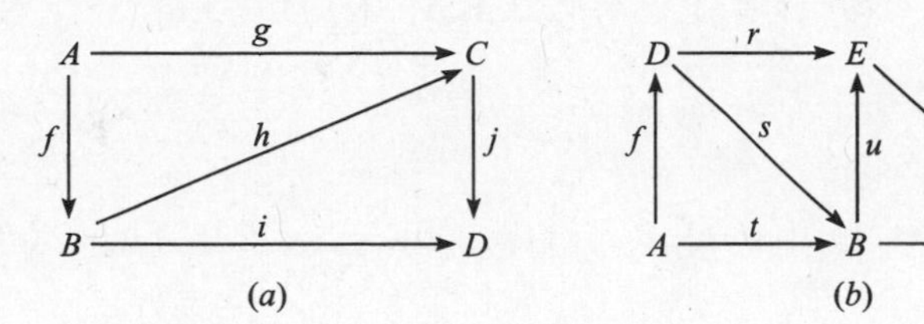


**Fig. 5-7**

**5.38.** Suppose the diagram in Fig. 5-7($b$) is commutative. Which functions are equal?

## ASSOCIATED SET FUNCTIONS

**5.39.** Let $A = \{1, 2, 3, 4, 5\}$, and let $f\colon A \to A$ be defined by

$$f = \{(1,3),\ (2,2),\ (3,5),\ (4,3),\ (5,2)\}$$

Find: ($a$) $f[\{1,2,5\}]$; ($b$) $f^{-1}[\{3,4,5\}]$; ($c$) $f^{-1}[\{1,4\}]$.

**5.40.** Consider the function $f\colon R \to R$ defined by $f(x) = |x|$. Find:

($a$) $f^{-1}[\{7\}]$; ($b$) $f^{-1}[\{-5\}]$; ($c$) $f^{-1}[\{x : x \leq 0\}]$; ($d$) $f^{-1}[[2,3]] = f^{-1}[\{x : 2 \leq x \leq 3\}]$.

**5.41.** Suppose $f\colon S \to T$ is onto. Prove that the associated set function $f\colon \mathscr{P}(S) \to \mathscr{P}(T)$ is also onto.

**5.42.** Let $f\colon S \to T$ and let $A$ and $B$ be subsets of $S$. Prove: ($a$) $f[A \cap B] \subseteq f[A] \cap f[B]$; ($b$) $f[A \backslash B] \supseteq f[A] \backslash f[B]$.

**5.43.** Let $f\colon S \to T$ and let $A$ and $B$ be subsets of $T$. Prove:

($a$) $f^{-1}[A \cap B] = f^{-1}[A] \cap f^{-1}[B]$; ($b$) $f^{-1}[A \backslash B] = f^{-1}[A] \backslash f^{-1}[B]$. (Compare with Problem 5.42.)

## EXTENSIONS, RESTRICTIONS, CHOICE FUNCTIONS

**5.44.** Let $f$ be the following function with domain $D = \{1, 3, 5, 7\}$:

$$f = \{(1,6),\ (3,4),\ (5,2),\ (7,4)\}$$

For what values of $x$ and $y$ will the following functions be extensions of $f$:

($a$) $g_1 = \{(1,6),\ (2,2),\ (3,x),\ (4,1),\ (5,y),\ (6,2),\ (7,4)\}$
($b$) $g_2 = \{(1,x),\ (2,4),\ (3,4),\ (4,2),\ (5,2),\ (6,2),\ (7,y)\}$
($c$) $g_3 = \{(1,6),\ (2,x),\ (3,4),\ (4,7),\ (5,y),\ (6,3),\ (7,4)\}$
($d$) $g_4 = \{(1,6),\ (2,5),\ (3,1),\ (4,3),(5,x),\ (6,8),\ (7,y)\}$

**5.45.** Consider the following subsets of $B = \{1, 2, 3, 4, 5\}$:

$$A_1 = \{1,2,3\},\quad A_2 = \{2,4\},\quad A_3 = \{5\},\quad A_4 = \{1,3,4,5\}$$

State whether or not each of the following functions from $\{A_1, A_2, A_3, A_4\}$ into $B$ is a choice function:

($a$) $f_1 = \{(A_1,1),\ (A_2,2),\ (A_3,3),\ (A_4,4)\}$
($b$) $f_2 = \{(A_1,2),\ (A_2,2),\ (A_3,5),\ (A_4,5)\}$
($c$) $f_3 = \{(A_1,3),\ (A_2,1),\ (A_3,5),\ (A_4,3)\}$
($d$) $f_4 = \{(A_1,3),\ (A_2,4),\ (A_3,5),\ (A_4,1)\}$

**5.46.** Let $f$ be the following function with domain $D = \{1, 2, 3, 4, 5, 6\}$:

$$f = \{(1,4),\ (2,5),\ (3,6),\ (4,5),\ (5,4),\ (6,3)\}$$

Find the restriction of $f$ to: (*a*) $\{1, 3, 5\}$, (*b*) $\{2, 3, 4, 5\}$, (*c*) $\{1, 2, 5, 6\}$.

## CHARACTERISTIC FUNCTIONS

**5.47.** Let $\mathbf{U} = \{a, b, c, d, e\}$, and let $A = \{a, b, c\}$, $B = \{c, d\}$, and $C = \{a, d, e\}$. Find: (*a*) $\chi_A$; (*b*) $\chi_B$; (*c*) $\chi_C$.

**5.48.** Let $\mathbf{U} = \{a, b, c, d\}$. Each of the following functions from $\mathbf{U}$ into $\{0, 1\}$ is the characteristic function of a subset of $\mathbf{U}$. Find each subset.

(*a*) $\{(a,1),\ (b,0),\ (c,0),\ (d,1)\}$, (*c*) $\{(a,0),\ (b,0),\ (c,0),\ (d,0)\}$,
(*b*) $\{(a,0),\ (b,1),\ (c,0),\ (d,0)\}$, (*d*) $\{(a,1),\ (b,1),\ (c,0),\ (d,1)\}$.

**5.49.** Let $A$ and $B$ be subsets of a universal set $\mathbf{U}$. Prove: (*a*) $\chi_{A \cup B} = \chi_A + \chi_B - \chi_{A \cap B}$; (*b*) $\chi_{A \setminus B} = \chi_A - \chi_{A \cap B}$.

## FUNDAMENTAL FACTORIZATION

**5.50.** Let $A = \{\text{Marc, Erik, Audrey, Britt, Emily}\}$. Find $A/f$ and $f(A)$ where $f\colon A \to \mathbf{P}$ is defined by:

(*a*) $f(a) =$ number of letters in $a$; (*b*) $f(a) =$ number of distinct letters in $a$

**5.51.** Suppose $f\colon \mathbf{R} \to \mathbf{R}$ is defined by $f(x) = |x|/x$ when $x \neq 0$ and $f(0) = 0$. Find $\mathbf{R}/f$ and $f(\mathbf{R})$.

## MISCELLANEOUS PROBLEMS

**5.52.** Let $f\colon D \to \mathbf{R}$ and $g : D \to \mathbf{R}$ for some domain $D$. Define $(f + g) : D \to \mathbf{R}$ and $(fg) : D \to \mathbf{R}$ by

$$(f + g)(x) = f(x) + g(x) \qquad \text{and} \qquad (fg)(x) = f(x)g(x)$$

(Note $fg$ is not the composition of $f$ and $g$.)

(*a*) Let $0_D : D \to \mathbf{R}$ be the zero function, i.e., for every $x \in D$, $0_D(x) = 0$. Prove that, for any function $f\colon D \to \mathbf{R}$,

$$f + 0_D = 0_D + f = f \qquad \text{and} \qquad g = f \cdot 0_D = 0_D \cdot f = 0_D$$

(*b*) Consider the following functions on $D = \{1, 2, 3\}$:

$$f = \{(1,3),\ (2,5),\ (3,8)\} \qquad \text{and} \qquad g = \{(1,5),\ (2,-3),\ (3,4)\}$$

Find $f + g$ and $fg$.

**5.53.** Find $f(a)$ where: (*a*) $a = 4$ and $f(x) = 2x^4 - 5x^3 - 9x^2 + 7$;
(*b*) $a = 7$ and $f(x) = x^5 - 8x^4 + 6x^3 + 9x^2 - 7x - 27$.

**5.54.** Consider $n$ distinct sets $A_1, A_2, \ldots, A_n$ in a universal set $\mathbf{U}$. Prove:

(*a*) There are $2^n$ fundamental products of the $n$ sets.
(*b*) Any two such fundamental products are disjoint.
(*c*) $\mathbf{U}$ is the union of all the fundamental products.

## Answers to Supplementary Problems

**5.28.** (*a*) $\bigcup \mathscr{A} = \{1, 2, 3, \ldots, 89\}$; (*b*) $\bigcap \mathscr{A} = \{1, 3\}$

**5.29.** (*a*) (1) $A_{14}$; (2) $A_{24}$; (3) $A_3$; (4) $A_{12}$

**5.30.** (*a*) $(4, 6]$; (*b*) $(6, 8]$; (*c*) $(4, 21]$; (*d*) $(s, s+3]$; (*e*) $(s, s+16]$; (*f*) $\mathbf{R}$

**5.31.** (*a*) $\{0\}$; (*b*) $\varnothing$; (*c*) $\{0\}$

**5.33.** (*a*) 1, −8, 27, −64, 125, −216; (*b*) 1/3, 4/51, 9/7, 16/9, 25/11, 36/13; (*c*) 1, 6, 9, 8, 25, 10

**5.34.** (*a*) 1, 6, 21, 58, 141, 318; (*b*) 1, 2, 8, 28, 100, 356

**5.35.** (*a*) 962; (*b*) $a_0 + a_1x + a_2x^2 + a_3x^3 + a_4x^4$; (*c*) 44; (*d*) 15

**5.36.** (*a*) $\bar{x} = \dfrac{\sum x_i f_i}{\sum f_i}$; (*b*) $c_{ij} = \sum_{k=1}^{n} a_{ik} b_{kj}$

**5.37.** Three: $i \circ f$, $j \circ g$, and $j \circ h \circ f$

**5.38.** $t = s \circ f$, $r = u \circ s$, $r \circ f = u \circ s \circ f = u \circ t$, $w = v \circ u$, $w \circ s = v \circ r = v \circ u \circ s$,
$w \circ t = w \circ s \circ f = v \circ u \circ t = v \circ u \circ s \circ f = v \circ r \circ f$

**5.39.** (*a*) $\{2, 3\}$; (*b*) $\{1, 4, 3\}$; (*c*) $\varnothing$

**5.40.** (*a*) $\{7, -7\}$; (*b*) $\varnothing$; (*c*) $\{0\}$; (*d*) $[-3, -2] \cup [2, 3]$

**5.44.** (*a*) $x = 4$, $y = 2$; (*b*) $x = 6$, $y = 4$; (*c*) $x$ any value, $y = 2$; (*d*) $(3, 1)$ means no extension.

**5.45.** (*a*) No; (*b*) yes; (*c*) no; (*d*) yes

**5.46.** (*a*) $\{(1, 4), (3, 6), (5, 4)\}$; (*b*) $\{(2, 5), (3, 6), (4, 5), (5, 4)\}$; (*c*) $\{(1, 4), (2, 5), (5, 4), (6, 3)\}$

**5.47.** (*a*) $\chi_A = \{(a, 1), (b, 1), (c, 1), (d, 0), (e, 0)\}$
(*b*) $\chi_B = \{(a, 0), (b, 0), (c, 1), (d, 1), (e, 0)\}$
(*c*) $\chi_C = \{(a, 1), (b, 0), (c, 0), (d, 1), (e, 1)\}$

**5.48.** (*a*) $\{a, d\}$; (*b*) $\{b\}$; (*c*) $\varnothing$; (*d*) $\{a, b, d\}$

**5.50.** (*a*) $A/f = [\{\text{Marc, Erik}\}, \{\text{Britt, Emily}\}, \{\text{Audrey}\}]$, $f(a) = \{4, 5, 6\}$
(*b*) $A/f = [\{\text{Marc, Erik, Britt}\}, \{\text{Emily}\}, \{\text{Audrey}\}]$, $f(A) = \{4, 5, 6\}$

**5.51.** $\mathbf{R}/f = \{(-\infty, 0), \{0\}, (0, \infty)\}$; $f(\mathbf{R}) = \{-1, 0, 1\}$

**5.52.** (*b*) $f + g = \{(1, 8), (2, 2), (3, 12)\}$; $fg = \{(1, 15), (2, -15), (3, 32)\}$

**5.53.** (*a*) $f(4) = 19$; (*b*) $f(7) = 22$

# Chapter 6

# Cardinal Numbers

## 6.1 INTRODUCTION

It is natural to ask whether or not two sets have the same number of elements. For finite sets the answer can be found by simply counting the number of elements. For example, each of the sets

$$\{a, b, c, d\}, \qquad \{2, 3, 5, 7\}, \qquad \{x, y, z, t\}$$

has four elements. Thus these sets have the same number of elements. However, it is not always necessary to know the number of elements in two finite sets before we know that they have the same number of elements. For example, if each chair in a room is occupied by exactly one person and there is no one standing, then clearly there are "just as many" people as there are chairs in the room.

The above simple notion, that two sets have "the same number of elements" if their elements can be "paired-off", can also apply to infinite sets. In fact, it has the following startling results:

(*a*) Infinite sets need not have the "same number of elements"; some are "more infinite" than others.
(*b*) There are "just as many" even integers as there are integers, and "just as many" rational numbers **Q** as positive integers **P**.
(*c*) There are "more" points on the real line **R** than there are positive integers **P**; and there are "more" curves in the plane $\mathbf{R}^2$ than there are points in the plane.

This chapter will investigate and prove the above results. First we will formally define when two sets, finite or infinite, have the same number of elements or, in other words, the same cardinality. Lastly, we define addition and multiplication for these "cardinal numbers", and show that many of their properties reflect corresponding properties of sets.

We remark that, at one time, all infinite sets were considered to have the same number of elements. The German mathematician Georg Cantor (1845–1918) gave the above alternative definition which revolutionized the entire theory of sets.

## 6.2 ONE-TO-ONE CORRESPONDENCE, EQUIPOTENT SETS

Recall that a one-to-one correspondence between sets $A$ and $B$ is a function $f: A \to B$ which is bijective, that is, which is one-to-one and onto. In such a case, each element $a \in A$ is paired with a unique element $b \in B$ given by $b = f(a)$. We sometimes write

$$a \leftrightarrow b$$

to denote such a pairing.

**Remark**: Frequently, a child counts the objects of a set by forming a one-to-one correspondence between the objects and his fingers. An adult counts the objects of a set by forming a one-to-one correspondence between the objects and the set

$$\{1, 2, 3, \ldots, n\}$$

In fact, if one is asked the question:

"How many days are there until next Saturday?"

the response is often to actually pair the remaining days with one's fingers.

The following definition applies.

**Definition 6.1**: Sets $A$ and $B$ are said to have the *same cardinality* or the *same number of elements*, or to be *equipotent*, written

$$A \approx B$$

if there is a function $f: A \to B$ which is bijective, that is, both one-to-one and onto.

Recall that such a function $f$ is said to define a *one-to-one correspondence* between $A$ and $B$.

Since the identity function is bijective, and the composition and inverse of bijective functions are bijective, we immediately obtain the following theorem:

**Theorem 6.1**: The relation $\approx$ of being equipotent is an equivalence relation in any collection of sets. That is:

(i) $A \approx A$ for any set $A$.

(ii) If $A \approx B$, then $B \approx A$.

(iii) If $A \approx B$ and $B \approx C$, then $A \approx C$.

**EXAMPLE 6.1**

(*a*) Let $A$ and $B$ be sets with exactly three elements, say,

$$A = \{2, 3, 5\}, \qquad \text{and } B = \{\text{Marc, Erik, Audrey}\}$$

Then clearly we can find a one-to-one correspondence between $A$ and $B$. For example, we can label the elements of $A$ as the first element, the second element, and the third element, and label $B$ similarly. Then the rule which pairs the first elements of $A$ and $B$, pairs the second elements of $A$ and $B$, and pairs the third elements of $A$ and $B$, that is, the function $f: A \to B$ defined by

$$f(2) = \text{Marc}, \qquad f(3) = \text{Erik}, \qquad f(5) = \text{Audrey}$$

is one-to-one and onto. Thus $A$ and $B$ are equipotent.

The same idea may be used to show that any two finite sets with the same number of elements are equipotent.

(*b*) Let $A = \{a, b, c, d\}$ and $B = \{1, 2, 3\}$. Then $A$ and $B$ are not equipotent. For suppose there were a rule for pairing the elements of $A$ and $B$. If there were four or more pairs, then an element of $B$ would be used twice, and if there were three or fewer pairs then some element of $A$ would not be used. In other words, since $A$ has more elements than $B$, any function $f: A \to B$ must assign at least two elements of $A$ to the same element of $B$, and hence $f$ would not be one-to-one.

In a similar way, we can see that any two finite sets with different numbers of elements are not equipotent.

(*c*) Let $\mathbf{I} = [0, 1]$, the closed unit interval, and let $S$ be any other closed interval, say $S = [a, b]$ where $a < b$. The function $f: \mathbf{I} \to S$ defined by

$$f(x) = (b - a)x + a$$

is one-to-one and onto. Thus $\mathbf{I}$ and $S$ have the same cardinality. Therefore, by Theorem 6.1, any two closed intervals have the same cardinality.

(*d*) Consider the set $\mathbf{P} = \{1, 2, 3, \ldots\}$ of positive integers and the set $E = \{2, 4, 6, \ldots\}$ of even positive integers. The following defines a one-to-one correspondence between $\mathbf{P}$ and $E$:

$$\begin{array}{cccccccccc} \mathbf{P} = \{1, & 2, & 3, & 4, & 5, & . & . & .\} \\ \downarrow & \downarrow & \downarrow & \downarrow & \downarrow & \downarrow & \downarrow & \downarrow \\ E = \{2, & 4, & 6, & 8, & 10, & . & . & .\} \end{array}$$

In other words, the function $f: \mathbf{P} \to E$ defined by $f(n) = 2n$ is one-to-one and onto. Thus $\mathbf{P}$ and $E$ have the same cardinality.

More generally, if $K = \{0, k, 2k, 3k, \ldots\}$ is the set of multiples of a positive integer $k$, then $f: \mathbf{P} \to K$ defined by $f(n) = kn$ is a one-to-one correspondence between $\mathbf{P}$ and $K$. Therefore $\mathbf{P}$ and $K$ have the same cardinality.

Parts (*a*) and (*b*) of the above Example 6.1 show that finite sets are equipotent if and only if they contain the same number of elements. Thus, for finite sets, Definition 6.1 corresponds to the usual meaning of two sets containing the same number of elements.

On the other hand, Example 6.1(*d*) shows that the infinite set **P** has the same cardinality as a proper subset of itself. This property is characteristic of infinite sets. In fact, we state this observation formally.

**Definition 6.2**: A set $S$ is *infinite* if it has the same cardinality as a proper subset of itself. Otherwise $S$ is *finite*.

Familiar examples of infinite sets are the counting numbers (positive integers) **P**, the natural numbers (nonnegative integers) **N**, the integers **Z**, the rational numbers **Q**, and the real numbers **R**.

There might be a temptation to think that all infinite sets have the same cardinality; but we will show later that this is definitely not true.

We conclude this section with the following example, which tells us that any two sets have the same cardinality, respectively, to two disjoint sets.

**EXAMPLE 6.2** Consider any two sets $A$ and $B$. Let $A' = A \times \{1\}$ and $B' = B \times \{2\}$. Then

$$A \approx A' \qquad \text{and} \qquad B \approx B'$$

For example, the functions

$$f(a) = (a, 1),\ a \in A \qquad \text{and} \qquad g(b) = (b, 2),\ b \in B$$

are each bijective. Although $A$ and $B$ need not be disjoint, the sets $A'$ and $B'$ are disjoint, i.e.,

$$A' \cap B' = \varnothing$$

Specifically, each ordered pair in $A'$ has 1 as a second component, whereas each ordered pair in $B'$ has 2 as a second component.

## 6.3 DENUMERABLE AND COUNTABLE SETS

The reader is familiar with the set $\mathbf{P} = \{1, 2, 3, \ldots\}$ of counting numbers or positive integers. The following definitions apply.

**Definition 6.3**: A set $D$ is said to be *denumerable* or *countably infinite* if $D$ has the same cardinality as **P**.

**Definition 6.4**: A set is *countable* if it is finite or denumerable, and a set is *nondenumerable* if it is not countable.

Thus a set $S$ is nondenumerable if $S$ is infinite and $S$ does not have the same cardinality as **P**.

**EXAMPLE 6.3**

(*a*) Any infinite sequence

$$a_1, a_2, a_3, \ldots$$

of distinct elements is countably infinite, for a sequence is essentially a function $f(n) = a_n$ whose domain is **P**. So if the $a_n$ are distinct, the function is one-to-one and onto. Thus each of the following sets is countably infinite:

$$\{1, 1/2, 1/3, \ldots, 1/n, \ldots\}$$
$$\{1, -2, 3, -4, \ldots (-1)^{n-1}n, \ldots\}$$
$$\{(1, 1), (4, 8), (9, 27), \ldots, (n^2, n^3), \ldots\}$$

(*b*) Consider the product set $\mathbf{P} \times \mathbf{P}$ as exhibited in Fig. 6-1. The set $\mathbf{P} \times \mathbf{P}$ can be written as an infinite sequence as follows:

$$\{(1,1),\ (2,1),\ (1,2),\ (1,3),\ (2,2),\ldots\}$$

This sequence is determined by "following the arrows" in Fig. 6-1. Thus $\mathbf{P} \times \mathbf{P}$ is countably infinite for the reasons stated in (*a*).

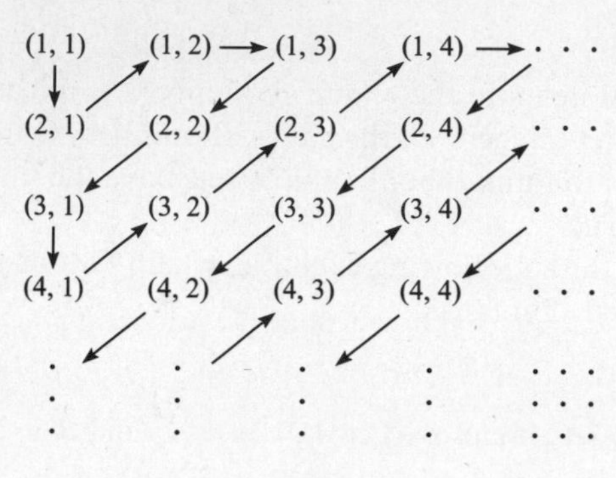


**Fig. 6-1**

(*c*) Recall that $\mathbf{N} = \{0, 1, 2, \ldots\} = \mathbf{P} \cup \{0\}$ is the set of natural numbers or nonnegative integers. Now each positive integer $a \in \mathbf{P}$ can be written uniquely in the form

$$a = 2^r(2s+1)$$

where $r, s \in \mathbf{N}$. Consider the function $f\colon \mathbf{P} \to \mathbf{N} \times \mathbf{N}$ defined by

$$f(a) = (r, s)$$

where $r$ and $s$ are as above. Then $f$ is one-to-one and onto. Thus $\mathbf{N} \times \mathbf{N}$ is denumerable (countably infinite) or, in other words, $\mathbf{N} \times \mathbf{N}$ has the same cardinality as $\mathbf{P}$. Note that $\mathbf{P} \times \mathbf{P}$ is a subset of $\mathbf{N} \times \mathbf{N}$.

The following theorems apply.

**Theorem 6.2**: Every infinite set contains a subset which is denumerable.

**Theorem 6.3**: A subset of a denumerable set is finite or denumerable.

**Corollary 6.4**: A subset of a countable set is countable.

**Theorem 6.5**: Let $A_1, A_2, A_3, \ldots$ be a sequence of pairwise disjoint denumerable sets. Then the union

$$A_1 \cup A_2 \cup A_3 \cup \cdots = \cup(A_i : i \in \mathbf{P})$$

is denumerable.

**Corollary 6.6**: A countable union of countable sets is countable.

Observe that Corollary 6.6 tells us that if each of the sets $A_1, A_2, A_3, \ldots$ is countable then the union

$$A_1 \cup A_2 \cup A_3 \cup \cdots$$

is also countable.

The next theorem gives a very important, and not entirely obvious, example of a denumerable (countably infinite) set.

**Theorem 6.7:** The set $\mathbf{Q}$ of rational numbers is denumerable.

*Proof*: Note that $\mathbf{Q} = \mathbf{Q}^+ \cup \{0\} \cup \mathbf{Q}^-$ where $\mathbf{Q}^+$ and $\mathbf{Q}^-$ denote, respectively, the sets of positive and negative rational numbers. Let $f\colon \mathbf{Q}^+ \to \mathbf{P} \times \mathbf{P}$ be defined by

$$f(p/q) = (p, q)$$

where $p/q$ is any element of $\mathbf{Q}^+$ expressed as the ratio of two relatively prime positive integers. Then $f$ is one-to-one and so $\mathbf{Q}^+$ has the same cardinality as a subset of $\mathbf{P} \times \mathbf{P}$. By Example 6.3($b$), $\mathbf{P} \times \mathbf{P}$ is denumerable; hence, by Theorem 6.3, the infinite set $\mathbf{Q}^+$ is denumerable. Similarly $\mathbf{Q}^-$ is denumerable. Thus the set $\mathbf{Q}$ of rational numbers, the union of $\mathbf{Q}^+, \{0\}$, and $\mathbf{Q}^-$, is also denumerable.

**Remark:** Theorem 6.7 tells us that there are just as many rational numbers as there are positive integers, that is, that $\mathbf{Q}$ has the same cardinality as $\mathbf{P}$.

## 6.4 REAL NUMBERS R AND THE POWER OF THE CONTINUUM

Not every infinite set is countable. The next theorem (proved in Problem 6.15) gives a specific and extremely important example of such a set.

**Theorem 6.8:** The unit interval $\mathbf{I} = [0, 1]$ is nondenumerable.

Observe that this theorem also tells us that infinite sets need not have the same cardinality.

The following definition applies.

**Definition 6.5:** A set $A$ is said to have the *power of the continuum* if $A$ has the same cardinality as the unit interval $\mathbf{I} = [0, 1]$.

Besides the unit interval $\mathbf{I}$, all the other intervals also have the power of the continuum. There are several such kinds of intervals. Specifically, if $a$ and $b$ are real numbers with $a < b$, then we define:

| | |
|---|---|
| closed interval: | $[a, b] = \{x \in \mathbf{R} : a \leq x \leq b\}$ |
| open interval: | $(a, b) = \{x \in \mathbf{R} : a < x < b\}$ |
| half-open intervals: | $[a, b) = \{x \in \mathbf{R} : a \leq x < b\}$ |
| | $(a, b] = \{x \in \mathbf{R} : a < x \leq b\}$ |

Example 6.1($c$) shows that any closed interval $[a, b]$ has the power of the continuum. Problem 6.3 shows that any open or half-open interval also has the power of the continuum.

### Real Numbers R

Lastly, we note that the set $\mathbf{R}$ of real numbers also has the power of the continuum. Specifically, consider the function $f\colon \mathbf{R} \to D$ where $D = (-1, 1)$ and $f$ is defined by

$$f(x) = \frac{x}{1 + |x|}$$

Figure 6-2 is the graph of this function. Clearly the values of $f$ belong to $(-1, 1)$ since $|x| < 1 + |x|$. It is not difficult to show that $f$ is both one-to-one and onto. Thus the set $\mathbf{R}$ of real numbers has the same cardinality as the open interval $D = (-1, 1)$, and hence $\mathbf{R}$ has the power of the continuum.

**Remark:** Some texts define a set $A$ to have the power of the continuum if it has the same cardinality as $\mathbf{R}$ rather than the unit interval $\mathbf{I}$. By the above remark, both definitions are equivalent. The use here of $\mathbf{I}$ rather than $\mathbf{R}$ is motivated by Theorem 6.8.

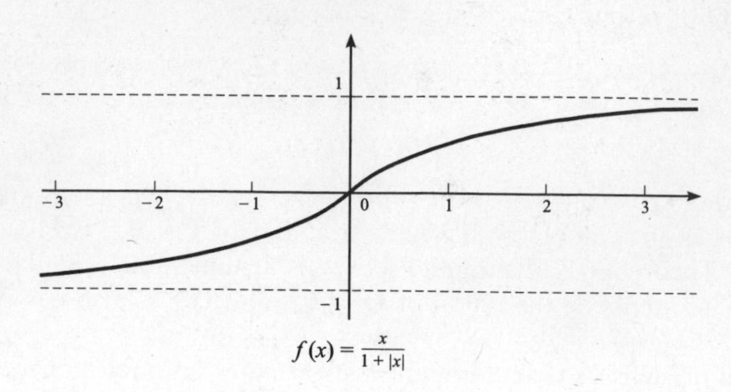


**Fig. 6-2**

## 6.5 CARDINAL NUMBERS

Frequently, we want to know the "size" of a given set without necessarily comparing it to another set. For finite sets, there is no difficulty. For example, the set $A = \{a, b, c\}$ has 3 elements. Any other set with 3 elements is equipotent to $A$. On the other hand, for infinite sets it is not sufficient to just say that the set has infinitely many elements since not all infinite sets are equipotent. To solve this problem, we introduce the concept of a cardinal number.

Each set $A$ is assigned a symbol in such a way that two sets $A$ and $B$ are assigned the same symbol if and only if they are equipotent. This symbol is called the *cardinality* or *cardinal number* of $A$, and it is denoted by

$$|A|, \qquad n(A), \qquad \text{or} \qquad \text{card}(A)$$

We will use $|A|$. Thus:

$$|A| = |B| \quad \text{if and only if} \quad A \approx B$$

One may also view a cardinal number as the equivalence class of all sets which are equipotent.

### Finite Cardinal Numbers

The obvious symbols are used for the cardinal numbers of finite sets. That is, 0 is assigned to the empty set $\varnothing$, and $n$ is assigned to the set $\{1, 2, \ldots, n\}$. Thus:

$$|A| = n \quad \text{if and only if} \quad A \approx \{1, 2, \ldots, n\}$$

Alternatively, the symbols $0, 1, 2, 3, \ldots$ are assigned, respectively, to the sets

$$\varnothing,\ \{\varnothing\},\ [\varnothing,\ \{\varnothing\}], [\varnothing,\ \{\varnothing\},\ [\varnothing,\ \{\varnothing\}]], \ldots$$

Although the natural number $n$ and the cardinal number $n$ are technically different things, there is no conflict using the same symbol in these two roles. The cardinal numbers of finite sets are called *finite cardinal numbers*.

### Transfinite Cardinal Numbers, $\aleph_0$ and c

Cardinal numbers of infinite sets are called *infinite* or *transfinite cardinal numbers*.

The cardinal number of the infinite set **P** of positive integers is

$$\aleph_0$$

which is read aleph-nought. This notation was introduced by Cantor. (The symbol $\aleph$ is the first letter aleph of the Hebrew alphabet.) Thus:

$$\boxed{|A| = \aleph_0 \qquad \text{if and only if} \qquad A \approx \mathbf{P}}$$

In particular, we have $|\mathbf{Z}| = \aleph_0$ and $|\mathbf{Q}| = \aleph_0$. (The significance of 0 in $\aleph_0$ is discussed in Chapter 8.)

The cardinal number of the unit interval $\mathbf{I} = [0, 1]$ is denoted by:

$$\mathbf{c}$$

and it is called the *power of the continuum*. Thus:

$$\boxed{|A| = \mathbf{c} \qquad \text{if and only if} \qquad A \approx \mathbf{I}}$$

In particular, we have $|\mathbf{R}| = \mathbf{c}$, and the cardinal number of any interval is $\mathbf{c}$.

The following statements follow directly from the above definitions:

(*a*) $A$ is denumerable or countably infinite means $|A| = \aleph_0$.
(*b*) $A$ is countable means $|A|$ is finite or $|A| = \aleph_0$.
(*c*) $A$ has the power of the continuum means $|A| = \mathbf{c}$.

## 6.6 ORDERING OF CARDINAL NUMBERS

One frequently wants to compare the size of two sets. This is done by means of an inequality relation which is defined for cardinal numbers as follows.

**Definition 6.6**: Let $A$ and $B$ be sets. We say that

$$|A| \le |B|$$

if $A$ has the same cardinality as a subset of $B$ or, equivalently, if there exists a one-to-one (injective) function $f: A \to B$.

As expected, $|A| \le |B|$ is read:

"The cardinal number of $A$ is less than or equal to the cardinal number of $B$."

As usual with the symbol $\le$, we have the following addition notation:

$$|A| < |B| \quad \text{means} \quad |A| \le |B| \quad \text{but } |A| \ne |B|$$
$$|A| \ge |B| \quad \text{means} \quad |B| \le |A|$$
$$|A| > |B| \quad \text{means} \quad |B| < |A|$$

Again, as usual, the symbols $<$, $\ge$, $>$ are read "less than", "greater than or equal to", and "greater than", respectively.

We emphasize that the above relations between cardinal numbers are well defined, that is, the relations are independent of the particular sets involved. Namely, if $A \approx A'$ and $B \approx B'$, then

$$|A| \le |B| \text{ if and only if } |A'| \le |B'| \quad \text{and} \quad |A| < |B| \text{ if and only if } |A'| < |B'|$$

**EXAMPLE 6.4**

(*a*) Let $A$ be a proper subset of a finite set $B$. Clearly, $|A| \leq |B|$. Since $A$ is a proper subset of $B$, where $A$ and $B$ are finite, we know that $|A| \neq |B|$. Thus $|A| < |B|$. In other words, for finite cardinals $m$ and $n$, we have $m < n$ as cardinal numbers if and only if $m < n$ as nonnegative integers. Accordingly, the inequality relation $\leq$ for cardinal numbers is an extension of the inequality relation $\leq$ for nonnegative integers.

(*b*) Let $n$ be a finite cardinal. Then $n < \aleph_0$ since any finite set $A$ is equipotent to a subset of $\mathbf{P}$ and $|A| \neq |\mathbf{P}|$. Thus we may write

$$0 < 1 < 2 < \cdots < \aleph_0$$

(*c*) Consider the set $\mathbf{P}$ of positive integers and the unit interval $\mathbf{I}$, that is, consider the sets

$$\mathbf{P} = \{1, 2, 3, \ldots\} \quad \text{and} \quad \mathbf{I} = \{x \in \mathbf{R} : 0 \leq x \leq 1\}$$

The function $f: \mathbf{P} \to \mathbf{I}$ defined by $f(n) = 1/n$ is one-to-one. Therefore, $|\mathbf{P}| \leq |\mathbf{I}|$. On the other hand, by Theorem 6.7, $|\mathbf{P}| \neq |\mathbf{I}|$. Therefore, $\aleph_0 = |\mathbf{P}| < |\mathbf{I}| = \mathbf{c}$. Accordingly, we may now write

$$0 < 1 < 2 < \cdots < \aleph_0 < \mathbf{c}$$

(*d*) Let $A$ be any infinite set. By Theorem 6.2, $A$ contains a subset which is denumerable. Accordingly, for any infinite set $A$, we always have $\aleph_0 \leq |A|$.

## Cantor's Theorem

The only transfinite cardinal numbers we have seen are $\aleph_0$ and $\mathbf{c}$. It is natural to ask if there are any others. The answer is yes. In fact, Cantor's theorem, which follows, tells us that the cardinal number of the power set $\mathscr{P}(A)$ of any set $A$ is larger than the cardinal number of the set $A$ itself; namely:

**Theorem 6.9 (Cantor):** For any set $A$, we have $|A| < |\mathscr{P}(A)|$.

This important theorem is proved in Problem 6.18.

**Notation:** If $\alpha = |A|$, then we let $2^\alpha = |\mathscr{P}(A)|$. This no doubt comes from the fact that if a finite set $A$ has $n$ elements then $\mathscr{P}(A)$ has $2^n$ elements.

Accordingly, Cantor's theorem may be restated as follows.

**Theorem 6.9 (Cantor):** For any cardinal number $\alpha$, we have $\alpha < 2^\alpha$.

## Schroeder–Bernstein Theorem, Law of Trichotomy

Note first that the relation $\leq$ for cardinal numbers is reflexive and transitive. That is:

(i) For any set $A$, we have $|A| = |A|$.
(ii) If $|A| \leq |B|$ and $|B| \leq |C|$, then $|A| \leq |C|$.

The second property (transitivity) comes from the fact that if $f: A \to B$ and $g: B \to C$ are both one-to-one, then the composition $g \circ f: A \to C$ is also one-to-one.

Since we have used the familiar $\leq$ notation, we would hope that the relation $\leq$ for cardinal numbers possesses other commonly used properties of the relation $\leq$ for the real numbers $\mathbf{R}$ and the integers $\mathbf{Z}$. One such property follows:

If $a$ and $b$ are real numbers such that $a \leq b$ and $b \leq a$, then $a = b$.

This property certainly holds for finite cardinal numbers. If $A$ is a proper subset of a finite set $B$, then $|A| < |B|$. Therefore, for finite sets $A$ and $B$, the only way that we can have $|A| \leq |B|$ and $|B \leq |A|$ is that $A$ and $B$ have the same number of elements, that is, that $|A| = |B|$.

On the other hand, it is possible for a proper subset of an infinite set to have as many elements as the entire set. For example, consider the infinite sets

$$E = \{2, 4, 6, \ldots\} \qquad \text{and} \qquad \mathbf{P} = \{1, 2, 3, \ldots\}$$

As illustrated in Example 6.1($d$), the subset $E$ does have the same cardinality as $\mathbf{P}$. Accordingly, the above property for infinite cardinal numbers is not obvious. But it is still indeed true in view of the celebrated Schroeder–Bernstein theorem which follows.

**Theorem 6.10 (Schroeder–Bernstein):** If $|A| \leq |B|$ and $|B| \leq |A|$, then $|A| = |B|$.

In other words, if $\alpha$ and $\beta$ are cardinal numbers such that $\alpha \leq \beta$ and $\beta \leq \alpha$, then $\alpha = \beta$. This important theorem, proved in Problem 6.19, can be stated in the following equivalent form.

**Theorem 6.11:** Let $X$, $Y$, $X_1$ be sets such that $X \supseteq Y \supseteq X_1$ and $X \approx X_1$. Then $X \approx Y$.

Another familiar property of the relation $\leq$ for the real numbers $\mathbf{R}$, called the law of trichotomy, is the following:

> If $a$ and $b$ are real numbers, then exactly one of the following is true:
> $$a < b, \qquad a = b, \qquad a > b$$

It is clear that the above property holds for finite cardinal numbers. Again, it is not obvious that it holds for infinite cardinal numbers. The fact that it does is the content of the next theorem.

**Theorem 6.12 (Law of Trichotomy):** For any two sets $A$ and $B$, exactly one of the following is true:

$$|A| < |B|, \qquad |A| = |B|, \qquad |A| > |B|$$

In other words, if $\alpha$ and $\beta$ are cardinal numbers, then either $\alpha < \beta$, $\alpha = \beta$, or $\alpha > \beta$. The proof of this theorem uses transfinite induction which is discussed in Chapter 9; hence the proof will be postponed until then.

### Continuum Hypothesis

By Cantor's theorem, $\aleph_0 < 2^{\aleph_0}$ and, as noted previously, $\aleph_0 < \mathbf{c}$. The next theorem (proved in Problem 6.20) tells us the relationship between $2^{\aleph_0}$ and $\mathbf{c}$.

**Theorem 6.13:** $2^{\aleph_0} = \mathbf{c}$.

It is natural to ask if there exists a cardinal number $\beta$ which lies "between" $\aleph_0$ and $\mathbf{c}$. Originally, Cantor supported the conjecture, which is known as the continuum hypothesis, that the answer to the above question is in the negative. Specifically:

**Continuum Hypothesis:** There exists no cardinal number $\beta$ such that

$$\aleph_0 < \beta < \mathbf{c}$$

In 1963 it was shown by Paul Cohen that the continuum hypothesis is independent of our axioms of set theory in somewhat the same sense that Euclid's fifth postulate on parallel lines is independent of the other axioms of geometry.

## 6.7 CARDINAL ARITHMETIC

The collection of all cardinal numbers can be considered to be a superset of the finite cardinal numbers (nonnegative integers)

$$0, 1, 2, 3, \ldots$$

This section shows how certain arithmetic operations on the finite cardinals can be extended to all the cardinal numbers.

### Cardinal Addition and Multiplication

Addition and multiplication of the counting numbers **N** are sometimes treated from the point of view of set theory. The interpretation of $2 + 3 = 5$, for example, is given by the picture in Fig. 6-3. Namely, the union of two disjoint sets, one having two elements and the other having three elements, is a set with five elements. This idea leads to a completely general definition of addition of cardinal numbers.

$$(x\,x) + (x\,x\,x) = (x\,x \quad x\,x\,x)$$

**Fig. 6-3**

**Definition 6.7:** Let $\alpha$ and $\beta$ be cardinal numbers and let $A$ and $B$ be disjoint sets with $\alpha = |A|$ and $\beta = |B|$. Then the *sum* of $\alpha$ and $\beta$ is denoted and defined by

$$\alpha + \beta = |(A \cup B)|$$

Two comments are appropriate with this definition. First of all, the addition of cardinal numbers is well-defined. That is, if $A'$ and $B'$ are also disjoint sets with cardinality $\alpha$ and $\beta$ respectively, then

$$|(A' \cup B')| = |(A \cup B)|$$

Second, if $A$ and $B$ are any two sets, then $A \times (1)$ and $B \times \{2\}$ are disjoint. Accordingly, there is no difficulty in finding disjoint sets with given cardinalities.

**EXAMPLE 6.5**

(*a*) Let $m$ and $n$ be finite cardinal numbers. Then $m + n$ corresponds to the usual addition in **N**.

(*b*) Let $n$ be a finite cardinal number. Then $n + \aleph_0 = \aleph_0$ since

$$n + \aleph_0 = |\{1, 2, \ldots, n\} \cup \{n+1, n+2, \ldots\}| = \aleph_0$$

(*c*) $\aleph_0 + \aleph_0 = \aleph_0$ since

$$\aleph_0 + \aleph_0 = |\{2, 4, 6, \ldots\} \cup \{1, 3, 5, \ldots\}| = \aleph_0$$

(*d*) $\mathbf{c} + \mathbf{c} = \mathbf{c}$ since

$$\mathbf{c} + \mathbf{c} = |[0, \tfrac{1}{2}] \cup (\tfrac{1}{2}, 1]| = \mathbf{c}$$

The definition of cardinal multiplication follows.

**Definition 6.8:** Let $\alpha$ and $\beta$ be cardinal numbers and let $A$ and $B$ be sets with $\alpha = |A|$ and $\beta = |B|$. Then the *product* of $\alpha$ and $\beta$ is denoted and defined by

$$\alpha\beta = |A \times B|$$

As with addition, multiplication of cardinal numbers is well-defined. (Observe that, in the definition of cardinal multiplication, $A$ and $B$ need not be disjoint.)

**EXAMPLE 6.6**

(*a*) Let $m$ and $n$ be finite cardinal numbers. Then $mn$ corresponds to the usual multiplication in **N**.

(*b*) Since $\mathbf{N} \times \mathbf{N}$ is countably infinite, $\aleph_0 \aleph_0 = \aleph_0$.

(*c*) Theorem 6.15 below tells us that the cartesian plane $\mathbf{R}^2$ has the same cardinality as **R**. That is, $\mathbf{cc} = \mathbf{c}$.

Table 6-1 lists properties of the addition and multiplication of cardinal numbers and gives the corresponding properties of sets under union and cartesian product. We state this result formally.

**Theorem 6.14**: The addition and multiplication of cardinal numbers satisfy the properties in Table 6-1.

**Table 6-1**

| Cardinal numbers | Sets |
|---|---|
| (1) $(\alpha+\beta)+\gamma = \alpha+(\beta+\gamma)$ | (1) $(A \cup B) \cup C = A \cup (B \cup C)$ |
| (2) $\alpha+\beta = \beta+\alpha$ | (2) $A \cup B = B \cup A$ |
| (3) $(\alpha\beta)\gamma = \alpha(\beta\gamma)$ | (3) $(A \times B) \times C \approx A \times (B \times C)$ |
| (4) $\alpha\beta = \beta\alpha$ | (4) $A \times B \approx B \times A$ |
| (5) $\alpha(\beta+\gamma) = \alpha\beta+\alpha\gamma$ | (5) $A \times (B \cup C) = (A \times B) \cup (A \times C)$ |
| (6) If $\alpha \leq \beta$, then $\alpha+\gamma \leq \beta+\gamma$ | (6) If $A \subseteq B$, then $(A \cup C) \subseteq (B \cup C)$ |
| (7) If $\alpha \leq \beta$, then $\alpha\gamma \leq \beta\gamma$ | (7) If $A \subseteq B$, then $(A \times C) \subseteq (B \times C)$ |

We emphasize that not every property of addition and multiplication of finite cardinals holds for cardinal numbers in general. For example, cancellation holds for finite cardinal numbers, that is,

(i) If $a+b = a+c$, then $b = c$.
(ii) If $ab = ac$ and $a \neq 0$, then $b = c$.

On the other hand, using Example 6.5 and Example 6.6, we have

(i) $\aleph_0 + \aleph_0 = \aleph_0 = \aleph_0 + 1$, but $\aleph_0 \neq 1$.
(ii) $\aleph_0 \aleph_0 = \aleph_0 = \aleph_0 1$, but $\aleph_0 \neq 1$.

Accordingly, the cancellation law is not true for the operations of addition and multiplication of infinite cardinal numbers.

On the other hand, the addition and multiplication of infinite cardinal numbers turn out to be very simple. We state the following theorem whose proof lies beyond the scope of this text.

**Theorem 6.15**: Let $\alpha$ and $\beta$ be nonzero cardinal numbers such that $\beta$ is infinite and $\alpha \leq \beta$. Then

$$\alpha+\beta = \alpha\beta = \beta$$

That is, given two nonzero cardinal numbers, at least one of which is infinite, their sum or product is simply the larger of the two. Examples 6.5 and 6.6 verify some instances of the theorem.

### Exponents and Cardinal Numbers

First we note that if $A$ and $B$ are sets, then

$$A^B$$

denotes the set of all functions from $B$ (the exponent) into $A$. This notation comes from the fact that if $A$ and $B$ are finite sets, say, $|A| = m$ and $|B| = n$, then there are $m^n$ functions from $B$ into $A$. This is illustrated in the next example, where $|A| = 2$ and $|B| = 3$.

**EXAMPLE 6.7** Let $A = \{1, 2\}$ and $B = \{x, y, z\}$. Then $A^B$ consists of exactly eight functions, which follow:

$$\{(x,1),(y,1),(z,1)\}, \quad \{(x,1),(y,1),(z,2)\}, \quad \{(x,1),(y,2),(z,1)\}, \quad \{(x,1),(y,2),(z,2)\},$$
$$\{(x,2),(y,1),(z,1)\}, \quad \{(x,2),(y,1),(z,2)\}, \quad \{(x,2),(y,2),(z,1)\}, \quad \{(x,2),(y,2),(z,2)\}$$

That is, there are 2 choices for $x$, 2 choices for $y$, and 2 choices for $z$, and hence there are $2^3 = 8$ functions altogether.

Exponents are introduced into the arithmetic of cardinal numbers in the next definition and, as illustrated above, this definition agrees with the case when $A$ and $B$ are finite sets.

**Definition 6.9**: Let $\alpha$ and $\beta$ be cardinal numbers and let $A$ and $B$ be sets with $\alpha = |A|$ and $\beta = |B|$. Then *$\alpha$ to the power $\beta$* is denoted and defined by

$$\alpha^\beta = |A^B|$$

**Remark**: Previously, if $\alpha = |A|$, then we used the exponent notation $2^\alpha = |\mathscr{P}(A)|$ where $\mathscr{P}(A)$ is the power set (collection of all subsets) of a set $A$. We note that there is a one-to-one correspondence between the subsets $X$ of $A$ and functions $f\colon A \to \{0, 1\}$ as follows:

$$f(a) = \begin{cases} 1 & \text{if } a \in X \\ 0 & \text{if } a \notin X \end{cases}$$

Thus there is no contradiction between the two notations.

The following familiar rules for working with exponents continue to hold.

**Theorem 6.16**: Let $\alpha, \beta, \gamma$ be cardinal numbers. Then:

(1) $(\alpha\beta)^\gamma = \alpha^\gamma\beta^\gamma$. (3) $(\alpha^\beta)^\gamma = \alpha^{\beta\gamma}$.
(2) $\alpha^\beta\alpha^\gamma = \alpha^{\beta+\gamma}$. (4) If $\alpha \le \beta$, then $\alpha^\gamma \le \beta^\gamma$.

**EXAMPLE 6.8** Using the rules for exponentiation we can make the following calculations:

(*a*) $\mathbf{c}^{\aleph_0} = (2^{\aleph_0})^{\aleph_0} = 2^{\aleph_0\aleph_0} = 2^{\aleph_0} = \mathbf{c}$.

(*b*) $\mathbf{c}^{\mathbf{c}} = (2^{\aleph_0})^{\mathbf{c}} = 2^{\aleph_0\mathbf{c}} = 2^{\mathbf{c}}$.

# Solved Problems

## EQUIPOTENT SETS, DENUMERABLE SETS, CONTINUUM

**6.1.** Consider the following concentric circles:

$$C_1 = \{(x, y) : x^2 + y^2 = a^2\}, \qquad C_2 = \{(x, y) : x^2 + y^2 = b^2\}$$

where, say, $0 < a < b$. Establish, geometrically, a one-to-one correspondence between $C_1$ and $C_2$.

Let $x \in C_2$. Consider the function $f\colon C_2 \to C_1$ where $f(x)$ is the point of intersection of the radius from the center of $C_2$, (and $C_1$) to $x$ and $C_1$, as shown in Fig. 6-4. Note that $f$ is both one-to-one and onto. Thus $f$ defines a one-to-one correspondence between $C_1$ and $C_2$.

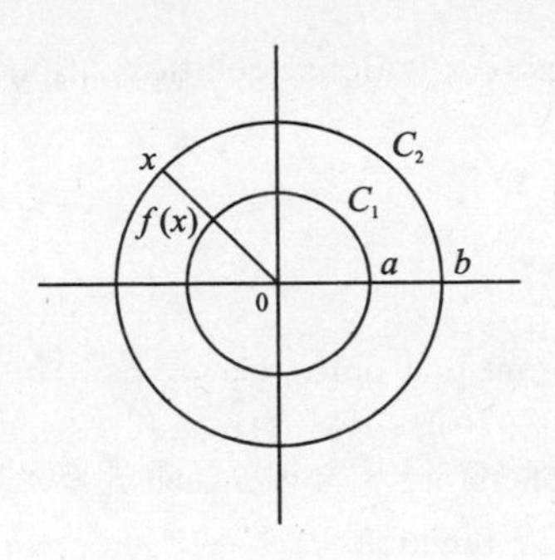


Fig. 6-4

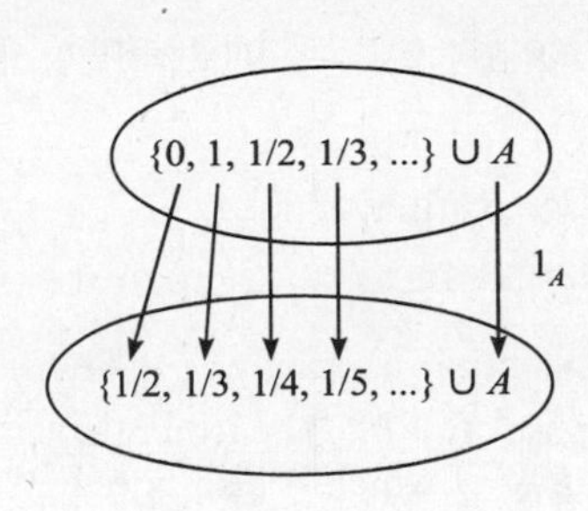


Fig. 6-5

**6.2.** Prove: (*a*) $[0,1] \approx (0,1)$; (*b*) $[0,1] \approx [0,1)$; (*c*) $[0,1] \approx (0,1]$.

(*a*) Note that

$$[0,1] = \{0, 1, 1/2, 1/3, \ldots\} \cup A$$
$$(0,1) = \{1/2, 1/3, 1/4, \ldots\} \cup A$$

where

$$A = [0,1] \setminus \{0, 1, 1/2, 1/3, \ldots\} = (0,1) \setminus \{1/2, 1/3, \ldots\}$$

Consider the function $f\colon [0,1] \to (0,1)$ defined by the diagram in Fig. 6-5. That is,

$$f(x) = \begin{cases} 1/2 & \text{if } x = 0 \\ 1/(n+2) & \text{if } x = 1/n, n \in \mathbf{P} \\ x & \text{if } x \neq 0, 1/n, n \in \mathbf{P} \end{cases}$$

The function $f$ is one-to-one and onto. Consequently, $[0,1] \approx (0,1)$.

(*b*) The function $f\colon [0,1] \to [0,1)$ defined by

$$f(x) = \begin{cases} 1/(n+1) & \text{if } x = 1/n, n \in \mathbf{P} \\ x & \text{if } x \neq 1/n, n \in \mathbf{P} \end{cases}$$

is one-to-one and onto. [It is similar to the function in part (*a*).] Hence $[0,1] \approx [0,1)$.

(*c*) Let $f\colon [0,1) \to (0,1]$ be the function defined by $f(x) = 1 - x$. Then $f$ is one-to-one and onto and, therefore, $[0,1) \approx (0,1]$. By part (*b*) and Theorem 6.1, we have $[0,1] \approx (0,1]$.

**6.3.** Prove that each of the following intervals (where $a < b$) has the power of the continuum, i.e., has cardinality $\mathbf{c}$:

$$(1)\ [a,b], \qquad (2)\ (a,b), \qquad (3)\ [a,b), \qquad (4)\ (a,b]$$

The formula $f(x) = a + (b-a)x$ defines a bijective mapping between each pair of sets:

(1) $[0,1)$ and $[a,b]$  (3) $[0,1)$ and $[a,b)$
(2) $(0,1)$ and $(a,b)$  (4) $(0,1]$ and $(a,b]$

Thus, by Theorem 6.1 and Problem 6.2, every interval has the same cardinality as the unit interval $\mathbf{I} = [0,1]$, that is, has the power of the continuum.

**6.4.** Prove Theorem 6.1: The relation $A \approx B$ in sets is an equivalence relation. Specifically:

(1) $A \approx A$ for any set $A$.

(2) If $A \approx B$, then $B \approx A$.

(3) If $A \approx B$ and $B \approx C$, then $A \approx C$.

(1) The identity function $1_A : A \to A$ is bijective (one-to-one and onto); hence $A \approx A$.

(2) Suppose $A \approx B$. Then there exists a bijective function $f: A \to B$. Hence $f$ has an inverse function $f^{-1}: B \to A$ which is also bijective. Hence $B \approx A$. Therefore, if $A \approx B$ then $B \approx A$.

(3) Suppose $A \approx B$ and $B \approx C$. Then there exist bijective functions $f: A \to B$ and $g: B \to C$. Then the composition function $g \circ f: A \to C$ is also bijective. Hence $A \approx B$. Therefore, if $A \approx B$ and $B \approx C$, then $A \approx C$.

**6.5.** Prove Theorem 6.2: Every infinite set $A$ contains a subset $D$ which is denumerable.

Let $f: \mathscr{P}(A) \to A$ be a choice function. Consider the following sequence:

$$\begin{aligned} a_1 &= f(A) \\ a_2 &= f(A \backslash \{a_1\}) \\ a_3 &= f(A \backslash \{a_1, a_2\}) \\ &\cdots\cdots\cdots \\ a_n &= f(A \backslash \{a_1, a_2, \ldots, a_{n-1}\}) \\ &\cdots\cdots\cdots \end{aligned}$$

Since $A$ is infinite, $A \backslash \{a_1, a_2, \ldots, a_{n-1}\}$ is not empty for every $n \in \mathbf{P}$. Furthermore, since $f$ is a choice function,

$$a_n \neq a_i \qquad \text{for} \qquad i < n$$

Thus the $a_n$ are distinct and, therefore, $D = \{a_1, a_2, \ldots\}$ is a denumerable subset of $A$.

Essentially, the choice function $f$ "chooses" an element $a_1 \in A$, then chooses an element $a_2$ from the elements which "remain" in $A$, and so on. Since $A$ is infinite, the set of elements which "remain" in $A$ is nonempty.

**6.6.** Prove: (*a*) For any sets $A$ and $B$, $A \times B \approx B \times A$.

(*b*) For any sets $A, B, C$,

$$(A \times B) \times C \approx A \times B \times C \approx A \times (B \times C)$$

(*c*) If $A \approx C$ and $B \approx D$, then $A \times B \approx C \times D$.

(*a*) Let $f: A \times B \to B \times A$ be defined by

$$f((a, b)) = (b, a)$$

Clearly $f$ is bijective. Hence $A \times B \approx B \times A$.

(*b*) Let $f: (A \times B) \times C \to A \times B \times C$ be defined by

$$f((a, b), c) = (a, b, c)$$

Then $f$ is bijective. Hence $(A \times B) \times C \approx A \times B \times C$. Similarly, $A \times (B \times C) \approx A \times B \times C$. Thus

$$(A \times B) \times C \approx A \times B \times C \approx A \times (B \times C)$$

(c) Let $f: A \to C$ and $g: B \to D$ be one-to-one correspondences. Define $h: A \times B \to C \times D$ by

$$h(a, b) = (f(a), g(b))$$

One can easily check that $h$ is one-to-one and onto. Hence $A \times B \approx C \times D$.

**6.7.** Prove: Let $X$ be any set and let $C(X)$ be the family of characteristic functions of $X$, that is, the family of functions $f: X \to \{0, 1\}$. Then $\mathcal{P}(X) \approx C(X)$ where $\mathcal{P}(X)$ is the power set of $X$, i.e., the collection of subsets of $X$.

Let $A$ be any subset of $X$, i.e., let $A \in \mathcal{P}(X)$. Let $f: \mathcal{P}(X) \to C(X)$ be defined by

$$f(A) = \chi_A$$

that is, $f$ maps each subset $A$ of $X$ into the characteristic function $\chi_A$ of $A$ (relative to $X$). [Recall $\chi_A : X \to \{0, 1\}$ is defined by $f(x) = 1$ if and only if $x \in A$.] Then $f$ is both one-to-one and onto. Hence $\mathcal{P}(X) \approx C(X)$.

**6.8.** Suppose $A$ is an infinite set and $F$ is a finite subset of $A$. Show that $A \backslash F \approx A$. In other words, removing a finite number of elements from an infinite set does not change its cardinality.

Suppose $F = \{a_1, a_2, \ldots, a_n\}$. Choose a denumerable subset $D = \{a_1, a_2, \ldots, a_n, a_{n+1}, \ldots\}$ of $A$ so that the first $n$ elements of $D$ are the elements of $F$. Let $g: A \to A \backslash F$ be defined by

$$g(a) = a \text{ if } a \notin D \quad \text{and} \quad g(a_k) = a_{k+n} \text{ if } a \in D$$

Then $g$ is one-to-one correspondence between $A$ and $A \backslash F$. Thus $A \approx A \backslash F$.

**6.9.** Prove Theorem 6.3: A subset of a denumerable set is either finite or denumerable.

Consider any denumerable set, say,

$$A = \{a_1, a_2, a_3, \ldots\} \tag{1}$$

Let $B$ be a subset of $A$. If $B = \varnothing$, then $B$ is finite. Suppose $B \neq 0$. Let $b_1$ be the first element in the sequence in (1) such that $b_1 \in B$; let $b_2$ be the first element which follows $b_1$ in the sequence in (1) such that $b_2 \in B$; and so on. Then $B = \{b_1, b_2, \ldots\}$. If the sequence $b_1, b_2, \ldots$ ends, then $B$ is finite. Otherwise $B$ is denumerable.

**6.10.** Prove: A countable union of finite sets is countable.

Let $\mathscr{C} = \{S_i : i \in \mathbf{P}\}$ be a countable collection of finite sets, and let $C = \cup_i S_i$. If $C$ is empty, then $C$ is countable. Suppose $C \neq \varnothing$. Define $A_1 = S_1$, $A_2 = S_2 \backslash S_1$, $A_3 = S_3 \backslash S_2$, and so on. Then the sets $A_i$ are finite and pairwise disjoint. Say,

$$A_1 = \{a_{11}, a_{12}, \ldots, a_{1n_1}\}, \qquad A_2 = \{a_{21}, a_{22}, \ldots a_{2n_2}\}, \ldots$$

Then the union $B = \cup_i A_i$ can be written as a sequence as follows:

$$B = \{a_{11}, a_{12}, \ldots, a_{1n_1}, a_{21}, a_{22}, \ldots, a_{2n_2}, \ldots\}$$

That is, first we write down the elements of $A_1$, then the elements of $A_2$, and so on. Formally, define $f: D \to \mathbf{P}$ as follows:

$$f(a_{ij}) = n_1 + n_2 + \cdots + n_{i-1+j}$$

Then $f$ is bijective. Hence $B$ is countable. However, $B$ is also the union of the sets in $\mathscr{C}$; that is, $B = C$. Therefore, $C$ is countable, as claimed.

**6.11.** Prove Theorem 6.5: Let $A_1, A_2, A_3, \ldots$ be a sequence of pairwise disjoint denumerable sets. Then the union $S = \cup_i A_i$ is denumerable.

Suppose

$$A_1 = \{a_{11}, a_{12}, a_{13}, \ldots\}, \qquad A_2 = \{a_{21}, a_{22}, a_{23}, \ldots\}, \ \ldots$$

Define $D_n = \{a_{ij} : i + j = n, n > 1\}$. For example,

$$D_2 = \{a_{11}\}, \qquad D_3 = \{a_{12,}, a_{21}\}, \qquad D_4 = \{a_{13}, a_{22}, a_{31}\}, \ \ldots$$

Note that each $D_n$ is finite. In fact, $D_n$ has $n - 1$ elements. By Problem 6.10, $T = \bigcup(D_j : j > 1)$ is countable. On the other hand, the union of the finite $D$'s is the same as the union of the $A$'s, that is, $T = S$. Thus $S$ is countable.

**6.12.** Show that $\mathbf{R} \approx \mathbf{R}^+$. (The sets of positive and negative real numbers are denoted, respectively, by $\mathbf{R}^+$ and $\mathbf{R}^-$.)

The function $f(x) = x/(1 + |x|)$ is a one-to-one correspondence between $\mathbf{R}^-$ and the open interval $(-1, 0)$. Hence the function $h$ defined by

$$h(x) = \begin{cases} \dfrac{x}{1 + |x|} + 1 & \text{if } x < 0 \\ x + 1 & \text{if } x \geq 0 \end{cases}$$

is a one-to-one correspondence between $\mathbf{R}$ and $\mathbf{R}^+$. Hence $\mathbf{R} \approx \mathbf{R}^+$.

**6.13.** Suppose $A$ is any uncountable set and $B$ is a denumerable subset of $A$. Show that $A \backslash B \approx A$. In other words, removing a denumerable set from an uncountable set does not change its cardinality.

Suppose $B = \{b_1, b_2, b_3, \ldots\}$. The set $A \backslash B$ is infinite (indeed uncountable) and contains a denumerable subset, say, $D = \{d_1, d_2, d_3, \ldots\}$. Let $A^* = A \backslash (B \cup D)$. Then $A$ and $A \backslash B$ are the following disjoint unions,

$$A = A^* \cup D \cup B = A^* \cup \{d_1, d_2, d_3, \ldots\} \cup \{b_1, b_2, b_3, \ldots\}$$
$$A \backslash B = A^* \cup D = A^* \cup \{d_1, d_2, d_3, \ldots\}$$

Define $f\colon A \to A \backslash B$ as in Fig. 6-6, that is,

$$f(a) = a \qquad \text{if } a \in A^*$$
$$f(d_n) = d_{2n} - 1 \qquad n \in \mathbf{P}$$
$$f(b_n) = d_{2n} \qquad n \in \mathbf{P}$$

Then $f$ is one-to-one and onto; hence $A \backslash B \approx A$.

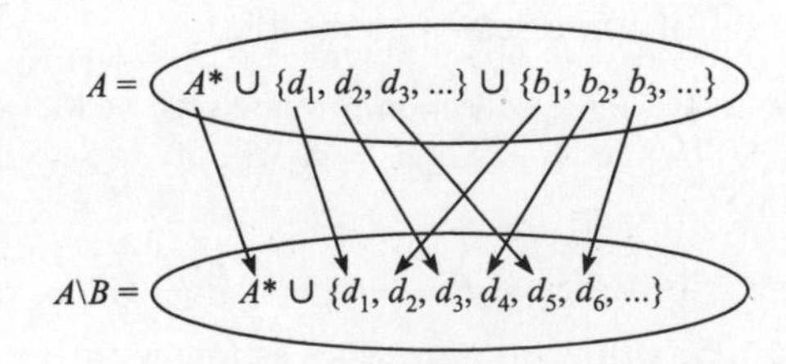


**Fig. 6-6**

**6.14.** Prove: The plane $\mathbf{R}^2$ is not the union of a countable number of lines.

Let $\mathscr{L}$ be any countable collection of lines. Since there are $\mathbf{c}$ vertical lines and $\mathscr{L}$ is countable, there is a vertical line $T$ such that $T \notin \mathscr{L}$. Now each line in $\mathscr{L}$ can intersect $T$ in at most one point. Thus there are only a countable number of points in $T$ which lie on lines in $\mathscr{L}$. Hence there is a point $p \in T \subseteq \mathbf{R}^2$ which does not line on any line in $\mathscr{L}$.

**6.15.** Prove Theorem 6.8: The unit interval $\mathbf{I} = [0, 1]$ is not denumerable.

**Method 1:** Assume $\mathbf{I}$ is denumerable. Then

$$\mathbf{I} = \{x_1, x_2, x_3, \ldots\}$$

that is, the elements of $\mathbf{I}$ can be written in a sequence.

Now each element in $\mathbf{I}$ can be written in the form of an infinite decimal as follows:

$$x_1 = 0.a_{11}a_{12}a_{13} \cdots a_{1n} \cdots$$
$$x_2 = 0.a_{21}a_{22}a_{23} \cdots a_{2n} \cdots$$
$$\cdots\cdots\cdots\cdots\cdots\cdots\cdots\cdots$$
$$x_n = 0.a_{n1}a_{n2}a_{n3} \cdots a_{nn} \cdots$$
$$\cdots\cdots\cdots\cdots\cdots\cdots\cdots\cdots$$

where $a_{ij} \in \{0, 1, \ldots, 9\}$ and where each decimal contains an infinite number of nonzero elements. Thus we write 1 as 0.999... and, for those numbers which can be written in the form of a decimal in two ways, for example,

$$1/2 = 0.5000\ldots = 0.4999\ldots$$

(in one of them there is an infinite number of nines and in the other all except a finite set of digits are zeros), we write the infinite decimal in which an infinite number of nines appear.

Now construct the real number

$$y = 0.b_1b_2b_3 \cdots b_n \cdots$$

which will belong to $\mathbf{I}$, in the following way:

Choose $b_1$ so $b_1 \neq a_{11}$ and $b_1 \neq 0$. Choose $b_2$ so $b_2 \neq a_{22}$ and $b_2 \neq 0$. And so on.

Note $y \neq x_1$ since $b_1 \neq a_{11}$ (and $b_1 \neq 0$); $y \neq x_2$ since $b_2 \neq a_{22}$ (and $b_2 \neq 0$), and so on. That is, $y \neq x_n$ for all $n \in \mathbf{P}$. Thus $y \notin \mathbf{I}$, which contradicts the fact that $y \in \mathbf{I}$. Thus the assumption that $\mathbf{I}$ is denumerable has led to a contradiction. Consequently, $\mathbf{I}$ is nondenumerable.

**Method 2:** [This second proof of Theorem 6.8 uses Problem 6.17(*b*).]

Assume $\mathbf{I}$ is denumerable. Then, as above,

$$\mathbf{I} = \{x_1, x_2, x_3, \ldots\}$$

that is, the elements of $\mathbf{I}$ can be written in a sequence.

Now construct a sequence of closed intervals $I_1, I_2, \ldots$ as follows. Consider the following three closed subintervals of $[0, 1]$:

$$[0, 1/3], \qquad [1/3, 2/3], \qquad [2/3, 1] \tag{1}$$

where each has length $1/3$. Now $x_1$ cannot belong to all three intervals. (If $x_1$ is one of the endpoints, then it could belong to two of the intervals, but not all three.) Let $I_1 = [a_1, b_1]$, be one of the intervals in (*1*) such that $x_1 \notin I_1$. Now consider the following three closed subintervals of $I_1 = [a_1, b_1]$:

$$[a_1, a_1 + 1/9], \qquad [a_1 + 1/9,\ a_1 + 2/9], \qquad [a_1 + 2/9,\ b_1] \tag{2}$$

where each has length $1/9$. Similarly, let $I_2$ be one of the intervals in (*2*) with the property that $x_2$ does not belong to $I_2$. Continue in this manner. Thus we obtain a sequence of closed intervals,

$$I_1 \supseteq I_2 \supseteq I_3 \supseteq \cdots \tag{3}$$

such that $x_n \notin I_n$ for all $n \in \mathbf{P}$.

By the above property of real numbers, there exists a real number $y \in \mathbf{I} = [0, 1]$ such that $y$ belongs to every interval in (*3*). But since

$$y \in \mathbf{I} = \{x_1, x_2, x_3, \ldots\}$$

we must have $y = x_m$ for some $m \in \mathbf{P}$. By our construction $y = x_m \notin I_m$, which contradicts the fact that $y$ belongs to every interval in (*3*). Thus our assumption that $\mathbf{I}$ is denumerable has led to a contradiction. Accordingly, $\mathbf{I}$ is nondenumerable.

**6.16.** Prove that $\mathbf{R}^2 \approx \mathbf{R}$ and, more generally, that $\mathbf{R}^n \approx \mathbf{R}$.

Since $\mathbf{R} \approx S = (0, 1)$, it suffices to show that the open unit square

$$S^2 = \{(x, y) : 0 < x < 1,\ 0 < y < 1) = (0, 1) \times (0, 1)$$

has the same cardinality as $S = (0, 1)$. Any point $(x, y) \in S$ can be written in the decimal form

$$(x, y) = (0.d_1 d_2 d_3, \cdots, 0.e_1 e_2 e_3 \cdots)$$

where each decimal expansion contains an infinite number of nonzero digits (e.g., for 1/2 write 0.4999... instead of 0.5000... 0. The function

$$f(x, y) = 0.d_1 e_1 d_2 e_2 d_3 e_3 \cdots$$

is one-to-one by the uniqueness of decimal expansions. Furthermore, the function $g: S \to S^2$ defined by $g(x) = (x, 1/2)$ is one-to-one. Accordingly, by the Schroeder–Bernstein Theorem 6.10, $S^2 \approx S$. Thus $\mathbf{R}^2 \approx \mathbf{R}$.

Therefore, $\mathbf{R}^3 \approx \mathbf{R}^2 \times \mathbf{R} \approx \mathbf{R} \times \mathbf{R} \approx \mathbf{R}$. Similarly, by induction, $\mathbf{R}^n \approx \mathbf{R}$.

**6.17.** A sequence $I_1, I_2, \ldots$ of intervals is said to be "nested" if $I_1 \supseteq I_2 \supseteq \ldots$.

(*a*) Give an example of a nested sequence of open intervals $I_k$ whose intersection is empty.

(*b*) Prove the Nested Interval Property of the real numbers $\mathbf{R}$: A nested sequence $I_1 = [a_1, b_1]$, $I_2 = [a_2, b_2], \ldots$ of closed intervals is not empty.

(*a*) Let $I_k = (0, 1/k)$. Then $\bigcap(I_k : k \in P) = \varnothing$. [This follows from the fact that, for any $\mathbf{c} > 0$ there exists a $k$ such that $1/k < \mathbf{c}$.]

(*b*) Let $A = \{a_1, a_2, \ldots\}$. Since the intervals are nested, $A$ is bounded and every $b_k$ is an upper bound of $A$. By the completion property of $\mathbf{R}$, $y = \sup(A)$ exists. Thus, for every $k$, $a_k \le y \le b_k$. Thus $y$ belongs to every interval, and hence $\bigcap_k I_k \neq \varnothing$.

## CARDINAL NUMBERS AND THE INEQUALITY OF CARDINAL NUMBERS

**6.18.** Prove Cantor's Theorem 6.9: For any set $A$, we have $|A| < |\mathscr{P}(A)|$.

The function $g: A \to \mathscr{P}(A)$ which sends each element $a \in A$ into the set consisting of $a$ alone, i.e., which is defined by $g(a) = \{a\}$, is one-to-one. Thus $|A| \le |\mathscr{P}(A)|$.

If we now show that $|A| \neq |\mathscr{P}(A)|$, then the theorem will follow. Suppose the contrary, that is, suppose $|A| = |\mathscr{P}(A)|$ and that $f: A \to \mathscr{P}(A)$ is one-to-one and onto. Let $a \in A$ be called a "bad" element if $a$ is not a member of the set which is its image, i.e., if $a \notin f(a)$. Now let $B$ be the set of "bad" elements. That is,

$$B = \{x : x \in A, x \notin f(x)\}$$

Now $B$ is a subset of $A$, that is, $B \in \mathscr{P}(A)$. Since $f: A \to \mathscr{P}(A)$ is onto, there exists an element $b \in A$ such that $f(b) = B$. Is $b$ a "bad" element or a "good" element? If $b \in B$ then, by definition of $B$, $b \notin f(b) = B$, which is impossible. Likewise, if $b \notin B$, then $b \in f(b) = B$, which is also impossible. Thus the original assumption, that $|A| = |\mathscr{P}(A)|$, has led to a contradiction. Hence the assumption is false, and so the theorem is true.

**6.19.** Prove Theorem 6.11 (which is an equivalent formulation of the Schroeder–Bernstein theorem 6.10): Let $X, Y, X_1$ be sets such that $X \supseteq Y \supseteq X_1$ and $X \approx X_1$. Then $X \approx Y$.

Since $X \approx X_1$, there exists a one-to-one correspondence (bijection) $f: X \to X_1$. Since $X \supseteq Y$, the restriction of $f$ to $Y$, which we also denote by $f$, is also one-to-one. Let $f(Y) = Y_1$. Then $Y$ and $Y_1$ are equipotent,

$$X \supseteq Y \supseteq X_1 \supseteq Y_1$$

and $f: Y \to Y_1$ is bijective. But now $Y \supseteq X_1 \supseteq Y_1$ and $Y \approx Y_1$. For similar reasons, $X_1$ and $f(X_1) = X_2$ are equipotent,

$$X \supseteq Y \supseteq X_1 \supseteq Y_1 \supseteq X_2$$

and $f: X_1 \to X_2$ is bijective. Accordingly, there exist equipotent sets $X, X_1, X_2, \ldots$ and equipotent sets $Y, Y_1, Y_2, \ldots$ such that

$$X \supseteq Y \supseteq X_1 \supseteq Y_1 \supseteq X_2 \supseteq Y_2 \supseteq X_3 \supseteq Y_3 \supseteq \ldots$$

and $f: X_k \to X_{k+1}$ and $f: Y_k \to Y_{k+1}$ are bijective.
Let

$$B = X \cap Y \cap X_1 \cap Y_1 \cap X_2 \cap Y_2 \cap \ldots$$

Then

$$X = (X \backslash Y) \cup (Y \backslash X_1) \cup (X_1 \backslash Y_1) \cup \cdots \cup B$$
$$Y = (Y \backslash X_1) \cup (X_1 \backslash Y_1) \cup (Y_1 \backslash X_2) \cup \cdots \cup B$$

Furthermore, $X \backslash Y, X_1 \backslash Y_1, X_2 \backslash Y_2, \ldots$ are equipotent. In fact, the function

$$f: (X_k \backslash Y_k) \to (X_{k+1} \backslash Y_{k+1})$$

is one-to-one and onto.

Consider the function $g: X \to Y$ defined by the diagram in Fig. 6-7. That is,

$$g(x) = \begin{cases} f(x) & \text{if } x \in X_k \backslash Y_k \text{ or } x \in X \backslash Y \\ x & \text{if } x \in Y_k \backslash X_k \text{ or } x \in B \end{cases}$$

Then $g$ is one-to-one and onto. Therefore $X \approx Y$.

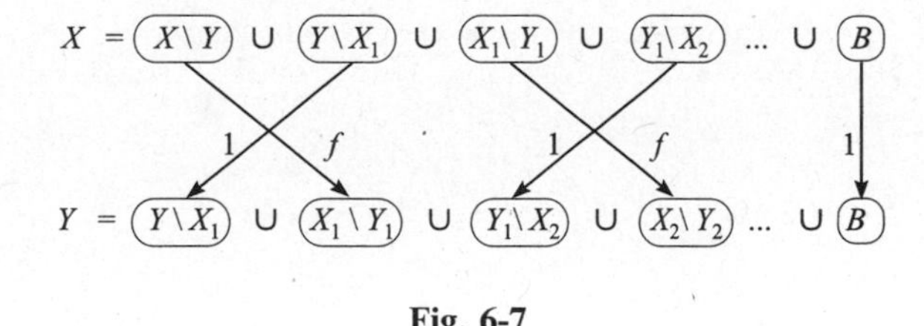


**Fig. 6-7**

**6.20.** Prove Theorem 6.13: $\mathbf{c} = 2^{\aleph_0}$.

Let $\mathbf{R}$ be the set of real numbers and let $\mathscr{P}(\mathbf{Q})$ be the power set of the set $\mathbf{Q}$ of rational numbers, i.e., the family of subsets of $\mathbf{Q}$. Furthermore, let the function $f: \mathbf{R} \to \mathscr{P}(\mathbf{Q})$ be defined by

$$f(a) = \{x : x \in \mathbf{Q}, x < a\}$$

That is, $f$ maps each real number $a$ into the set of rational numbers less than $a$. We shall show that $f$ is one-to-one. Let $a, b \in \mathbf{R}$, $a \neq b$ and, say, $a < b$. By a property of the real numbers, there exists a rational number $r$ such that

$$a < r < b$$

Then $r \in f(b)$ and $r \notin f(a)$; hence $f(b) \neq f(a)$. Therefore, $f$ is one-to-one. Thus $|\mathbf{R}| \leq |\mathscr{P}(\mathbf{Q})|$. Since $|\mathbf{R}| = \mathbf{c}$ and $|\mathbf{Q}| = \aleph_0$, we have

$$\mathbf{c} \leq 2^{\aleph_0}$$

Now let $C(\mathbf{P})$ be the family of characteristic functions $f: \mathbf{P} \to \{0, 1\}$ which, as proven in Problem 6.8, is equivalent to $\mathscr{P}(\mathbf{P})$. Here $\mathbf{P} = \{1, 2, \ldots\}$. Let $\mathbf{I} = [0, 1]$, the closed unit interval, and let the function $F : C(\mathbf{P}) \to \mathbf{I}$ be defined by

$$F(f) = 0.f(1)f(2)f(3)\cdots$$

an infinite decimal consisting of zeros or ones. Suppose $f, g \in C(\mathbf{P})$ and $f \neq g$. Then the decimals would be different, and so $F(f) \neq F(g)$. Accordingly, $F$ is one-to-one. Therefore,

$$|\mathscr{P}(\mathbf{Q})| = |C(\mathbf{P})| \leq |\mathbf{I}|$$

Since $|\mathbf{Q}| = \aleph_0$ and $|\mathbf{I}| = \mathbf{c}$, we have

$$2^{\aleph_0} \leq \mathbf{c}$$

Both inequalities give us

$$\mathbf{c} = 2^{\aleph_0}$$

**6.21.** Let $S = (0, 1)$, the open unit interval, and let $T$ be the set of real numbers in $S$ which have an infinite number of threes in their decimal expansion. Show that $|T| = |S|$.

Let $x \in S$ and suppose $x = 0.d_1d_2d_3 \cdots d_n \cdots$. Let the function $f$: $S \to T$ be defined by

$$f(x) = 0.d_1 3 d_2 3 d_3 3 \cdots 3 d_n 3 \cdots$$

Then $f$ is one-to-one and hence $|S| \leq |T|$. Since $T$ is a subset of $S$, we have $|T| \leq |S|$. By the Schroeder–Bernstein theorem, $|T| = |S|$.

**6.22.** Let $S$ denote the open unit interval $(0, 1)$, and let $S^\omega$ denote the set of all denumerable sequences $(x_1, x_2, x_3, \ldots)$ where $x_i \in S$. (*a*) Prove $|S^\omega| \approx |S|$. (*b*) Prove the set $\mathbf{R}^\omega$ of all denumerable sequences of real numbers has cardinality $\mathbf{c}$.

(*a*) Let $(x_1, x_2, x_3, \ldots) \in S^\omega$. Consider the decimal expansions:

$$x_1 = 0.d_{11}d_{12}d_{13}d_{14}\cdots$$
$$x_2 = 0.d_{21}d_{22}d_{23}d_{24}\cdots$$
$$x_3 = 0.d_{31}d_{32}d_{33}d_{34}\cdots$$

And so on

Associate the sequence $(x_1, x_2, x_3, \ldots)$ with the decimal number

$$0.d_{11} : d_{21}d_{12} : d_{13}d_{22}d_{31} : \cdots$$

where the subscripts in the successive blocks of digits $d_{11}, d_{22}d_{12}, d_{13}d_{22}d_{31}, \ldots$ are obtained by "following the arrows" in Fig. 6-1. (This procedure was used to show that $\mathbf{P} \times \mathbf{P}$ is countable.) This association defines a one-to-one function from $S^\omega$ into $S$. The function $g : S \to S^\omega$ defined by $f(x) = (x, x, x, \ldots)$ is also one-to-one. By the Schroeder–Bernstein theorem $|S^\omega| \approx |S|$.

(*b*) Since $\mathbf{R} \approx S$, it follows that $|\mathbf{R}^\omega| = |S^\omega| = |S| = \mathbf{c}$.

## CARDINAL ARITHMETIC

**6.23.** Let $A_1, A_2, A_3, A_4$ be any sets. Define sets $B_1, B_2, B_3, B_4$ such that

$$|A_1| + |A_2| + |A_3| + |A_4| = |B_1 \cup B_2 \cup B_3 \cup B_4|$$

Let $B_1 = A_1 \times \{1\}$, $B_2 = A_2 \times \{2\}$, $B_3 = A_3 \times \{3\}$, $B_4 = A_4 \times \{4\}$. Then $B_k \approx A_k$ for $k = 1, 2, 3, 4$. Also, the $B_k$ are disjoint, that is, $B_i \cap B_j = \varnothing$ if $i \neq j$. Consequently, the above will be true.

**6.24.** Let $\{A_i : i \in I\}$ be any family of sets. Define a family of sets $\{B_i : i \in I\}$ such that $B_i \approx A_i$, for $i \in I$, and $B_i \cap B_j = \varnothing$ for $i \neq j$.

Let $B_i = A_i \times \{i\}$. Then the family $\{B_i : i \in I\}$ has the required properties.

**6.25.** Prove Theorem 6.14: The addition and multiplication of cardinal numbers satisfy the properties in Table 6-1. That is, for cardinal numbers $\alpha, \beta, \gamma$:

(1) $(\alpha + \beta) + \gamma = \alpha + (\beta + \gamma)$

(2) $\alpha + \beta = \beta + \alpha$

(3) $(\alpha\beta)\gamma = \alpha(\beta\gamma)$

(4) $\alpha\beta = \beta\alpha$

(5) $\alpha(\beta + \gamma) = \alpha\beta + \alpha\gamma$

(6) If $\alpha \leq \beta$, then $\alpha + \gamma \leq \beta + \gamma$

(7) If $\alpha \leq \beta$, then $\alpha\gamma \leq \beta\gamma$

Let $A, B, C$ be pairwise disjoint sets such that $\alpha = |A|$, $\beta = |B|$, $\gamma = |C|$.

(1) We have:

$$(\alpha + \beta) + \gamma = |A \cup B| + |C| = |(A \cup B) \cup C|$$
$$\alpha + (\beta + \gamma) = |A| + |B \cup C| = |A \cup (B \cup C)|$$

However, the union of sets is associative, i.e., $(A \cup B) \cup C = A \cup (B \cup C)$. Hence

$$(\alpha + \beta) + \gamma = \alpha + (\beta + \gamma)$$

(2) Since $A \cup B = B \cup A$, we have

$$\alpha + \beta = |A \cup B| = |B \cup A| = \beta + \alpha$$

(3) We have:

$$(\alpha\beta)\gamma = |A \times B||C| = |(A \times B) \times C|$$
$$\alpha(\beta\gamma) = |A||B \times C| = |A \times (B \times C)|$$

However, by Problem 6.6(*b*), $(A \times B) \times C \approx A \times (B \times C)$. Hence

$$(\alpha\beta)\gamma = \alpha(\beta\gamma)$$

(4) By Problem 6.6(*a*), $A \times B \approx B \times A$; hence

$$\alpha\beta = |A \times B| = |B \times A| = \beta\alpha$$

(5) Note first that $B \cap C = \varnothing$ implies $(A \times B) \cap (A \times C) = \varnothing$. Then:

$$\alpha(\beta + \gamma) = |A||B \cup C| = |A \times (B \cup C)|$$
$$\alpha\beta + \alpha\gamma = |A \times B| + |A \times C| = |(A \times B) \cup (A \times C)|$$

However, $A \times (B \cup C) = (A \times B) \cup (A \times C)$. Therefore,

$$\alpha(\beta + \gamma) = \alpha\beta + \alpha\gamma$$

(6) Suppose $\alpha \leq \beta$. Then there exists a one-to-one mapping $f: A \to B$. Let $g : A \cup C \to B \cup C$ be defined by

$$g(x) = \begin{cases} f(x) & \text{if } x \in A \\ x & \text{if } x \in C \end{cases}$$

Then $g$ is one-to-one. Accordingly, $|A \cup C| \leq |B \cup C|$ and so

$$\alpha + \gamma \leq \beta + \gamma$$

(7) Suppose $\alpha \leq \beta$. Then there exists a one-to-one mapping $f: A \to B$. Let $g : A \times C \to B \times C$ be defined by

$$g(a, c) = (f(a), c)$$

Then $g$ is one-to-one. Accordingly, $|A \times C| \leq |B \times C|$ and so

$$\alpha\gamma \leq \beta\gamma$$

**6.26.** Prove: $\aleph_0 \mathbf{c} = \mathbf{c}$.

Consider the integers $\mathbf{Z} = \{\ldots, -2, -1, 0, 1, 2, \ldots\}$ and the half-open interval $A = [0, 1)$. Furthermore, let $f\colon \mathbf{Z} \times A \to \mathbf{R}$ be defined by

$$f(n, a) = n + a$$

In other words, $f(\{n\} \times [0, 1))$ is mapped onto $[n, n+1)$. Then $f$ is a one-to-one correspondence between $\mathbf{Z} \times A$ and $\mathbf{R}$. Since $|\mathbf{Z}| = \aleph_0$ and $|A| = |\mathbf{R}| = \mathbf{c}$, we have

$$\aleph_0 \mathbf{c} = |\mathbf{Z} \times A| = |\mathbf{R}| = \mathbf{c}$$

**6.27.** Prove: Let $\alpha$ be any infinite cardinal number. Then $\aleph_0 + \alpha = \alpha$.

We have shown that $\aleph_0 + \aleph_0 = \aleph_0$. Suppose $\alpha$ is uncountable, and $\alpha = |A|$. By Problem 6.13, $A \backslash B \approx A$ where $B$ is a denumerable subset of $A$. Recall $A = (A \backslash B) \cup B$ and the union is disjoint. Hence

$$\alpha = |A| = |(A \backslash B) \cup B| = |A \backslash B| + |B| = \alpha + \aleph_0 = \aleph_0 + \alpha$$

## MISCELLANEOUS PROBLEMS

**6.28.** Prove: The set $\mathscr{P}$ of all polynomials

$$p(x) = a_0 + a_1 x + a_2 x^2 + a_m x^m \tag{1}$$

with integral coefficients, that is, where $a_0, a_1, \ldots, a_m$ are integers, is denumerable.

For each pair of nonnegative integers $(n, m)$, let $P(n, m)$ be the set of polynomials in (*1*) of degree $m$ in which

$$|a_0| + |a_1| + \cdots + |a_m| = n$$

Note that $P(n, m)$ is finite. Therefore

$$\mathscr{P} = \bigcup(P(n, m) : (n, m) \in \mathbf{N} \times \mathbf{N})$$

is countable since it is a countable family of countable sets. But $\mathscr{P}$ is not finite; hence $\mathscr{P}$ is denumerable.

**6.29.** A real number $r$ is called an *algebraic* number if $r$ is a solution to a polynomial equation

$$p(x) = a_0 + a_1 x + a_2 x^2 + a_m x^m = 0$$

with integral coefficients. Prove the set $A$ of algebraic numbers is denumerable.

By the preceding Problem 6.28, that the set $E$ of polynomial equations is denumerable:

$$E = \{p_1(x) = 0,\ p_2(x) = 0,\ p_3(x) = 0, \ldots\}$$

Define

$$A_k = \{x : x \text{ is a solution of } p_k(x) = 0\}$$

Since a polynomial of degree $n$ can have at most $n$ roots, each $A_k$ is finite. Therefore

$$A = \bigcup\{A_k : k \in \mathbf{P}\}$$

is a countable family of countable sets. Accordingly, $A$ is countable and, since $A$ is not finite, $A$ is denumerable.

**6.30.** Explicitly exhibit $\aleph_0$ pairwise-disjoint denumerable subsets of $\mathbf{P} = \{1, 2, 3, \ldots\}$.

Let $p$ and $q$ be distinct prime numbers. The sets

$$S_p = \{p, p^2, p^3, \ldots\} \qquad \text{and} \qquad S_q = \{q, q^2, q^3, \ldots\}$$

are pairwise disjoint. One can show that the set $\{p_1, p_2, p_3, \ldots\}$ of prime numbers is an infinite subset of $\mathbf{P}$ and hence has cardinality $\aleph_0$. Thus the family $\{S_{p_1}, S_{p_2}, S_{p_3}, \ldots\}$ has the desired properties.

## Supplementary Problems

### EQUIPOTENT SETS, COUNTABLE SETS, CONTINUUM

**6.31.** The set $\mathbf{Z}$ of integers can be put into a one-to-one correspondence with $\mathbf{P} = \{1, 2, 3, \ldots\}$ as follows:

$$\begin{array}{cccccccc} 1 & 2 & 3 & 4 & 5 & 6 & 7 & \ldots \\ \downarrow & \downarrow & \downarrow & \downarrow & \downarrow & \downarrow & \downarrow & \\ 0 & 1 & -1 & 2 & -2 & 3 & -3 & \ldots \end{array}$$

Find a formula for the function $f\colon \mathbf{P} \to \mathbf{Z}$ which gives the above correspondence between $\mathbf{P}$ and $\mathbf{Z}$.

**6.32.** $\mathbf{P} \times \mathbf{P}$ was written as a sequence by considering the diagram in Fig. 6-1. This is not the only way to write $\mathbf{P} \times \mathbf{P}$ as a sequence. Write $\mathbf{P} \times \mathbf{P}$ as a sequence in two other ways by drawing appropriate diagrams.

**6.33.** Prove that the set $S$ of rational points in the plane $\mathbf{R}^2$ is denumerable. [A point $p = (x, y)$ in $\mathbf{R}^2$ is rational if $x$ and $y$ are rational.]

**6.34.** Let $S$ be the set of rational points in the plane $\mathbf{R}^2$. Show that $S$ can be partitioned into two sets $V$ and $H$ such that the intersection of $V$ with any vertical line is finite and the intersection of $H$ with any horizontal line is finite.

**6.35.** Let $\mathscr{A} = \{A_i : i \in I\}$ be a set of pairwise disjoint intervals in the line $\mathbf{R}$. Show that $\mathscr{A}$ is countable.

**6.36.** Let $\mathscr{B} = \{B_i : i \in I\}$ be a set of pairwise disjoint circles in the plane $\mathbf{R}^2$. Show that $\mathscr{B}$ is countable.

**6.37.** A function $f\colon \mathbf{P} \to \mathbf{P}$ is said to have finite support if $f(n) = 0$ for all but a finite number of $n$. Show that the set of all such functions is denumerable.

**6.38.** A real number $x$ is called *transcendental* if $x$ is not algebraic, i.e., if $x$ is not a solution to a polynomial equation

$$p(x) = a_0 + a_1 x + a_2 x^2 + \cdots + a_n x^n = 0$$

with integral coefficients. (See Problem 6.29.) For example, $\pi$ and $e$ are transcendental numbers. Prove that the set $T$ of transcendental numbers has the power of the continuum.

**6.39.** Recall that a *permutation* of $\mathbf{P} = \{1, 2, 3, \ldots\}$ is a bijective function $\sigma : \mathbf{P} \to \mathbf{P}$. Show that the set PERM($\mathbf{P}$) of all permutations of $\mathbf{P}$ has the power of the continuum.

### CARDINAL NUMBERS, CARDINAL ARITHMETIC

**6.40.** Suppose $\alpha$ and $\beta$ are cardinal numbers such that $\alpha \leq \beta$. Show that there exists a set $S$ with a subset $A$ such that $\alpha = |A|$ and $\beta = |S|$.

**6.41.** Show that Theorems 6.10 and 6.11 are equivalent. (Hence each proves the Schroeder–Bernstein theorem.)

**6.42.** Prove $\mathbf{c}^{\aleph_0} = \mathbf{c}$.

**6.43.** Show that there are only $\mathbf{c}$ continuous functions from $\mathbf{R}$ into $\mathbf{R}$. (Assume that if $f$ and $g$ are such continuous functions and $f(q) = g(q)$ for all rational numbers $q$ in $\mathbf{R}$, then $f = g$, that is, $f(x) = g(x)$ for all $x$ in $\mathbf{R}$.)

**6.44.** Prove Theorem 6.16(2): Let $\alpha, \beta, \gamma$ be cardinal numbers. Then $\alpha^\beta \alpha^\gamma = \alpha^{\beta+\gamma}$.

**6.45.** Let $\alpha, \beta, \gamma$ be cardinal numbers such that $\alpha \leq \beta$. Prove: (*a*) $\alpha^\gamma \leq \beta^\gamma$, (*b*) $\gamma^\alpha \leq \gamma^\beta$.

**6.46.** Show that the cardinal inequality relations are well defined; that is, if $A \approx A'$ and $B \approx B'$, show that:

(*a*) $|A| \leq |B|$ if and only if $|A'| \leq |B'|$. (*b*) $|A| < |B|$ if and only if $|A'| < |B'|$.

**6.47.** Show that cardinal addition and multiplication are well defined, that is:

(*a*) *Cardinal Addition*: If $A \approx A'$ and $B \approx B'$, where $A$ and $B$ are disjoint and $A'$ and $B'$ are disjoint, show that $|A \cup B| = |A' \cup B'|$.

(*b*) *Cardinal Multiplication*: If $A \approx A'$ and $B \approx B'$, show that $|A \times B| = |A' \times B'|$.

**6.48.** Let $\mathscr{C}$ be the collection of all circles in the plane $\mathbf{R}^2$. Show that $\mathscr{C}$ has cardinality **c**.

**MISCELLANEOUS PROBLEMS**

**6.49.** (Heine–Borel Property of the real numbers **R**.) Let $\mathscr{C} = \{I_k : k \in K\}$ be a collection of open intervals which covers a closed interval $A = [a, b]$. Show that $\mathscr{C}$ contains a finite subcover of $A$, that is, a finite subcollection of $\mathscr{C}$ is a cover of $A$. [A collection $\{I_k : k \in K\}$ of intervals is called a "cover" of a set $A$ if $A \subseteq \bigcup_k I_k$.]

# Answers to Supplementary Problems

**6.31.** The following function $f$: **P** → **P** has the required property:

$$f(n) = \begin{cases} -n/2 + 1/2 & \text{if } n \text{ is odd} \\ n/2 & \text{if } n \text{ is even} \end{cases}$$

**6.32.** Each diagram in Fig. 6-8 shows that $\mathbf{P} \times \mathbf{P}$ can be written as an infinite sequence of distinct elements as follows:

(*a*) $\mathbf{P} \times \mathbf{P} = \{(1,1), (2,1), (2,2), (1,2), (1,3), (2,3), \ldots\}$
(*b*) $\mathbf{P} \times \mathbf{P} = \{(1,1), (1,2), (2,1), (1,3), (2,2), (3,1), (1,4), \ldots\}$

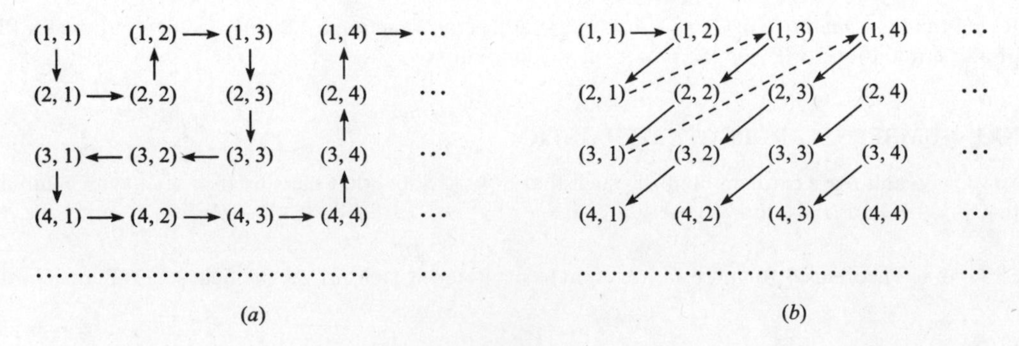


**Fig. 6-8**

**6.33.** $|\mathbf{Q} \times \mathbf{Q}| = |\mathbf{P} \times \mathbf{P}| = |\mathbf{P}| = \aleph_0$

**6.35.** *Hint*: Each interval contains a distinct rational number.

**6.36.** *Hint*: Each circle contains a distinct rational point in $\mathbf{R}^2$.

**6.38.** *Hint*: **R** is the union of the algebraic and transcendental numbers.

**6.42.** *Hint*: Use Problem 6.22

**6.43.** *Hint*: Use Problem 6.22 or 6.42.

**6.44.** *Hint*: Let $\alpha = |A|, \beta = |B|, \gamma = |C|$ where $B$ and $C$ are disjoint. Let $D = B \cup C$. Then $\beta + \gamma = |B \cup C| = |D|$. Associate with each function $f$: $D \to A$ the pair $f_1$: $B \to A$ and $f_2$: $C \to A$ where $f_1 = f|_B$ and $f_2 = f|_C$. Show that the map $F(f) = (f_1, f_2)$ is bijective.

**6.45.** *Hint*: Let $\alpha = |A|, \beta = |B|, \gamma = |C|$ where we can assume $A \subseteq B$ since $\alpha \leq \beta$.

(*a*) For each function $f$: $C \to A$ associate the function $f'$: $C \to B$ defined by $f'(x) = f(x)$. Show that the map $F(f) = g$ is one-to-one.

(*b*) For each function $f$: $A \to C$ associate a function $f'$: $B \to C$ which extends $f$, i.e., for each $a \in A$, $f'(a) = f(a)$. Show that the map $F(f) = f'$ is one-to-one.

**6.48.** Since each circle in $\mathscr{C}$ is determined by its center $(x, y)$ and radius $r$, $\mathscr{C} \approx \mathbf{R} \times \mathbf{R} \times \mathbf{R}^+ \approx \mathbf{R}$.

**6.49.** Suppose no finite subcollection of $\mathscr{C}$ is a cover of $A$. Let $p_1$ be the midpoint of the interval $A = A_1 = [a_1, b_1]$. At least one of $[a_1, p_1]$ and $[p_1, b_1]$ cannot be covered by a finite subcollection of $\mathscr{C}$ or else the whole interval $A_1$ will be, and let $A_2 = [a_2, b_2]$ be that subinterval. Similarly, let $p_2$ be the midpoint of the interval $A_2 = [a_2, b_2]$, and let $A_3 = [a_3, b_3]$ be one of the two intervals $[a_2, p_2]$ and $[p_2, b_2]$ which cannot be covered by a finite subcollection of $\mathscr{C}$, and so on. Thus we have a sequence $A_1, A_2, \ldots$ of nested closed intervals, and each cannot be covered by a finite subcollection of $\mathscr{C}$. Furthermore, $\lim_{n\to\infty} d_n = 0$ where $d_n = b_n - a_n$ is the length of $A_n$. By Problem 6.17(*b*), there exists a real number $y$ in every $A_k$. Since $\mathscr{C}$ is a cover of $A$, $y$ belongs to some element of $\mathscr{C}$, say $y \in I_j$ where $I_j = (c, d)$. Let $e$ be the distance from $y$ to the closest endpoint of $I_j$. Then there exists $d_j$ such that $d_j < e$. This means $A_j \subseteq I_j$. This contradicts the fact that $A_j$ cannot be covered by a finite subcollection of $\mathscr{C}$. Thus the original assumption that no finite subcollection of $\mathscr{C}$ covers $A$ leads to a contradiction, and so a finite subcollection of $\mathscr{C}$ covers $A$.

# Chapter 7

# Ordered Sets and Lattices

## 7.1 INTRODUCTION

Order and precedence relationships appear in many different places in mathematics and computer science. This chapter makes these notions precise. We also define a lattice, which is a special kind of an ordered set.

## 7.2 ORDERED SETS

Suppose $R$ is a relation on a set $S$ satisfying the following three properties:

[$\mathbf{O}_1$] (***Reflexive***): For any $a \in S$, we have $a\,R\,a$.

[$\mathbf{O}_2$] (***Antisymmetric***): If $a\,R\,b$ and $b\,R\,a$, then $a = b$.

[$\mathbf{O}_3$] (***Transitive***): If $a\,R\,b$ and $b\,R\,c$, then $a\,R\,c$.

Then $R$ is called a *partial order* or, simply an *order* relation, and $R$ is said to define a *partial ordering* of $S$. The set $S$ with the partial ordering $R$ is called a *partially ordered set* or, simply, an *ordered set*. (Sometimes the term *poset* is used for partially ordered set.)

The most familiar order relation, called the *usual order*, is the relation $\leq$ (read "less than or equal") on the positive integers **P** or, more generally, on any subset of the real numbers **R**. For this reason, a partial ordering relation is frequently denoted by

$$\precsim$$

With this notation, the above three properties of a partial order appear in the following usual form:

[$\mathbf{O}_1$] (***Reflexive***): For any $a \in S$, we have $a \precsim a$.

[$\mathbf{O}_2$] (***Antisymmetric***): If $a \precsim b$ and $b \precsim a$, then $a = b$.

[$\mathbf{O}_3$] (***Transitive***): If $a \precsim b$ and $b \precsim c$, then $a \precsim c$.

Although an ordered set consists of two things, a set $S$ and the partial ordering $\precsim$, one usually simply writes $S$ to denote the ordered sets as long as the partial ordering is fixed in the context of the discussion; otherwise the ordered set is denoted by the pair $(S, \precsim)$.

Suppose $S$ is an ordered set. Then the statement

$a \precsim b$ is read "*a precedes b*"

In this context we also write:

$a \prec b$ means $a \precsim b$ and $a \neq b$; read "$a$ strictly precedes $b$".

$b \succsim a$ means $a \precsim b$; read "$b$ succeeds $a$".

$b \succ a$ means $a \prec b$; read "$b$ strictly succeeds $a$".

$\not\precsim$, $\not\prec$, $\not\succsim$ and $\not\succ$ are self-explanatory.

When there is no ambiguity, the symbols $\leq$, $<$, $>$, $\geq$ are frequently used instead of $\precsim$, $\prec$, $\succ$, and $\succsim$, respectively.

**EXAMPLE 7.1**

(*a*) Let $\mathscr{S}$ be any collection of sets. The relation $\subseteq$ of set inclusion is a partial ordering of $\mathscr{S}$. Specifically, $A \subseteq A$ for any set $A$; if $A \subseteq B$ and $B \subseteq A$ then $A = B$; and if $A \subseteq B$ and $B \subseteq C$ then $A \subseteq C$.

(*b*) Consider the set **P** of positive integers. We say "$a$ divides $b$", written $a|b$, if there exists an integer $c$ such that $ac = b$. For example, $2|4$, $3|12$, $7|21$, and so on. This relation of divisibility is a partial ordering of **P**.

(*c*) The relation "|" of divisibility is not an ordering of the set **Z** of integers. Specifically, the relation is not antisymmetric. For instance, $2|-2$ and $-2|2$, but $2 \neq -2$.

(*d*) Consider the set **Z** of integers. Define $a\,R\,b$ if there is a positive integer $r$ such that $b = a^r$. For instance, $2\,R\,8$ since $8 = 2^3$. One can show (Problem 7.8) that $R$ is a partial ordering of **Z**.

### Dual Order

Let $\precsim$ be any partial ordering of a set $S$. The relation $\succsim$, that is, $a$ succeeds $b$, is also a partial ordering of $S$; it is called the *dual order*. Observe that $a \precsim b$ if and only if $b \succsim a$; hence the dual order $\succsim$ is the inverse of the relation $\precsim$, that is $\succsim\; = \;\precsim^{-1}$.

### Ordered Subsets

Let $A$ be a subset of an ordered set $S$, and suppose $a, b \in A$. Then the order in $S$ induces an order in $A$ in the following natural way:

$$\boxed{a \precsim b \text{ as elements of } A \text{ whenever } a \precsim b \text{ as elements of } S}$$

More precisely, if $R$ is a partial ordering of $S$, then the relation

$$R_A = R \cap (A \times A)$$

is a partial ordering of $A$ called the *induced order* on $A$ or the *restriction* of $R$ to $A$. The subset $A$ with the induced order is called an *ordered subset* of $S$. Unless otherwise stated or implied, any subset of an ordered set $S$ will be treated as an ordered subset of $S$.

### Quasi-order

Suppose $\prec$ is a relation on a set $S$ satisfying the following two properties:

[$Q_1$] (***Irreflexive***): For any $a \in A$, we have $a \nprec a$.
[$Q_2$] (***Transitive***): If $a \prec b$, and $b \prec c$, then $a \prec c$.

Then $\prec$ is called a *quasi-order* on $S$.

There is a close relationship between partial orders and quasi-orders. Specifically, if $\precsim$ is a partial order on a set $S$ and we define $a \prec b$ to mean $a \precsim b$ but $a \neq b$, then $\prec$ is a quasi-order on $S$. Conversely, if $\prec$ is a quasi-order on a set $S$ and we define $a \precsim b$ to mean $a \prec b$ or $a = b$, then $\precsim$ is a partial order on $S$. This allows us to switch back and forth between a partial order and its corresponding quasi-order using whichever is more convenient.

### Comparability

Suppose $a$ and $b$ are distinct elements in a partially ordered set $S$. We say $a$ and $b$ are *comparable* if

$$a \prec b \qquad \text{or} \qquad b \prec a$$

that is, if one of them precedes the other. Thus $a$ and $b$ are *noncomparable*, written

$$a \| b$$

if $a \nprec b$ and $b \nprec a$.

### Linearly Ordered Sets

The word "partial" is used in defining a partially ordered set $S$ since some of the elements of $S$ need not be comparable. Suppose, on the other hand, every pair of elements of $S$ are comparable. Then $S$ is said to be *linearly* or *totally ordered.* Although an ordered set $S$ may not be linearly ordered, it is still possible for a subset $A$ of $S$ to be linearly ordered. Such a linearly ordered subset $A$ of an ordered set $S$ is called a *chain* in $S$. Clearly, every subset of a linearly ordered set $S$ must also be linearly ordered.

**EXAMPLE 7.2**

(*a*) Consider the set **P** of positive integers ordered by divisibility. Then 21 and 7 are comparable since $7|21$. On the other hand, 3 and 5 are noncomparable since neither $3|5$ nor $5|3$. Thus **P** is not linearly ordered by divisibility. Observe that $A = \{2, 6, 12, 36\}$ is a chain (linearly ordered subset) in **P** since $2|6$, $6|12$, and $12|36$.

(*b*) The set **P** of positive integers with the usual order $\leq$ (less than or equal) is linearly ordered and hence every ordered subset of **P** is also linearly ordered.

(*c*) The power set $\mathscr{P}(A)$ of a set $A$ with 2 or more elements is not linearly ordered by set inclusion. For instance, suppose $a$ and $b$ belong to $A$. Then $\{a\}$ and $\{b\}$ are noncomparable. Observe that the empty set $\varnothing$, $\{a\}$, and $A$ do form a chain in $\mathscr{P}(A)$ since $\varnothing \subseteq \{a\} \subseteq A$. Similarly, $\varnothing$, $\{b\}$, and $A$ form a chain in $\mathscr{P}(A)$.

## 7.3 SET CONSTRUCTIONS AND ORDER

This section discusses different ways of defining an order on a set which is constructed from ordered sets.

### Product Sets and Order

There are a number of ways to define an order relation on the cartesian product of given ordered sets. Two of these ways follow:

(*a*) ***Product Order***: Suppose $S$ and $T$ are ordered scts. Then the following, is an order relation on the product set $S \times T$, called the *product order*:

$$(a, b) \precsim (a', b') \qquad \text{if } a \leq a' \text{ and } b \leq b'$$

Problem 7.15 shows that this relationship does satisfy the necessary axioms of an order.

(*b*) ***Lexicographical Order***: Suppose $S$ and $T$ are linearly ordered sets. Then the following is an order relation on the product set $S \times T$, called the *lexicographical* or *dictionary order*:

$$(a, b) \prec (a', b') \begin{cases} \text{if } a < a', \\ \text{or if } a = a' \text{ and } b < b' \end{cases}$$

This order can be extended to $S_1 \times S_2 \times \cdots \times S_n$ as follows:

$$(a_1, a_2, \ldots, a_n) \prec (a_1', a_2', \ldots, a_n')$$
$$\text{if } a_1 = a_1', a_2 = a_2', \ldots, a_{k-1} = a_{k-1}', \text{ but } a_k < a_k'$$

Note that the lexicographical order is also linear.

### Concatenation or Sum Order

Suppose $\{A_i : i \in I\}$ is a linearly ordered collection of disjoint linearly ordered sets; that is, the index set $I$ is linearly ordered, each set $A_i$ is linearly ordered, and $A_i \cap A_j = \varnothing$ when $i \neq j$. Then we assume, unless otherwise specified, the following linear order on the union $S = \bigcup_i A_i$, which we call the *concatenation order* or *usual order* or *sum order*:

$$x \prec y \begin{cases} \text{if } x \in A_i,\ y \in A_j, \text{ and } i < j \\ \text{or if } x, y \in A_i \text{ and } x < y \text{ as elements of } A_i \end{cases}$$

This order can sometimes be pictured by listing the elements of $A_i$ before the elements $A_j$ when $i < j$ and separating the sets by semicolons. For example, consider the sets

$$A = \{1, 3, 5, 7, \ldots\}, \qquad B = \{a, b, c\}, \qquad C = \{2, 4, 6, \ldots\}$$

where position in each set determines the linear order. Then the concatenation order on $S = A \cup B \cup C$ (where we assume the sets are ordered by the position in the union, i.e., $A < B < C$) may be pictured by writing

$$S = \{1, 3, 5, \ldots;\ a, b, c;\ 2, 4, 6, \ldots\}$$

Note that the order on $S' = B \cup A \cup C$ may be pictured by

$$S' = \{a, b, c;\ 1, 3, 5, \ldots;\ 2, 4, 6, \ldots\}$$

and this is not the same as the order on $S$.

### Kleene Closure and Order

Let $A$ be a nonempty linearly ordered set (sometimes called an *alphabet*). A *word* $w$ over $A$ is a finite sequence

$$w = a_1 a_2, \ldots, a_n$$

of elements of $A$. We will let $|w|$ denote the *length* $n$ of $w$. (The empty sequence, denoted by $\lambda$, is also a word and $|\lambda| = 0$.) The Kleene closure of $A$, denoted by $A^*$, is defined to be the collection of all such words over $A^*$. The following are two order relations on $A^*$.

(*a*) ***Alphabetical (Lexicographical) Order***: The reader is no doubt familiar with the usual alphabetical ordering of $A'$. That is:

(i) $\lambda < w$, where $\lambda$ is the empty word and $w$ is any nonempty word.

(ii) Suppose $u = au'$ and $v = bv'$ are distinct nonempty words where $a, b \in A$ and $u', v' \in A^*$. Then:

$$u \prec v \begin{cases} \text{if } a < b \\ \text{or if } a = b \text{ but } u' \prec v' \end{cases}$$

(*b*) ***Short-lex Order***: Here $A^*$ is ordered first by length, and then alphabetically. That is, for any distinct words $u, v$ in $A^*$:

$$u \prec v \begin{cases} \text{if } |u| < |v| \\ \text{or if } |u| = |v| \text{ but } u \text{ precedes } v \text{ alphabetically} \end{cases}$$

For example, "to" precedes "and" since |"to"| = 2 but |"and"| = 3. However, "an" precedes "to" since they have the same length, but "an" precedes "to" alphabetically. This order is also called the *free semigroup order*.

## 7.4 PARTIALLY ORDERED SETS AND HASSE DIAGRAMS

Let $S$ be a partially ordered set, and suppose $a, b \in S$. We say that $a$ is an *immediate predecessor* of $b$, or that $b$ is an *immediate successor* of $a$, or that $b$ is a *cover* of $a$, written

$$a << b$$

if $a < b$ but no element in $S$ lies between $a$ and $b$, that is, there exists no element $c$ in $S$ such that $a < c < b$.

Suppose $S$ is a finite partially ordered set. Then the order on $S$ is completely known once we know all pairs $a, b$ in $S$ such that $a << b$, that is, once we know the relation $<<$ on $S$. This follows from the fact that $x < y$ if and only if $x << y$ or there exist elements $a_1, a_2, \ldots, a_m$ in $S$ such that

$$x << a_1 << a_2 << \cdots << a_m << y$$

### Hasse Diagrams

The *Hasse diagram* of a finite partially ordered set $S$ is a graphical representation of $S$ as follows. The elements of $S$ are represented by points in the plane (called *vertices*), and there is a directed line segment (*arrow*) drawn from $a$ to $b$ (called an *edge*) whenever $a << b$ in $S$. Instead of drawing an arrow from $a$ to $b$, we sometimes place $b$ higher than $a$ and draw a line between them. It is then understood that movement upwards indicates succession. In the diagram thus created, $x < y$ if and only if there is a directed path (sequence of edges) from vertex $x$ to vertex $y$. Also, there can be no (directed) cycles in the diagram of $S$ since the order relation is antisymmetric.

The Hasse diagram of an ordered set $S$ is a picture of $S$; hence it is very useful in describing types of elements in $S$. Sometimes we define a partially ordered set by simply presenting its Hasse diagram.

**EXAMPLE 7.3**

(*a*) Let $A = \{1, 2, 3, 4, 6, 8, 9, 12, 18, 24\}$ be ordered by the relation "$x$ divides $y$". The Hasse diagram of $A$ appears in Fig. 7-1(*a*).

(*b*) Let $B = \{a, b, c, d, e\}$. The diagram in Fig. 7-1(*b*) defines a partial ordering on $B$ in a natural way. That is, $d \leq b$, $d \leq a$, $e \leq c$, and so on. Note that $b$ and $c$ are noncomparable.

(*c*) The diagram of a finite linearly ordered set consists of simply one path. For example, Fig. 7-1(*c*) is the diagram of such a set with five elements.

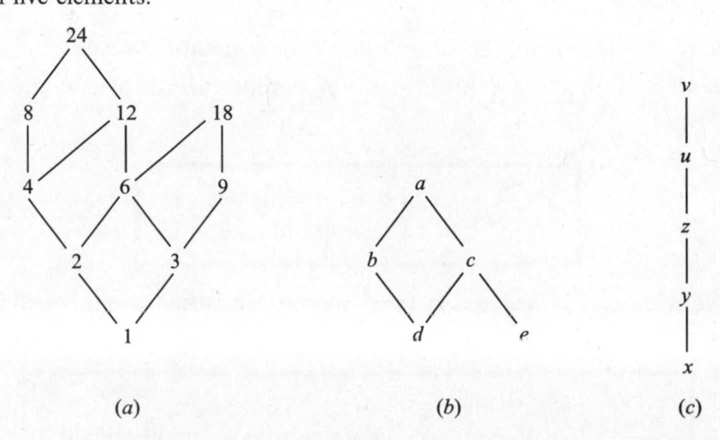


Fig. 7-1

**EXAMPLE 7.4** A *partition* of a positive integer $m$ is a set of positive integers whose sum is $m$. For instance, there are 7 partitions of $m = 5$ as follows:

5, 3-2, 2-2-1, 1-1-1-1-1, 4-1, 3-1-1, 2-1-1-1

We order the partitions of an integer $m$ as follows. A partition $P_1$ precedes a partition $P_2$ if the integers in $P_1$ can be added to obtain the integers in $P_2$ or, equivalently, if the integers in $P_2$ can be further subdivided to obtain the integers in $P_1$. For example,

2-2-1 precedes 3-2 and 4-1

since $2 + 1 = 3$ and $2 + 2 = 4$. On the other hand, 3-1-1 and 2-2-1 are noncomparable.

Figure 7-2 gives the Hasse diagram of the partitions of $m = 5$.

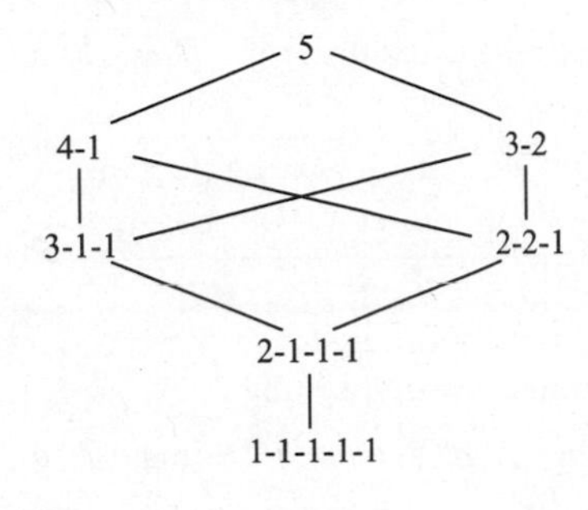


**Fig. 7-2**

## 7.5 MINIMAL AND MAXIMAL ELEMENTS, FIRST AND LAST ELEMENTS

Let $S$ be a partially ordered set. An element $a \in S$ is called a *minimal* element of $S$ if no element of $S$ strictly precedes (is less than) $a$; that is, if

$$x \leq a \text{ implies } x = a$$

Similarly, an element $b \in S$ is called a *maximal* element of $S$ if no element of $S$ strictly succeeds (is greater than) $b$; that is, if

$$x \geq b \text{ implies } x = b$$

Geometrically speaking, $a$ is a minimal element of $S$ if no edge enters $a$ (from below), and $b$ is a maximal element of $S$ if no edge leaves $b$ (in an upward direction). We note that $S$ can have more than one minimal and more than one maximal element.

If $S$ is infinite, then $S$ may have no minimal and no maximal element. For instance, the set **Z** of integers with the usual order $\leq$ has no minimal and no maximal element. On the other hand, if $S$ is finite, then $S$ has at least one minimal element and one maximal element.

An element $a \in S$ is called a *first* element of $S$ if

$$a \leq x$$

for every $x \in S$, that is, if $a$ precedes every other element in $S$. Similarly, an element $b \in S$ is called a *last* element of $S$ if

$$y \leq b$$

for every $y \in S$, that is, if $b$ succeeds every other element in $S$. We note that $S$ can have at most one first element which must be a minimal element of $S$, and $S$ can have at most one last element which must be a maximal element of $S$. Generally speaking, $S$ may have neither a first nor a last element, even when $S$ is finite.

Now suppose that $S$ is a linearly ordered set. If $S$ has a minimal element, then it must also be a first element; and if $S$ has a maximal element, then it must also be a last element. In particular, if $S$ is a finite linearly ordered set, then $S$ has both a first element and a last element.

**EXAMPLE 7.5** Consider the three partially ordered sets in Example 7.3 whose Hasse diagrams appear in Fig. 7-1.

(*a*) $A$ has two maximal elements, 18 and 24, and neither is a last element. $A$ has only one minimal element, 1, which is also a first element.

(*b*) $B$ has two minimal elements, $d$ and $e$, and neither is a first element. $B$ has only one maximal element $a$, which is also a last element.

(*c*) The linearly ordered set $\{x, y, z, u, v\}$ has one minimal element, $x$, which is a first element, and one maximal element, $v$, which is a last element.

**EXAMPLE 7.6**

(*a*) Consider the set $\mathbf{P} = \{1, 2, 3, \ldots\}$ with the usual order $\leq$. Then 1 is a first and only minimal element. $\mathbf{P}$ has no last and no maximal element.

(*b*) Let $A$ be any nonempty set and let $\mathscr{P}(A)$ be the power set of $A$ ordered by set inclusion. Then the empty set $\varnothing$ is a first element of $\mathscr{P}(A)$ since $\varnothing \subseteq X$ for any set $X$. Moreover, $A$ is a last element of $\mathscr{P}(A)$ since every set $Y$ in $\mathscr{P}(A)$ is a subset of $A$, that is, $Y \subseteq A$.

(*c*) Let $S = \{a_1, a_2, \ldots, a_m\}$ be a finite linearly ordered set. Then $S$ contains precisely one minimal element and precisely one maximal element, denoted respectively by

$$\min(a_1, a_2, \ldots, a_m) \qquad \text{and} \qquad \max(a_1, a_2, \ldots, a_m)$$

## 7.6 CONSISTENT ENUMERATION

Suppose $S$ is a finite partially ordered set. Frequently we want to assign a positive integer to each element of $S$ in such a way that the order is preserved. That is, we seek a function $f: S \to \mathbf{P}$ so that if $a < b$ then $f(a) < f(b)$. Such a function $f$ is called a *consistent enumeration* of $S$. The fact that this can be done is the content of the following theorem.

**Theorem 7.1**: There exists a consistent enumeration for any finite partially ordered set $S$.

We prove this theorem in Problem 7.17. In fact, we prove that if $S$ has $n$ elements then there exists a consistent enumeration $f: S \to \{1, 2, \ldots, n\}$.

We emphasize that such an enumeration need not be unique. For example, the following are two such enumerations for the ordered set in Fig. 7-1(*b*):

(i) $f(d) = 1, \quad f(e) = 2, \quad f(b) = 3, \quad f(c) = 4, \quad f(a) = 5$

(ii) $g(e) = 1, \quad g(d) = 2, \quad g(c) = 3, \quad g(b) = 4, \quad g(a) = 5$

On the other hand, the linearly ordered set in Fig. 7-1(*c*) admits only one consistent enumeration if we map the set into $\{1, 2, 3, 4, 5\}$. Specifically, we must assign:

$$h(x) = 1, \qquad h(y) = 2, \qquad h(z) = 3, \qquad h(u) = 4, \qquad h(v) = 5$$

## 7.7 SUPREMUM AND INFIMUM

Let $S$ be a partially ordered set, and let $A$ be a subset of $S$. An element $M$ in $S$ is called an *upper bound* of $A$ if $M$ succeeds every element of $A$, that is, for every $x \in A$, we have

$$x \leq M$$

If an upper bound of $A$ precedes every other upper bound of $A$, then it is called the *supremum* of $A$ and it is denoted by

$$\sup(A)$$

We also write $\sup(a_1, \ldots, a_n)$ instead of $\sup(A)$ when $A$ consists of the elements $a_1, \ldots, a_n$. We emphasize that there can be at most one $\sup(A)$; however, $\sup(A)$ may not exist.

Analogously, an element $m$ in $S$ is called a *lower bound* of a subset $A$ if $m$ precedes every element of $A$, that is, for every $y \in A$, we have

$$m \preceq y$$

If a lower bound of $A$ succeeds every other lower bound of $A$, then it is called the *infimum* of $A$ and it is denoted by

$$\inf(A)$$

We also write $\inf(a_1, \ldots, a_n)$ instead of $\inf(A)$ when $A$ consists of the elements $a_1, \ldots, a_n$. Similarly, there can be at most one $\inf(A)$ although $\inf(A)$ may not exist.

Some texts use the term *least upper bound* instead of supremum and then write $\text{lub}(A)$ instead of $\sup(A)$, and use the term *greatest lower bound* instead of infimum and then write $\text{glb}(A)$ instead of $\inf(A)$.

If $A$ has an upper bound we say $A$ is *bounded above*, and if $A$ has a lower bound we say $A$ is *bounded below*. In particular, $A$ is *bounded* if $A$ has an upper and lower bound.

**EXAMPLE 7.7**

(*a*) Let $S = \{a, b, c, d, e, f\}$ be ordered as pictured in Fig. 7-3(*a*), and let $A = \{b, c, d\}$. The upper bounds of $A$ are $e$ and $f$ since only $e$ and $f$ succeed every element in $A$. The lower bounds of $A$ are $a$ and $b$ since only $a$ and $b$ precede every element of $A$. Note $e$ and $f$ are noncomparable; hence $\sup(A)$ does not exist. However, $b$ also succeeds $a$, hence $\inf(A) = b$. Observe that $\inf(A) = b$ does belong to $A$.

(*b*) Let $S = \{1, 2, 3, \ldots, 8\}$ be ordered as pictured in Fig. 7-3(*b*), and let $A = \{4, 5, 7\}$. The upper bounds of $A$ are 1, 2, and 3, and the only lower bound is 8. Note that 7 is not a lower bound since 7 does not precede 4. Here $\sup(A) = 3$ since 3 precedes the other upper bounds 1 and 2, and $\inf(A) = 8$ since 8 is the only lower bound. Observe that neither $\inf(A) = 8$ nor $\sup(A) = 3$ belongs to $A$.

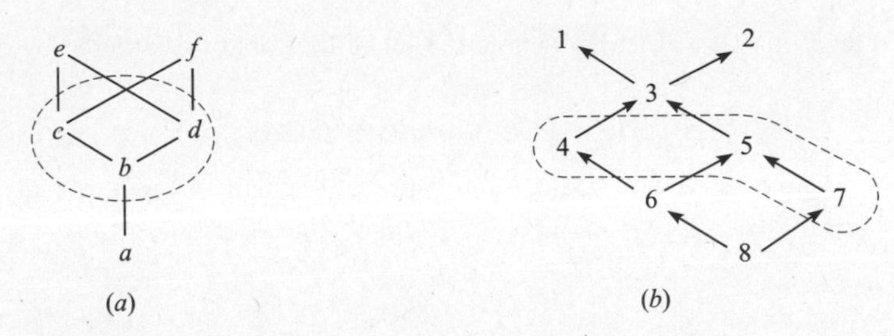


**Fig. 7-3**

(*c*) Consider the set $\mathbf{Q}$ of rational numbers, and its subset

$$B = \{x \in \mathbf{Q} : x > 0 \text{ and } 2 < x^2 < 3\}$$

that is, $B$ consists of those rational numbers which lie between $\sqrt{2}$ and $\sqrt{3}$ on the real line $\mathbf{R}$. Then $B$ has an infinite number of upper and lower bounds, but $\inf(B)$ and $\sup(B)$ do not exist. In other words, $B$ has no least upper bound and no greatest lower bound. Note that $\sqrt{2}$ and $\sqrt{3}$ do not belong to $\mathbf{Q}$ and cannot be considered as upper or lower bounds of $B$.

The above Example 7.7(*c*) points out one of the main differences between the real numbers **R** and the rational numbers **Q**. That is:

> **Completeness Axiom of the Real Numbers R:**
> Let $A$ be a nonempty subset of **R** and suppose $A$ has an upper bound. Then $A$ has a least upper bound, that is, $\sup(A)$ exists.

### Existence of $\sup(a, b)$ and $\inf(a, b)$

Let $S$ be an ordered set and let $a, b \in S$. If $S$ is linearly ordered, then $\sup(a, b)$ and $\inf(a, b)$ clearly exist. Specifically, if $a \leq b$, then $\sup(a, b) = b$ and $\inf(a, b) = a$. On the other hand, if $S$ is an arbitrary ordered set, then $\sup(a, b)$ and $\inf(a, b)$ need not exist. However, there are important examples of nonlinearly ordered sets where $\sup(a, b)$ and $\inf(a, b)$ do exist for every $a, b$ in the set.

**EXAMPLE 7.8**

(*a*) Consider the set $\mathbf{P} = \{1, 2, 3, \ldots\}$. The *greatest common divisor* of $a$ and $b$ in **P**, denoted by

$$\gcd(a, b)$$

is the largest integer which divides $a$ and $b$. The *least common multiple* of $a$ and $b$, denoted by

$$\operatorname{lcm}(a, b)$$

is the smallest integer divisible by both $a$ and $b$.

An important theorem in number theory says that every common divisor of $a$ and $b$ divides $\gcd(a, b)$. Also, one can prove that $\operatorname{lcm}(a, b)$ divides every multiple of $a$ and $b$.

Suppose **P** is ordered by divisibility. Then

$$\gcd(a, b) = \inf(a, b) \quad \text{and} \quad \operatorname{lcm}(a, b) = \sup(a, b)$$

In other words, $\inf(a, b)$ and $\sup(a, b)$ do exist for any pair $a, b$ of elements of **P** ordered by divisibility.

(*b*) For any positive integer $m$, we will let $\mathbf{D}_m$ denote the set of divisors of $m$ ordered by divisibility. The Hasse diagram of

$$\mathbf{D}_{36} = \{1, 2, 3, 4, 6, 9, 12, 18, 36\}$$

appears in Fig. 7-4. Again, $\inf(a, b) = \gcd(a, b)$ and $\sup(a, b) = \operatorname{lcm}(a, b)$ exist for any pair $a, b \in \mathbf{D}_m$.

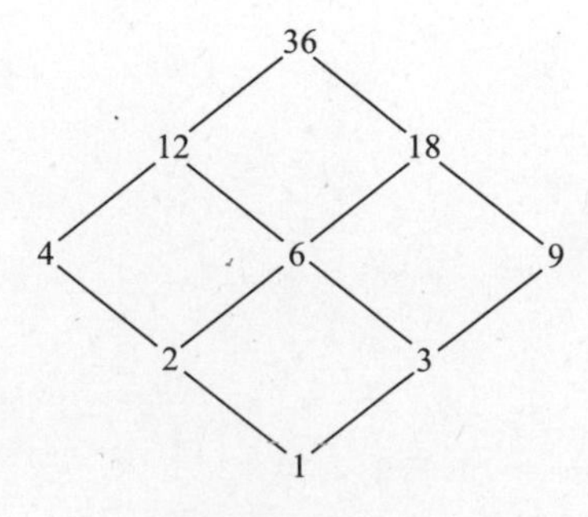


**Fig. 7-4**

(*c*) Let $S$ be a nonempty set with at least two elements, and let $\mathscr{P}(S)$ be the power set of $S$ ordered by set inclusion. Let $A$ and $B$ be any two elements of $\mathscr{P}(A)$, that is, let $A$ and $B$ be subsets of $S$. Then $\sup(A, B)$ and $\inf(A, B)$ do exist. Specifically, $\sup(A, B) = A \cup B$ and $\inf(A, B) = A \cap B$.

## 7.8 ISOMORPHIC (SIMILAR) ORDERED SETS

Suppose $X$ and $Y$ are partially ordered sets. A one-to-one (injective) function $f: X \to Y$ is called a *similarity mapping* from $X$ into $Y$ if $f$ preserves the order relation, that is, if the following condition holds for any pair $a, b \in X$:

$$a \leq b \text{ in } X \text{ if and only if } f(a) \leq f(b) \text{ in } Y$$

The above condition is equivalent to the following two conditions:

(1) If $a \leq b$ then $f(a) \leq f(b)$.
(2) If $a \| b$ (noncomparable), then $f(a) \| f(b)$.

Accordingly, if the underlying sets $X$ and $Y$ are both linearly ordered, then only (1) is needed for $f$ to be a similarity mapping.

Two ordered sets $X$ and $Y$ are said to be *order-isomorphic* or *isomorphic* or *similar*, written

$$X \simeq Y$$

if there exists a one-to-one correspondence (bijective mapping) $f: X \to Y$ which preserves the order relations, i.e., which is a similarity mapping. Such a function $f$ is then called an *order-isomorphism* or *isomorphism* from $X$ onto $Y$ or an *order-isomorphism* between $X$ and $Y$.

**EXAMPLE 7.9**

(*a*) Suppose $S = \{a, b, c, d\}$ is ordered by the diagram in Fig. 7-5(*a*) and suppose $T = \{1, 2, 6, 8\}$ is ordered by divisibility. Figure 7-5(*b*) is the Hasse diagram of the ordered set $T$. Then $S \simeq T$. In particular, the following function $f: S \to T$ is an isomorphism between $S$ and $T$:

$$f(a) = 6, \quad f(b) = 8, \quad f(c) = 2, \quad f(d) = 1$$

We note that the following function $g: S \to T$ is another isomorphism between $S$ and $T$:

$$g(a) = 8, \quad g(b) = 6, \quad g(c) = 2, \quad g(d) = 1$$

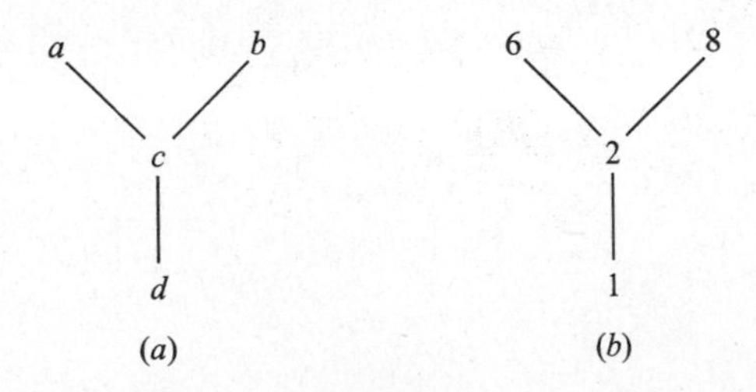


Fig. 7-5

(*b*) The set of positive integers $\mathbf{P} = \{1, 2, 3, \ldots\}$ is order-isomorphic to the set of even positive integers $E = \{2, 4, 6, \ldots\}$ since the function $f: \mathbf{P} \to E$ defined by $f(x) = 2x$ is an isomorphism between $\mathbf{P}$ and $E$.

(*c*) Consider the usual ordering $\leq$ of the positive integers $\mathbf{P} = \{1, 2, 3, \ldots\}$ and the negative integers $A = \{-1, -2, -3, \ldots\}$. Then $\mathbf{P}$ is not order-isomorphic to $A$. For if $f: \mathbf{P} \to A$ is an isomorphism then, for every $n \in \mathbf{P}$,

$$1 \leq n \quad \text{should imply} \quad f(1) \leq f(n)$$

for every $f(n) \in A$. Since $A$ has no first element, $f$ cannot exist.

The following theorems follow directly from the definition of order-isomorphic sets.

**Theorem 7.2**: Suppose $S$ is linearly ordered and $T \simeq S$. Then $T$ is linearly ordered.

**Theorem 7.3**: Suppose $f\colon S \to T$ is an order-isomorphism between ordered sets $S$ and $T$. Then $a \in S$ is a first, last, minimal, or maximal element of $S$ if and only if $f(a)$ is, respectively, a first, last, minimal, or maximal element of $T$.

**Theorem 7.4**: If $S$ is order-isomorphic to $T$, then $S$ is equipotent to $T$; that is, if $S \simeq T$ then $|S| = |T|$.

Example 7.9(*c*) shows that the converse of the above theorem is not true. That is, equipotent ordered sets need not be order-isomorphic.

**Theorem 7.5**: The relation of order-isomorphism between ordered sets is an equivalence relation. That is:

(i) $S \simeq S$, for any ordered set $S$.
(ii) If $S \simeq T$, then $T \simeq S$.
(iii) If $S \simeq T$ and $T \simeq U$, then $S \simeq U$.

## 7.9 ORDER TYPES OF LINEARLY ORDERED SETS

Consider a collection $\mathscr{S}$ of linearly ordered sets. Each set $A$ in $\mathscr{S}$ is assigned a symbol in such a way that two linearly ordered sets $A$ and $B$ in $\mathscr{S}$ are assigned the same symbol if and only if the sets are order-isomorphic. This symbol is called the *order type* of the sets. (One may view the order type as the equivalence class of all order-isomorphic sets in $\mathscr{S}$.) We emphasize that order type is only defined for linearly ordered sets, not ordered sets in general.

The order types of the following familiar sets (with the usual order) follow:

$$\omega = \text{order type of the set } \mathbf{P} \text{ of positive integers}$$
$$\pi = \text{order type of the set } \mathbf{Z} \text{ of integers}$$
$$\eta = \text{order type of the set } \mathbf{Q} \text{ of rational numbers}$$

Moreover, if $\zeta$ is the order type of a linearly ordered set $S$, then $\zeta^*$ will denote the order type of $S$ with the inverse order.

**EXAMPLE 7-10**

(*a*) Consider the following sets:

$$\mathbf{P} = \{1, 2, 3, \ldots\} \text{ of positive integers,}$$
$$E = \{2, 4, 6, \ldots\} \text{ of even positive integers,}$$
$$A = \{\ldots, -3, -2, -1\} \text{ of negative integers.}$$

The order type of set $E$ is $\omega$ since $E$ is order-isomorphic to $\mathbf{P}$, but the order type of the set $A$ is not $\omega$ since $A$ is not order-isomorphic to $\mathbf{P}$. However, the order type of $A$ is $\omega^*$ since $A$ is order-isomorphic to $\mathbf{P}$ with the inverse order.

(*b*) $\mathbf{P} = \{1, 2, 3, \ldots\}$ with the usual order is not order-isomorphic to $\mathbf{P}$ with the inverse order; hence $\omega \neq \omega^*$. On the other hand,

$$\mathbf{Z} = \{\ldots, -2, -1, 0, 1, 2, \ldots\}$$

with the usual order is order-isomorphic to $\mathbf{Z}$ with the inverse order; hence $\pi = \pi^*$.

## 7.10 LATTICES

Let $L$ be a nonempty set closed under two binary operations called *meet* and *join*, denoted respectively by $\wedge$ and $\vee$. Then $L$ is called a *lattice* if the following axioms hold where $a, b, c$ are any elements in $L$:

[$L_1$] ***Commutative law***:
(1*a*) $a \wedge b = b \wedge a$ (1*b*) $a \vee b = b \vee a$

[$L_2$] ***Associative law***:
(2*a*) $(a \wedge b) \wedge c = a \wedge (b \wedge c)$ (2*b*) $(a \vee b) \vee c = a \vee (b \vee c)$

[$L_3$] ***Absorption law***:
(3*a*) $a \wedge (a \vee b) = a$ (3*b*) $a \vee (a \wedge b) = a$

We will sometimes denote the lattice by $(L, \wedge, \vee)$ when we want to show which operations are involved.

### Duality and the Idempotent Law

The *dual* of any statement in a lattice $(L, \wedge, \vee)$ is defined to be the statement that is obtained by interchanging $\wedge$ and $\vee$. For example, the dual of

$$a \wedge (b \vee a) = a \vee a \quad \text{is} \quad a \vee (b \wedge a) = a \wedge a$$

Notice that the dual of each axiom of a lattice is also an axiom. Accordingly, the principle of duality holds; that is:

**Theorem 7.6 (Principle of Duality)**: The dual of any theorem in a lattice is also a theorem.

This follows from the fact that the dual theorem can be proven by using the dual of each step of the proof of the original theorem.

An important property of lattices follows directly from the absorption laws.

**Theorem 7.7 (Idempotent Law)**: (i) $a \wedge a = a$, (ii) $a \vee a = a$.

The proof of (i) requires only two lines:

$$\begin{aligned} a \wedge a &= a \wedge (a \vee (a \wedge b)) && \text{(using (3}b\text{))} \\ &= a && \text{(using (3}a\text{))} \end{aligned}$$

The proof of (ii) follows from the above principle of duality (or can be proved in a similar manner).

### Lattices and Order

Given a lattice $L$, we can define a partial order on $L$ as follows:

$$a \leq b \quad \text{if} \quad a \wedge b = a$$

Analogously, we could define

$$a \leq b \quad \text{if} \quad a \vee b = b$$

We state these results in a theorem.

**Theorem 7.8**: Let $L$ be a lattice. Then:

(i) $a \wedge b = a$ if and only if $a \vee b = b$.

(ii) The relation $a \leq b$ (defined by $a \wedge b = a$ or $a \vee b = b$) is a partial order on $L$.

Now that we have a partial order on any lattice $L$, we can picture $L$ by a diagram as was done for partially ordered sets in general.

**EXAMPLE 7.11** Let $C$ be a collection of sets closed under intersection and union. Then $(C, \cap, \cup)$ is a lattice. In this lattice, the partial order relation is the same as the set inclusion relation. Figure 7-6 shows the diagram of the lattice $L$ of all subsets of $\{a, b, c\}$.

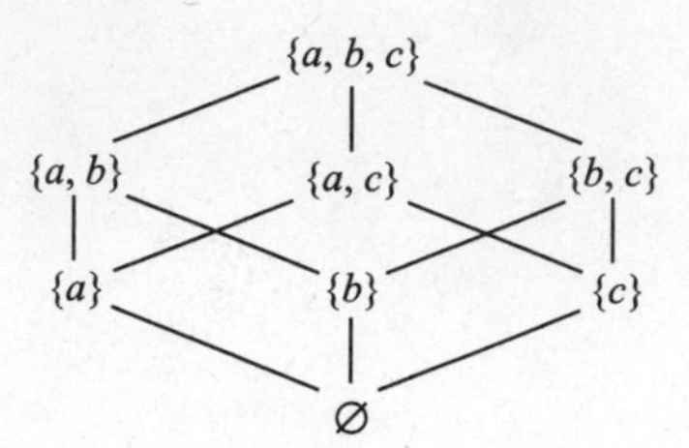


**Fig. 7-6**

We have shown how to define a partial order on a lattice $L$. The next theorem tells us when we can define a lattice on a partially ordered set $P$ such that the lattice will give back the original order on $P$.

**Theorem 7.9**: Let $P$ be a partially ordered set such that the $\inf(a, b)$ and $\sup(a, b)$ exist for any $a, b$ in $P$. Letting

$$a \wedge b = \inf(a, b) \qquad \text{and} \qquad a \vee b = \sup(a, b)$$

we have that $(P, \wedge, \vee)$ is a lattice. Furthermore, the partial order on $P$ induced by the lattice is the same as the original partial order on $P$.

The converse of the above theorem is also true. That is, let $L$ be a lattice and let $\leq$ be the induced partial order on $L$. Then $\inf(a, b)$ and $\sup(a, b)$ exist for any pair $a, b$ in $L$ and the lattice obtained from the ordered set $(L, \leq)$ is the original lattice. Accordingly, we have the following:

**Alternate Definition**: A lattice is a partially ordered set in which

$$a \wedge b = \inf(a, b) \qquad \text{and} \qquad a \vee b = \sup(a, b)$$

exist for any pair of elements $a$ and $b$.

We note first that any linearly ordered set is a lattice since $\inf(a, b) = a$ and $\sup(a, b) = b$ whenever $a \leq b$. By Example 7.8, the positive integers **P** and the set $\mathbf{D}_m$ of divisors of $m$ are lattices under the relation of divisibility.

### Sublattices, Isomorphic Lattices

Suppose $M$ is a nonempty subset of a lattice $L$. We say $M$ is a *sublattice* of $L$ if $M$ itself is a lattice (with respect to the operations of $L$). We note that $M$ is a sublattice of $L$ if and only if $M$ is closed under the operations of $\wedge$ and $\vee$ of $L$. For example, the set $\mathbf{D}_m$ of divisors of $m$ is a sublattice of the positive integers $N$ under divisibility.

Two lattices $L$ and $L'$ are said to be *isomorphic* if there is a one-to-one correspondence $f: L \to L'$ such that

$$f(a \wedge b) = f(a) \wedge f(b) \qquad \text{and} \qquad f(a \vee b) = f(a) \vee f(b)$$

for any elements $a, b$ in $L$.

## 7.11 BOUNDED, DISTRIBUTIVE, COMPLEMENTED LATTICES

This section discusses a number of different kinds of lattices, bounded, distributive, and complemented lattices. We also discuss a number of special kinds of elements in a lattice, join irreducible elements, atoms, and complements.

### Bounded Lattices

A lattice $L$ is said to have a *lower bound* 0 if for any element $x$ in $L$ we have $0 \precsim x$. Analogously, $L$ is said to have an *upper bound* $I$ if for any $x$ in $L$ we have $x \precsim I$. We say $L$ is *bounded* if $L$ has both a lower bound 0 and an upper bound $I$. In such a lattice we have the identities

$$a \vee I = I, \qquad a \wedge I = a, \qquad a \vee 0 = a, \qquad a \wedge 0 = 0$$

for any element $a$ in $L$.

The nonnegative integers with the usual ordering,

$$0 < 1 < 2 < 3 < 4 < \ldots$$

have 0 as a lower bound but have no upper bound. On the other hand, the lattice $P(\mathbf{U})$ of all subsets of any universal set $\mathbf{U}$ is a bounded lattice with $\mathbf{U}$ as an upper bound and the empty set $\varnothing$ as a lower bound.

Suppose $L = \{a_1; a_2, \ldots, a_n\}$ is a finite lattice. Then

$$a_1 \vee a_2 \vee \cdots \vee a_n \qquad \text{and} \qquad a_1 \wedge a_2 \wedge \cdots \wedge a_n$$

are upper and lower bounds for $L$, respectively. Thus we have

**Theorem 7.10**: Every finite lattice $L$ is bounded.

### Distributive Lattices

A lattice $L$ is said to be *distributive* if for any elements $a, b, c$ in $L$ we have the following:

[$L_4$] ***Distributive law***:

(4*a*) $a \wedge (b \vee c) = (a \wedge b) \vee (a \wedge c)$ (4*b*) $a \vee (b \wedge c) = (a \vee b) \wedge (a \vee c)$

Otherwise, $L$ is said to be *nondistributive*. We note that by the principle of duality the condition (4*a*) holds if and only if (4*b*) holds

Figure 7-7(*a*) is a nondistributive lattice since

$$a \vee (b \wedge c) = a \vee 0 = a$$

but

$$(a \vee b) \wedge (a \vee c) = I \wedge c = c$$

Figure 7-7(*b*) is also a nondistributive lattice. In fact, we have the following characterization of such lattices.

**Theorem 7.11**: A lattice $L$ is nondistributive if and only if it contains a sublattice isomorphic to Fig. 7-7(*a*) or (*b*).

The proof of this theorem lies beyond the scope of this text.

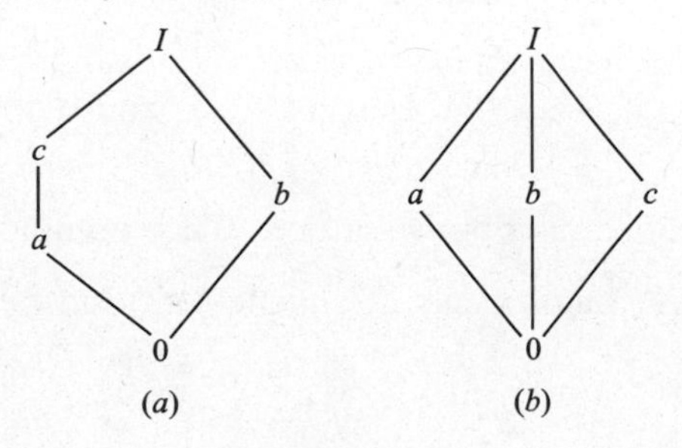


**Fig. 7-7**

### Join-Irreducible Elements, Atoms

Let $L$ be a lattice with a lower bound 0. An element $a$ in $L$ is said to be *join irreducible* if $a = x \vee y$ implies $a = x$ or $a = y$. (Prime numbers under multiplication have this property, i.e., if $p = ab$ then $p = a$

or $p = b$ where $p$ is prime.) Clearly 0 is join irreducible. If $a$ has at least two immediate predecessors, say $b_1$ and $b_2$ as in Fig. 7-8($a$), then $a = b_1 \vee b_2$, and so $a$ is not join irreducible. On the other hand, if $a$ has a unique immediate predecessor $c$, then $a \neq \sup(b_1, b_2) = b_1 \vee b_2$ for any other elements $b_1$ and $b_2$ because $c$ would lie between the $b$'s and $a$ as in Fig. 7-8($b$). In other words, $a \neq 0$ is join irreducible if and only if $a$ has a unique immediate predecessor. Those elements which immediately succeed 0, called *atoms*, are join irreducible. However, lattices can have other join-irreducible elements. For example, the element $c$ in Fig. 7-8($a$) is not an atom but is join irreducible since $a$ is its only immediate predecessor.

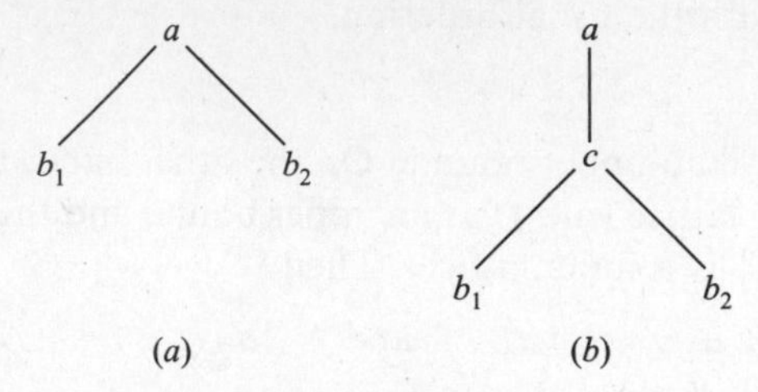


**Fig. 7-8**

If an element $a$ in a finite lattice $L$ is not join irreducible, then we can write $a = b_1 \vee b_2$. Then we can write $b_1$ and $b_2$ as the join of other elements if they are not join irreducible; and so on. Since $L$ is finite we finally have

$$a = d_1 \vee d_2 \vee \cdots \vee d_n$$

where the $d$'s are join irreducible. If $d_i$ precedes $d_j$ then $d_i \vee d_j = d_j$; so we can delete the $d_i$ from the expression. In other words, we can assume that the $d$'s are *irredundant*, i.e., no $d$ precedes any other $d$. We emphasize that such an expression need not be unique, e.g., $I = a \vee b$ and $I = b \vee c$ in both lattices in Fig. 7-7($b$). We now state the main theorem of this section (proved in Problem 7.39).

**Theorem 7.12**: Let $L$ be a finite distributive lattice. Then every $a$ in $L$ can be written uniquely (except for order) as the join of irredundant join-irreducible elements.

Actually this theorem can be generalized to lattices with *finite length*, i.e., where all linearly ordered subsets are finite. (Problem 7.34 gives an infinite lattice with finite length.)

### Complements

Let $L$ be a bounded lattice with lower bound 0 and upper bound $I$. Let $a$ be an element of $L$. An element $x$ in $L$ is called a *complement* of $a$ if

$$a \vee x = I \qquad \text{and} \qquad a \wedge x = 0$$

Complements need not exist and need not be unique. For example, the elements $a$ and $c$ are both complements of $b$ in Fig. 7-7($a$). Also, the elements $y, z$, and $u$ in the chain in Fig. 7-1 have no complements. We have the following result.

**Theorem 7.13**: Let $L$ be a bounded distributive lattice. Then complements are unique if they exist.

*Proof*: Suppose $x$ and $y$ are complements of any element $a$ in $L$. Then

$$a \vee x = I, \qquad a \vee y = I, \qquad a \wedge x = 0, \qquad a \wedge y = 0$$

Using distributivity,

$$x = x \vee 0 = x \vee (a \wedge y) = (x \vee a) \wedge (x \vee y) = I \wedge (x \vee y) = x \vee y$$

Similarly,

$$y = y \vee 0 = y \vee (a \wedge x) = (y \vee a) \wedge (y \vee x) = I \wedge (y \vee x) = y \vee x$$

Thus $x = x \vee y = y \vee x = y$ and the theorem is proved.

### Complemented Lattices

A lattice $L$ is said to be *complemented* if $L$ is bounded and every element in $L$ has a complement. Figure 7-7($b$) shows a complemented lattice where complements are not unique. On the other hand, the lattice $P(\mathbf{U})$ of all subsets of a universal set $\mathbf{U}$ is complemented, and each subset $A$ of $\mathbf{U}$ has the unique complement $A^c = \mathbf{U} \backslash A$.

**Theorem 7.14**: Let $L$ be a complemented lattice with unique complements. Then the join-irreducible elements of $L$, other than 0, are its atoms.

Combining this theorem and Theorems 7.12 and 7.13 we get an important result.

**Theorem 7.15**: Let $L$ be a finite complemented distributive lattice. Then every element $a$ in $L$ is the join of a unique set of atoms.

**Remark**: Some texts define a lattice $L$ to be complemented if each $a$ in $L$ has a unique complement. Theorem 7.14 is then stated differently.

# Solved Problems

## ORDERED SETS AND SUBSETS

**7.1.** Suppose the set $\mathbf{P} = \{1, 2, 3, \ldots\}$ of positive integers is ordered by divisibility. Insert the correct symbol, $<$, $>$, or $\|$ (not comparable), between each pair of numbers:

($a$) 2 ____ 8, ($b$) 18 ____ 24, ($c$) 9 ____ 3, ($d$) 5 ____ 15.

($a$) Since 2 divides 8, 2 precedes 8; hence $2 < 8$.

($b$) 18 does not divide 24, and 24 does not divide 18; hence $18 \| 24$.

($c$) Since 9 is divisible by 3, $9 > 3$.

($d$) Since 5 divides 15, $5 < 15$.

**7.2.** Let $\mathbf{P} = \{1, 2, 3, \ldots\}$ be ordered by divisibility. State whether each of the following is a chain (linearly ordered subset) in $\mathbf{P}$.

($a$) $A = \{24, 2, 6\}$ ($c$) $C = \{2, 8, 32, 4\}$ ($e$) $E = \{15, 5, 30\}$

($b$) $B = \{3, 15, 5\}$ ($d$) $D = \{7\}$ ($f$) $\mathbf{P} = \{1, 2, 3, \ldots\}$

($a$) Since 2 divides 6 which divides 24, $A$ is a chain in $\mathbf{P}$.

($b$) Since 3 and 5 are noncomparable, $B$ is not a chain in $\mathbf{P}$.

($c$) $C$ is a chain in $\mathbf{P}$ since $2 < 4 < 8 < 32$, that is, $2|4|8|32$ where $|$ means divides.

($d$) Any set consisting of one element is linearly ordered; hence $D$ is a chain in $\mathbf{P}$.

($e$) Here $5 < 15 < 30$; hence $E$ is a chain in $\mathbf{P}$.

($f$) $\mathbf{P}$ is not linearly ordered, e.g., 2 and 3 are noncomparable; hence $\mathbf{P}$ itself is not a chain in $\mathbf{P}$.

**7.3.** Let $A = \{1, 2, 3, 4, 5\}$ be ordered by the Hasse diagram in Fig. 7-9. Insert the correct symbol, $<$, $>$, or $\|$ (not comparable), between each pair of elements:

(*a*) 1 ____ 5, (*b*) 2 ____ 3, (*c*) 4 ____ 1, (*d*) 3 ____ 4

(*a*) Since there is a "path" (edges slanting upward) from 5 to 3 to 1, 5 precedes 1; hence $1 > 5$.

(*b*) There is no path from 2 to 3, or vice versa; hence $2\|3$.

(*c*) There is a path from 4 to 2 to 1; hence $4 < 1$.

(*d*) Neither $3 < 4$ nor $4 < 3$; hence $3\|4$.

**7.4.** Consider the ordered set $A$ in Fig. 7-9.

(*a*) Find all minimal and maximal elements of $A$.

(*b*) Does $A$ have a first element or a last element?

(*a*) No element strictly precedes 4 or 5, so 4 and 5 are minimal elements of $A$. No element strictly succeeds 1, so 1 is a maximal element of $A$.

(*b*) $A$ has no first element. Although 4 and 5 are minimal elements of $A$, neither precedes the other. However, 1 is a last element of $A$ since 1 succeeds every element of $A$.

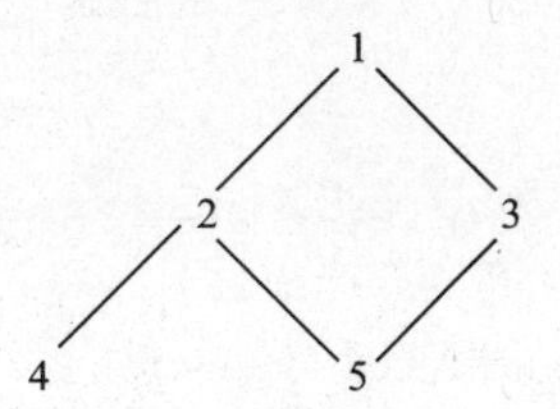


**Fig. 7-9**

**7.5.** Consider the ordered set $A$ in Fig. 7-9. For each $a \in A$, let $p(a)$ denote the set of predecessors of $a$, that is,

$$p(a) = \{x : x \leq a\}$$

Let $p(A)$ denote the collection of all predecessor sets of $A$, and let $p(A)$ be ordered by set inclusion. Draw the Hasse diagram of $p(A)$.

The elements of $p(A)$ follow:

$$p(1) = \{1, 2, 3, 4, 5\}, \quad p(2) = \{2, 4, 5\}, \quad p(3) = \{3, 5\}, \quad p(4) = \{4\}, \quad p(5) = \{5\}$$

Figure 7-10 gives the Hasse diagram of $p(A)$ ordered by set inclusion. [Observe that the diagrams of $A$ and $p(A)$ are identical except for the labeling of the vertices.]

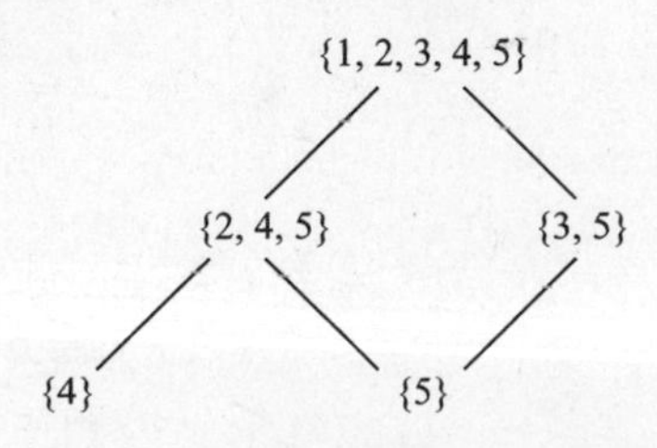


**Fig. 7-10**

**7.6.** Consider the ordered set $A$ in Fig. 7-9. Let $L(A)$ denote the collection of all chains (linearly ordered subsets) in $A$ with 2 or more elements, and let $L(A)$ be ordered by set inclusion. Draw the Hasse diagram of $L(A)$.

The elements of $L(A)$ are as follows:

$$\{1,2,4\}, \{1,2,5\}, \{1,3,5\}, \{1,2\}, \{1,4\}, \{1,3\}, \{1,5\}, \{2,4\}, \{2,5\}, \{3,5\}$$

(Note $\{2,5\}$ and $\{3,4\}$ are not linearly ordered, and there are no chains with four or more elements.) The diagram of $L(A)$ appears in Fig. 7-11.

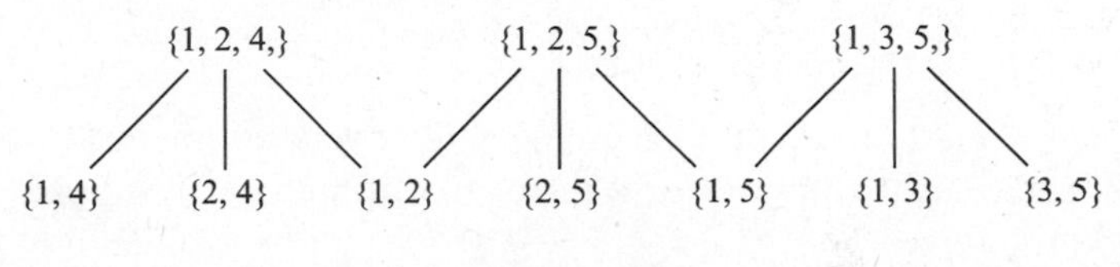


**Fig. 7-11**

**7.7.** Prerequisites in college is a familiar partial ordering of available classes. Define $A < B$ if class $A$ is a prerequisite for class $B$. Let $C$ be the set of mathematics classes and their prerequisites given in Fig. 7-12($a$).

($a$) Draw the Hasse diagram for the partial ordering of these classes.

($b$) Find all minimal and maximal elements of $C$.

($c$) Does $C$ have a first element or a last element?

($a$) Math 101 must be on the bottom of the diagram since it is the only course with no prerequisites. Since Math 201 and Math 250 only require Math 101, we have Math 101 $<<$ Math 201 and we have Math 101 $<<$ Math 250; hence draw a line slanting upward from Math 101 to Math 201 and one from Math 101 to Math 250. Continuing this process, we obtain the Hasse diagram in Fig. 7-12($b$).

| **Class** | **Prerequisites** |
|---|---|
| Math 101 | None |
| Math 201 | Math 101 |
| Math 250 | Math 101 |
| Math 251 | Math 250 |
| Math 340 | Math 201 |
| Math 341 | Math 340 |
| Math 450 | Math 201, Math 250 |
| Math 500 | Math 450, Math 251 |

($a$)

341 500
340 450 251
201 250
101

($b$)

**Fig. 7-12**

($b$) No element strictly precedes Math 101 so Math 101 is a minimal element of $C$. No element strictly succeeds Math 341 or Math 500, so each is a maximal element of $C$.

($c$) Math 101 is a first element of $C$ since it precedes every other element of $C$. However, $C$ has no last element. Although Math 341 and Math 500 are maximal elements, neither is a last element since neither precedes the other.

**7.8.** Consider the set **Z** of integers. Define $a\,R\,b$ by $b = a^r$ for some positive integer $r$. Show that $R$ is a partial order on **Z**, that is, show that $R$ is (*a*) reflexive, (*b*) antisymmetric, and (*c*) transitive.

(*a*) $R$ is reflexive since $a = a^1$.

(*b*) Suppose $a\,R\,b$ and $b\,R\,a$, say $b = a^r$ and $a = b^s$. Then $a = (a^r)^s = a^{rs}$. There are three possibilities: (i) $rs = 1$, (ii) $a = 1$, and (iii) $a = -1$. If $rs = 1$, then $r = 1$ and $s = 1$ and so $a = b$. If $a = 1$, then $b = 1^r = 1 = a$, and, similarly, if $b = 1$, then $a = 1$. Lastly, if $a = -1$, then $b = -1$ (since $b \neq 1$) and so $a = b$. In all three cases, $a = b$. Thus $R$ is antisymmetric.

(*c*) Suppose $a\,R\,b$ and $b\,R\,c$, say $b = a^r$ and $c = b^s$. Then $c = (a^r)^s = a^{rs}$ and hence $a\,R\,c$. Hence $R$ is transitive.

**7.9.** Consider the set $\mathbf{P} = \{1, 2, 3, \ldots\}$ of positive integers. Every number in **P** can be written uniquely as a product of a nonnegative power of 2 times an odd number. Suppose $a$ and $a'$ are positive integers such that

$$a = 2^r(2s+1) \qquad \text{and} \qquad a' = 2^{r'}(2s'+1)$$

where $r$ and $s$ are nonnegative integers. We define:

$$a < a' \begin{cases} \text{if } r < r' \\ \text{or if } r = r' \text{ but } s < s' \end{cases}$$

Insert the correct symbol, $<$ or $>$, between each of the following pairs of numbers:

(*a*) 5 ____ 14, (*b*) 6 ____ 9, (*c*) 26 ____ 12, (*d*) 20 ____ 30

The elements of **P** can be listed as in Fig. 7-13. The first row consists of the odd numbers, the second row of 2 times the odd numbers, the third row of $2^2 = 4$ times the odd numbers, and so on. Then $a < a'$ if $a$ is in a higher row than $a'$, or if $a$ and $a'$ are in the same row but $a$ comes before $a'$ in the row. Thus:

(*a*) $5 < 14$, (*b*) $6 > 9$, (*c*) $26 < 12$, (*d*) $20 > 30$.

| $r$ \ $s$ | 0 | 1 | 2 | 3 | 4 | 5 | 6 | 7 | |
|---|---|---|---|---|---|---|---|---|---|
| 0 | 1 | 3 | 5 | 7 | 9 | 11 | 13 | 15 | ... |
| 1 | 2 | 6 | 10 | 14 | 18 | 22 | 26 | 30 | ... |
| 2 | 4 | 12 | 20 | 28 | 36 | 44 | 52 | 60 | ... |
| | ⋮ | ⋮ | ⋮ | ⋮ | ⋮ | ⋮ | ⋮ | ⋮ | |

**Fig. 7-13**

**7.10.** Suppose $<$ is a quasi-order on a set $S$. Define:

$$a \leq b \qquad \text{if} \qquad a < b \qquad \text{or} \qquad a = b$$

Show that $a \leq b$ is a partial order on $S$.

We want to show that $\leq$ is (*a*) reflexive, (*b*) antisymmetric, and (*c*) transitive.

(*a*) Since $a = a$, we have $a \leq a$. Hence $\leq$ is reflexive.

(*b*) Suppose $a \leq b$ and $b \leq a$. Then either $a = b$ or else $a < b$ and $b < a$. Suppose $a < b$ and $b < a$. By transitivity of $<$, we have $a < a$. This contradicts the fact that $<$ is irreflexive. Thus $a = b$ and $\leq$ is antisymmetric.

(*c*) Suppose $a \leq b$ and $b \leq c$. There are four cases.

(1) Suppose $a = b$ and $b = c$. Then $a = c$ and $a \leq c$.

(2) Suppose $a = b$ and $b < c$. Then $a < c$ and so $a \leq c$.

(3) Suppose $a < b$ and $b = c$. Then $a < c$ and so $a \leq c$.

(4) Suppose $a < b$ and $b < c$. Since $<$ is transitive, $a < c$. Hence $a \leq c$.

In each case, $a \leq c$; hence $\leq$ is transitive.

## SET CONSTRUCTIONS AND ORDER

**7.11.** Suppose **P** has the usual order $\leq$. Consider the following pairs of elements of $\mathbf{P}^2 = \mathbf{P} \times \mathbf{P}$:

(*a*) (5, 7) ___ (7, 1) (*c*) (5, 5) ___ (4, 8) (*e*) (7, 9) ___ (4, 1)

(*b*) (4, 6) ___ (4, 2) (*d*) (1, 3) ___ (1, 7) (*f*) (7, 9) ___ (8, 2)

Insert the correct symbol, $<$, $>$, or $\|$ (not comparable), between each of the above pairs of elements of $\mathbf{P} \times \mathbf{P}$ when $\mathbf{P}^2$ is given (1) product order, (2) lexicographical order.

(1) Here $(a, b) \leq (a', b')$ provided $a \leq a'$ and $b \leq b'$. Hence $(a, b) < (a', b')$ if $a < a'$ and $b \leq b'$ or if $a \leq a'$ and $b < b'$. Thus:

(*a*) $\|$ since $5 < 7$ but $7 > 1$ (*c*) $\|$ since $5 > 4$ and $5 < 8$ (*e*) $>$ since $7 > 4$ and $9 > 1$

(*b*) $>$ since $4 \geq 4$ and $6 > 2$ (*d*) $<$ since $1 \leq 1$ and $3 < 7$ (*f*) $\|$ since $7 < 8$ and $9 > 2$

(2) Here $(a, b) < (a', b')$ if $a < a'$ or if $a = a'$ but $b < b'$. Thus:

(*a*) $<$ since $5 < 7$. (*c*) $>$ since $5 > 4$ (*e*) $>$ since $7 > 4$

(*b*) $>$ since $4 = 4$ and $6 > 2$ (*d*) $<$ since $1 = 1$ but $3 < 7$ (*f*) $<$ since $7 < 8$

**7.12.** Suppose the English alphabet $\mathbf{A} = \{\mathrm{a, b, c, \ldots, y, z}\}$ is given the *usual (alphabetical) order*. Consider the following two-letter words (viewed as elements of $\mathbf{A} \times \mathbf{A}$):

(*a*) cx ___ at (*c*) cx ___ cz (*e*) cx ___ dx

(*b*) cx ___ by (*d*) cx ___ rs (*f*) cx ___ cs

Insert the correct symbol, $<$, $>$, or $\|$ (not comparable), between each of the above two-letter words when $\mathbf{A}^2 = \mathbf{A} \times \mathbf{A}$ is given (1) the product order, (2) the lexicographical order.

(1) (*a*) $>$ since $\mathrm{c > a}$ and $\mathrm{x > t}$ (*c*) $<$ since $\mathrm{c \leq c}$ and $\mathrm{x < z}$ (*e*) $<$ since $\mathrm{c < d}$ and $\mathrm{x \leq x}$

(*b*) $\|$ since $\mathrm{c > b}$ but $\mathrm{x < y}$ (*d*) $\|$ since $\mathrm{c < r}$ but $\mathrm{x > s}$ (*f*) $>$ since $\mathrm{c \geq c}$ and $\mathrm{x > s}$

(2) (*a*) $>$ since $\mathrm{c > a}$ (*c*) $<$ since $\mathrm{c = c}$ and $\mathrm{x < z}$ (*e*) $<$ since $\mathrm{c < d}$

(*b*) $<$ since $\mathrm{c > b}$ (*d*) $<$ since $\mathrm{c < r}$ (*f*) $>$ since $\mathrm{c = c}$ and $\mathrm{x > s}$

**7.13.** Consider the set $\mathbf{P} = \{1, 2, 3, \ldots\}$ with the usual order, and the English alphabet $\mathbf{A} = \{\mathrm{a, b, c, \ldots, y, z}\}$ with the usual alphabetical order. Suppose $S = \mathbf{P} \cup \mathbf{A}$ and $T = \mathbf{A} \cup \mathbf{P}$ are each given the concatenation order:

$$S = \{1, 2, 3, \ldots; \mathrm{a, b, \ldots, z}\}, \qquad T = \{\mathrm{a, b, \ldots, z}; 1, 2, 3, \ldots\}$$

(Here $\mathbf{P} < \mathbf{A}$ in $S$ but $\mathbf{A} < \mathbf{P}$ in $T$.) (*a*) Insert the correct symbol, $<$ or $>$, between the pair "7 ___ y". (*b*) Which subsets of $S$ and of $T$ are chains? (*c*) Which elements in $S$ and which elements in $T$ have no immediate predecessors?

(*a*) We have $7 < \mathrm{y}$ when $7, \mathrm{y} \in S$, but $7 > \mathrm{y}$ when $7, \mathrm{y} \in T$.

(*b*) Since $S$ and $T$ are linearly ordered, every subset of $S$ and of $T$ are chains.

(*c*) In $S$, both 1 and a have no immediate predecessor. However, in $T$ only a has no immediate predecessor.

**7.14.** Consider the English alphabet $\mathbf{A} = \{\text{a}, \text{b}, \text{c}, \ldots, \text{y}, \text{z}\}$ with the *usual (alphabetical) order*. Recall that the Kleene closure $\mathbf{A}^*$ of $\mathbf{A}$ consists of all words in $\mathbf{A}$. Let $L$ be the following subset of $\mathbf{A}^*$:

$$L = \{\text{went, forget, to, medicine, me, toast, melt, for, we, arm}\}$$

Sort (arrange in order) $L$ where $\mathbf{A}^*$ is given (*a*) the short-lex (free semigroup) order, (*b*) the lexicographical order.

(*a*) First order the elements by length and then order them alphabetically to obtain:

me, to, we, arm, for, melt, went, toast, forget, medicine

(*b*) Use the usual alphabetical ordering to obtain:

arm, for, forget, me, medicine, melt, to, toast, we, went

**7.15.** Suppose $A$ and $B$ are ordered sets. Show that the product order on $A \times B$, defined by

$$(a, b) \precsim (c, d) \qquad \text{if } a \leq c \text{ and } b \leq d$$

is a partial ordering of $A \times B$.

We want to show that $\precsim$ is (*a*) reflexive, (*b*) antisymmetric, and (*c*) transitive.

(*a*) Since $a = a$ and $b = b$, we have $a \leq a$ and $b \leq b$. Hence $(a, b) \precsim (a, b)$ and $\precsim$ is reflexive.

(*b*) Suppose $(a, b) \precsim (c, d)$ and $(c, d) \precsim (a, b)$. Then

$$a \leq c \text{ and } b \leq d \qquad \text{and} \qquad c \leq a \text{ and } d \leq b$$

Thus $a = c$ and $b = d$. Hence $(a, b) = (c, d)$ and is antisymmetric.

(*c*) Suppose $(a, b) \precsim (c, d)$ and $(c, d) \precsim (e, f)$. Then

$$a \leq c \text{ and } b \leq d \qquad \text{and} \qquad c \leq e \text{ and } d \leq f$$

By transitivity of $\leq$, we have $a \leq e$ and $b \leq f$. Thus $(a, b) \precsim (e, f)$, and $\precsim$ is transitive.

## CONSISTENT ENUMERATIONS

**7.16.** Let $S = \{a, b, c, d, e\}$ be ordered as in Fig. 7-14. Find all possible consistent enumerations $f: S \to \{1, 2, 3, 4, 5\}$.

Since $a$ is the only minimal element $f(a) = 1$, and since $e$ is the only maximal element $f(e) = 5$. Also $f(b) = 2$ since $b$ is the only successor of $a$. The choices for $c$ and $d$ are $f(c) = 3$ and $f(d) = 4$ or vice versa. Thus there are two possible enumerations which follow:

$$f(a) = 1, \quad f(b) = 2, \quad f(c) = 3, \quad f(d) = 4, \quad f(e) = 5$$
$$f(a) = 1, \quad f(b) = 2, \quad f(c) = 4, \quad f(d) = 3, \quad f(e) = 5$$

We emphasize that we usually cannot recreate the original partial order from a given consistent enumeration.

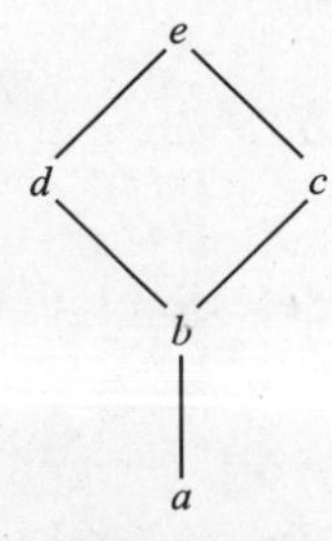


**Fig. 7-14**

**7.17.** Prove Theorem 7.1: Suppose $S$ is a finite partially ordered set with $n$ elements. Then there exists a consistent enumeration $f: S \to \{1, 2, \ldots, n\}$.

The proof is by induction on the number $n$ of elements in $S$. Suppose $n = 1$, say $S = \{s\}$. Then $f(s) = 1$ is a consistent enumeration of $S$. Now suppose $n > 1$ and the theorem holds for ordered sets with less than $n$ elements. Let $a$ in $S$ be a minimal element. [Such an element exists since $S$ is finite.] Let $T = S \setminus \{a\}$. Then $T$ is a finite poset with $n - 1$ elements and hence, by induction, $T$ admits a consistent enumeration; say $g: T \to \{1, 2, \ldots, n-1\}$. Define $f: S \to \{1, 2, \ldots, n\}$ by

$$f(x) = \begin{cases} 1 & \text{if } x = a \\ g(x) + 1 & \text{if } x \neq a \end{cases}$$

Then $f$ is the required consistent enumeration.

**7.18.** Suppose a student Ann wants to take all eight mathematics courses in Problem 7.7, but only one per semester.

(*a*) Which choice or choices does she have for her first and for her last (eighth) semester?

(*b*) Suppose she wants to take Math 250 in her first year (first or second semester) and Math 340 in her senior year (seventh or eighth semester). Find all possible ways that she can take the eight courses.

(*a*) By Fig. 7-12, Math 101 is the only minimal element and hence must be taken in the first semester, and Math 341 and 500 are the maximal elements and hence one of them must be taken in the last semester.

(*b*) Math 250 is not a minimal element and hence must be taken in the second semester, and Math 340 is not a maximal element so it must be taken in the seventh semester and Math 341 in the eighth semester. Also Math 500 must be taken in the sixth semester. The following give the three possible ways to take the eight courses:

$$[101, 250, 251, 201, 450, 500, 340, 341]$$
$$[101, 250, 201, 251, 450, 500, 340, 341]$$
$$[101, 250, 201, 450, 251, 500, 340, 341]$$

**7.19.** Suppose $\mathbf{P} = \{1, 2, 3, \ldots\}$ is ordered by divisibility "$|$". Find a consistent enumeration of $(\mathbf{P}, |)$ into $(\mathbf{P}, \leq)$.

The function $f: \mathbf{P} \to \mathbf{P}$ defined by $f(x) = x$ is a consistent enumeration since $a|b$ implies $a \leq b$.

**7.20.** Find a consistent enumeration of the real numbers $\mathbf{R}$ into $\mathbf{P}$.

Since $|\mathbf{R}| > |\mathbf{P}|$, there exists no one-to-one function from $\mathbf{R}$ into $\mathbf{P}$. Thus no consistent enumeration exists.

## UPPER AND LOWER BOUNDS, SUPREMUM AND INFIMUM

**7.21.** Let $S = \{a, b, c, d, e, f, g\}$ be ordered as in Fig. 7-15, and let $X = \{c, d, e\}$.

(*a*) Find the upper and lower bounds of $X$.

(*b*) Identify $\sup(X)$, the supremum of $X$, and $\inf(X)$, the infimum of $X$, if either exists.

(*a*) The elements $e$, $f$ and $g$ succeed every element of $X$; hence $e, f$, and $g$ are the upper bounds of $X$. The element $a$ precedes every element of $X$; hence it is the lower bound of $X$. Note that $b$ is not a lower bound since $b$ does not precede $c$; in fact, $b$ and $c$ are not comparable.

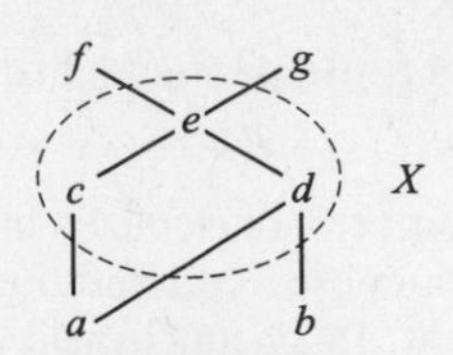


Fig. 7-15

(*b*) Since $e$ precedes both $f$ and $g$, we have $e = \sup(X)$. Likewise, since $a$ precedes (trivially) every lower bound of $X$, we have $a = \inf(X)$. Note that $\sup(X)$ belongs to $X$ but $\inf(X)$ does not belong to $X$.

**7.22.** Let $S = \{1, 2, 3, \ldots, 8\}$ be ordered as in Fig. 7-16, and let $A = \{2, 3, 6\}$.

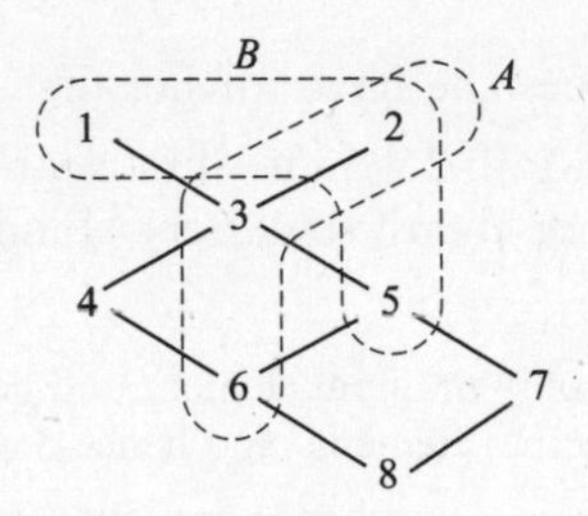


Fig. 7-16

(*a*) Find the upper and lower bounds of $A$.
(*b*) Identify $\sup(A)$ and $\inf(A)$ if either exists.

(*a*) The upper bound is 2, and the lower bounds are 6 and 8.
(*b*) Here $\sup(A) = 2$ and $\inf(A) = 6$.

**7.23.** Repeat Problem 7.22 for the subset $B = \{1, 2, 5\}$ of $S$.

(*a*) There is no upper bound for $B$ since no element succeeds both 1 and 2. The lower bounds are 6, 7, 8.
(*b*) Trivially, $\sup(A)$ does not exist since there are no upper bounds. Although $A$ has three lower bounds, $\inf(A)$ does not exist since no lower bound succeeds both 6 and 7.

**7.24.** Consider the set **Q** of rational numbers with the usual order $\leq$, and consider the subset $D$ of **Q** defined by

$$D = \{x \in \mathbf{Q} : 8 < x^3 < 15\}$$

(*a*) Is $D$ bounded above or below? (*b*) Do $\sup(D)$ and $\inf(D)$ exist?

(*a*) The subset $D$ is bounded both above and below. For example, 1 is a lower bound and 100 an upper bound.
(*b*) $\sup(D)$ does not exist. Suppose, on the contrary, $\sup(D) = x$. Since $\sqrt[3]{15}$ is irrational, $x > \sqrt[3]{15}$. However, there exists a rational number $y$ such that $\sqrt[3]{15} < y < x$. Thus $y$ is also an upper bound for $D$. This contradicts the assumption that $x = \sup(D)$. On the other hand, $\inf(D)$ does exist. Specifically, $\inf(D) = 2$.

**7.25.** Let $\mathscr{S}$ be a collection of sets ordered by set inclusion. Let $\mathscr{A} = \{A_i : i \in I\}$ be a subcollection of $\mathscr{S}$. Let $B = \bigcup_i A_i$. (*a*) Suppose $D$ is an upper bound of $\mathscr{A}$. Show that $B \subseteq D$. (*b*) Is $B$ an upper bound of $\mathscr{A}$?

(*a*) Let $x \in B$. Then there exists $j \in I$ such that $x \in A_j$. Since $D$ is an upper bound for $\mathscr{A}$, $A_j \subseteq D$. Hence $x \in D$. We have shown that $x \in B$ implies $x \in D$; hence $B \subseteq D$.

(*b*) Although $\mathscr{A} = \{A_i : i \in I\}$ is a subcollection of $\mathscr{S}$, it need not be true that $B = \bigcup_i A_i$ belongs to $\mathscr{S}$. Therefore, $B$ is an upper bound if and only if $B$ belongs to $\mathscr{S}$.

**7.26.** Given an example of a collection $\mathscr{S}$ of sets ordered by set inclusion, and a subcollection $\mathscr{A} = \{A_i : i \in I\}$ of $\mathscr{S}$ such that $B = \bigcup_i A_i$ is not an upper bound of $\mathscr{A}$.

Let $\mathscr{S}$ be the collection of all finite subsets of $\mathbf{P} = \{1, 2, 3, \ldots\}$ and let $\mathscr{A} = \{A_i\}$ be the subcollection of $\mathscr{S}$ consisting of sets with exactly two elements. Let $B = \bigcup_i A_i$. Then $B$ has an infinite number of elements and hence $B$ does not belong to $\mathscr{S}$. Thus $B$ is not an upper bound of $\mathscr{A}$ (in $\mathscr{S}$).

## ORDER-ISOMORPHIC SETS, SIMILARITY MAPPINGS

**7.27.** Suppose an ordered set $A$ is order-isomorphic to an ordered set $B$ and $f\colon A \to B$ is a similarity mapping. Are the following statements true or false?

(*a*) An element $a \in A$ immediately precedes an element $a' \in A$, that is, $a \prec a'$, if and only if $f(a) << f(a')$ in $B$.

(*b*) An element $a \in A$ has $r$ immediate successors in $A$ if and only if $f(a)$ has $r$ immediate successors in $B$.

(*c*) An element $a \in A$ has $r$ immediate predecessors in $A$ if and only if $f(a)$ has $r$ immediate predecessors in $B$.

All the statements are true; the order structure of $A$ is the same as the order structure of $B$.

**7.28.** Let $S$ be the ordered set in Fig. 7-14. Suppose $A = \{1, 2, 3, 4, 5\}$ is order-isomorphic to $S$ and suppose the following is a similarity mapping from $S$ onto $A$:

$$f = \{(a, 1), (b, 3), (c, 5), (d, 2), (e, 4)\}$$

Draw the Hasse diagram of $A$.

The similarity mapping $f$ preserves the order structure of $S$ and hence $f$ may be viewed simply as a relabeling of the vertices in the diagram of $S$. Thus Fig. 7-17 shows the Hasse diagram of $A$.

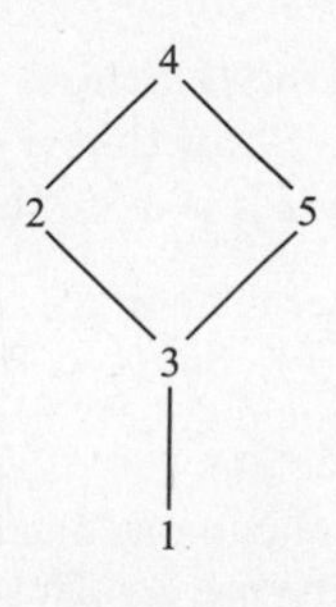


Fig. 7-17

**7.29.** Let $S = \{1, 2, 3, 4, 5, 6\}$ be ordered as in Fig. 7-18($a$).

($a$) Find the number $n$ of similarity mappings $f: S \to S$.

($b$) Is $S$ order-isomorphic to $S$ with the inverse ordering?

($a$) Since 1 and 2 are the minimal elements, there are only two possibilities for $f(1)$ and $f(2)$; that is, $f(1) = 1$ and $f(2) = 2$, or $f(1) = 2$ and $f(2) = 1$. Similarly, we must have $f(5) = 5$ and $f(6) = 6$, or $f(5) = 6$ and $f(6) = 5$. Furthermore, 3 precedes 4 and they both must succeed 1 and 2 and they both must precede 5 and 6. Thus we must have $f(3) = 3$ and $f(4) = 4$. In other words, $n = 4$. The four similarity mappings are listed in Fig. 7-18($b$).

($b$) $S$ with the inverse order is pictured in Fig. 7-18($c$), which may be obtained by inverting the original diagram which reverses the direction of the arrows. Clearly the diagrams are order-isomorphic. One such order-isomorphism between the sets follows:

$$f(1) = 5, \quad f(2) = 6, \quad f(3) = 4, \quad f(4) = 3, \quad f(5) = 1, \quad f(6) = 2$$

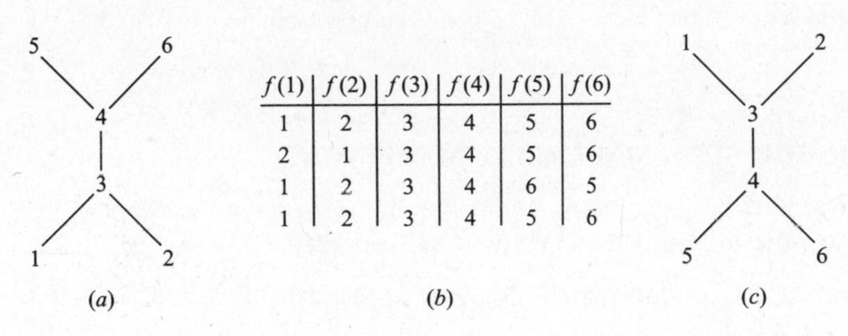

| $f(1)$ | $f(2)$ | $f(3)$ | $f(4)$ | $f(5)$ | $f(6)$ |
|---|---|---|---|---|---|
| 1 | 2 | 3 | 4 | 5 | 6 |
| 2 | 1 | 3 | 4 | 5 | 6 |
| 1 | 2 | 3 | 4 | 6 | 5 |
| 1 | 2 | 3 | 4 | 5 | 6 |

($a$) ($b$) ($c$)

**Fig. 7-18**

**7.30.** Consider $\mathbf{P} = \{1, 2, 3, \ldots\}$ and $\mathbf{A} = \{\text{a}, \text{b}, \text{c}, \ldots, \text{x}, \text{y}\}$ with the usual orders, and suppose $S = \mathbf{P} \cup \mathbf{A}$ and $T = \mathbf{A} \cup \mathbf{P}$ are each given the concatenation order

$$S = \{1, 2, 3, \ldots; \text{a}, \text{b}, \ldots, \text{z}\} \quad \text{and} \quad T = \{\text{a}, \text{b}, \ldots, \text{z}; 1, 2, 3, \ldots\}$$

Show that $S$ and $T$ are not order-isomorphic.

There are two elements, 1 and a, which have no predecessors in $S$, but there is only one element, a, which has no predecessor in $T$. Any order-isomorphism between sets must preserve the number of such elements. Thus $S$ is not order-isomorphic to $T$.

**7.31.** Let $A$ be an ordered set and, for each $a \in A$, let $p(a)$ denote the set of predecessors of $a$:

$$p(a) = \{x : x \leq a\}$$

(called the *predecessor set* of $a$). Let $p(A)$ denote the collection of all predecessor sets of the elements in $A$ ordered by set inclusion. Show that $A$ and $p(A)$ are isomorphic by showing that the map $f: A \to p(A)$, defined by $f(a) = p(a)$, is a similarity mapping of $A$ onto $p(A)$.

First we show that $f$ preserves the order relation of $A$. Suppose $a \leq b$. Let $x \in p(a)$. Then $x \leq a$, and hence $x \leq b$; so $x \in p(b)$. Thus $p(a) \subseteq p(b)$. Suppose $a \| b$ (noncomparable). Then $a \in p(a)$ but $a \notin p(b)$; hence $p(a) \not\subseteq p(b)$. Similarly, $b \in p(b)$ but $b \notin p(a)$; hence $p(b) \not\subseteq p(a)$. Therefore, $p(a) \| p(b)$. Thus $f$ preserves order.

We now need only show that $f$ is a one-to-one and onto. First we show that $f$ is an onto function. Suppose $y \in p(A)$. Then $y = p(a)$ for some $a \in A$. Thus $f(a) = p(a) = y$ so $f$ is a function from $A$ onto $p(A)$.

Next we show $f$ is one-to-one. Suppose $a \neq b$. Then $a < b$, $b > a$ or $a \| b$. In the first and third cases, $b \in p(b)$ but $b \notin p(a)$, and in the second case $a \in p(a)$ but $a \notin p(b)$. Accordingly, in all three cases, we have $p(a) \neq p(b)$. Therefore $f$ is one-to-one.

Consequently, $f$ is a similarity mapping of $A$ onto $p(A)$ and so $A \simeq p(A)$.

**7.32.** Consider the ordered set $A = \{a, b, c, d, e\}$ in Fig. 7-19($a$). Find the Hasse diagram of the collection $p(A)$ of predecessor sets of the elements of $A$ ordered by set inclusion.

The elements of $p(A)$ follow:

$$p(a) = \{a, c, d, e\}, \quad p(b) = \{b, c, d, e\}, \quad p(c) = \{c, d, e\}, \quad p(d) = \{d\}, \quad p(e) = \{e\}$$

Figure 7-19($b$) gives the diagram of $p(A)$ ordered by set inclusion. Observe that the two diagrams in Fig. 7-19 are identical except for the labeling of the vertices.

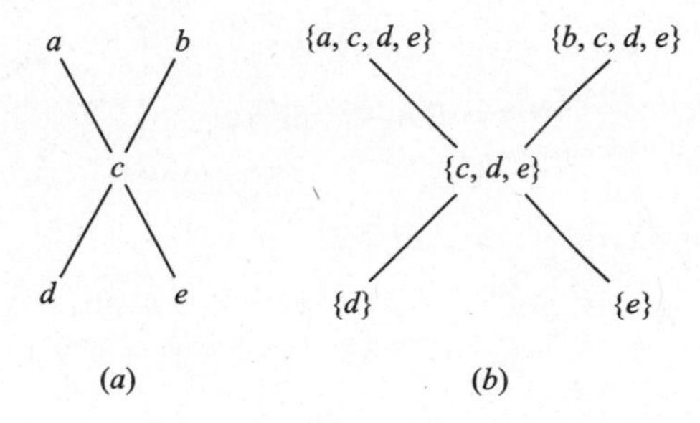


**Fig. 7-19**

## LATTICES

**7.33.** Write the dual of each statement:

($a$) $(a \wedge b) \vee c = (b \vee c) \wedge (c \vee a)$; ($b$) $(a \wedge b) \vee a = a \wedge (b \vee a)$

Replace $\vee$ by $\wedge$ and $\wedge$ by $\vee$ in each statement to obtain the dual statement:

($a$) $(a \vee b) \wedge c = (b \wedge c) \vee (c \wedge a)$

($b$) $(a \vee b) \wedge a = a \vee (b \wedge a)$

**7.34.** Give an example of an infinite lattice $L$ with finite length.

Let $L = \{0, 1, a_1, a_2, a_3, \ldots\}$ and let $L$ be ordered as in Fig. 7-20; that is, for each $n \in \mathbf{P}$ we have

$$0 < a_n < 1$$

Then $L$ has finite length since $L$ has no infinite linearly ordered subset.

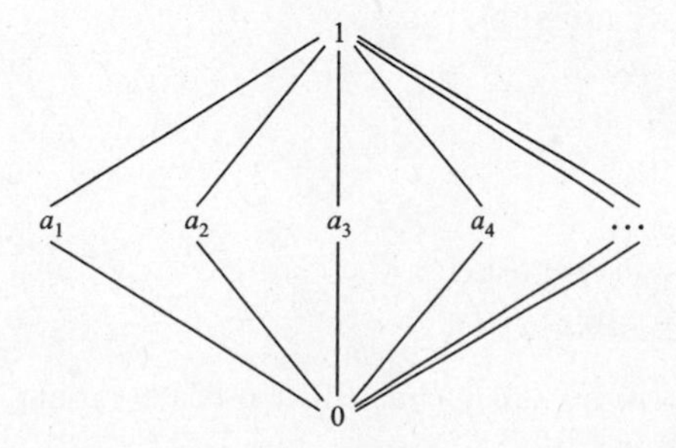


**Fig. 7-20**

**7.35.** Prove Theorem 7.8: Let $L$ be a lattice. Then: (i) $a \wedge b = a$ if and only if $a \vee b = b$. (ii) The relation $a \leq b$ (defined by $a \wedge b = a$ or $a \vee b = b$) is a partial order on $L$.

(*a*) Suppose $a \wedge b = a$. Using the absorption law in the first step we have:

$$b = b \vee (b \wedge a) = b \vee (a \wedge b) = b \vee a = a \vee b$$

Now suppose $a \vee b = b$. Again using the absorption law in the first step we have:

$$a = a \wedge (a \vee b) = a \wedge b$$

Thus $a \wedge b = a$ if and only if $a \vee b = b$.

(*b*) For any $a$ in $L$, we have $a \wedge a = a$ by idempotency. Hence $a \leq a$, and so $\leq$ is reflexive.

Suppose $a \leq b$ and $b \leq a$. Then $a \wedge b = a$ and $b \wedge a = b$. Therefore, $a = a \wedge b = b \wedge a = b$, and so $\leq$ is antisymmetric.

Lastly, suppose $a \leq b$ and $b \leq c$. Then $a \wedge b = a$ and $b \wedge c = b$. Thus

$$a \wedge c = (a \wedge b) \wedge c = a \wedge (b \wedge c) = a \wedge b = a$$

Therefore $a \leq c$, and so $\leq$ is transitive. Accordingly, $\leq$ is a partial order on $L$.

**7.36.** Which of the partially ordered sets in Fig. 7-21 are lattices?

A partially ordered set is a lattice if and only if $\sup(x, y)$ and $\inf(x, y)$ exist for each pair $x, y$ in the set. Only (*c*) is not a lattice since $\{a, b\}$ has three upper bounds, $c$, $d$, and $I$, and no one of them precedes the other two, i.e., $\sup(a, b)$ does not exist.

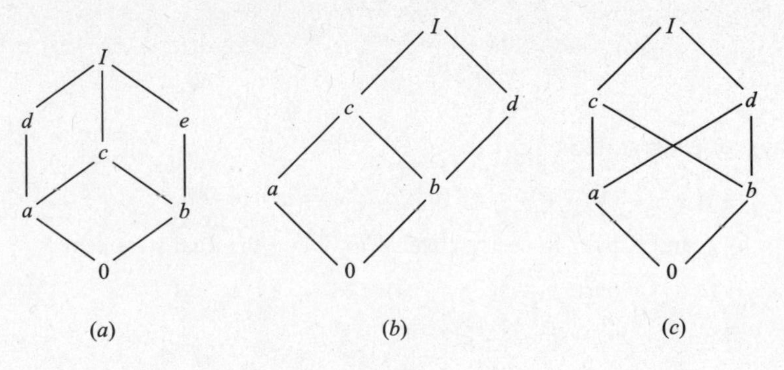


**Fig. 7-21**

**7.37.** Consider the lattice $L$ in Fig. 7-21(*a*).

(*a*) Which nonzero elements are join irreducible?

(*b*) Which elements are atoms?

(*c*) Which of the following are sublattices of $L$:

$$L_1 = \{0, a, b, I\} \qquad L_3 = \{a, c, d, I\}$$
$$L_2 = \{0, a, e, I\} \qquad L_4 = \{0, c, d, I\}$$

(*d*) Is $L$ distributive?

(*e*) Find complements, if they exist, for the elements $a$, $b$, and $c$.

(*f*) Is $L$ a complemented lattice?

(*a*) Those nonzero elements with a unique immediate predecessor are join irreducible. Hence $a, b, d$, and $e$ are join irreducible.

(*b*) Those elements which immediately succeed 0 are atoms, hence $a$ and $b$ are the atoms.

(*c*) A subset $L'$ is a sublattice if it is closed under $\wedge$ and $\vee$. $L_1$ is not a sublattice since $a \vee b = c$, which does not belong to $L_1$. The set $L_4$ is not a sublattice since $c \wedge d = a$ does not belong to $L_4$. The other two sets, $L_2$ and $L_3$, are sublattices.

(*d*) $L$ is not distributive since $M = \{0, a, d, e, I\}$ is a sublattice which is isomorphic to the nondistributive lattice in Fig. 7-7(*a*).

(*e*) We have $a \wedge e = 0$ and $a \vee e = I$, so $a$ and $e$ are complements. Also $b$ and $d$ are complements. However, $c$ has no complement.

(*f*) $L$ is not a complemented lattice since $c$ has no complement.

**7.38.** Consider the lattice $M$ in Fig. 7-21(*b*).

(*a*) Find the nonzero join-irreducible elements and atoms of $M$.

(*b*) Is $M$ distributive?

(*c*) Is $M$ complemented?

(*a*) The nonzero elements with a unique predecessor are $a$, $b$, and $d$, and of these three only $a$ and $b$ are atoms since their unique predecessor is 0.

(*b*) $M$ is distributive since $M$ does not have a sublattice which is isomorphic to one of the lattices in Fig. 7-7.

(*c*) $M$ is not complemented since $b$ has no complement. Note $a$ is the only solution to $b \wedge x = 0$ but $b \vee a = c \neq I$.

**7.39.** Prove Theorem 7.12: Let $L$ be a finite distributive lattice. Then every $a$ in $L$ can be written uniquely (except for order) as the join of irredundant join-irreducible elements.

Since $L$ is infinite we can write $a$ as the join of irredundant join-irreducible elements as discussed in Section 7.11. Thus we need only prove uniqueness. Suppose

$$a = b_1 \vee b_2 \vee \cdots \vee b_r = c_1 \vee c_2 \vee \cdots \vee c_s$$

where the $b$'s are irredundant and join irreducible and the $c$'s are irredundant and irreducible. For any given $i$ we have

$$b_i \leq (b_1 \vee b_2 \vee \cdots \vee b_r) = (c_1 \vee c_2 \vee \cdots \vee c_s)$$

Hence

$$b_i = b_i \wedge (c_1 \vee c_2 \vee \cdots \vee c_s) = (b_i \wedge c_1) \vee (b_1 \wedge c_2) \vee \cdots \vee (b_i \wedge c_s)$$

Since $b_i$ is join irreducible, there exists a $j$ such that $b_i = b_i \wedge c_j$, and so $b_i \leq c_j$. By a similar argument, for $c_j$ there exists a $b_k$ such that $c_j \leq b_k$. Therefore

$$b_i \leq c_j \leq b_k$$

which gives $b_i = c_j = b_k$ since the $b$'s are irredundant. Accordingly, the $b$'s and $c$'s may be paired off. Thus the representation for $a$ is unique except for order.

**7.40.** Prove Theorem 7.14: Let $L$ be a complemented lattice with unique complements. Then the join-irreducible elements of $L$, other than 0, are its atoms.

Suppose $a$ is join irreducible and is not an atom. Then $a$ has a unique immediate predecessor $b \neq 0$. Let $b'$ be the complement of $b$. Since $b \neq 0$ we have $b' \neq I$. If $a$ precedes $b'$, then $b \leq a \leq b'$, and so $b \wedge b' = b'$, which is impossible since $b \wedge b' = I$. Thus $a$ does not precede $b'$, and so $a \wedge b'$ must strictly precede $a$. Since $b$ is the unique immediate predecessor of $a$, we also have that $a \wedge b'$ precedes $b$ as in Fig. 7-22. But $a \wedge b'$ precedes $b'$. Hence

$$a \wedge b' \leq \inf(b, b') = b \wedge b' = 0$$

Thus $a \wedge b' = 0$. Since $a \vee b = a$, we also have that

$$a \vee b' = (a \vee b) \vee b' = a \vee (b \vee b') = a \vee I = I$$

Therefore $b'$ is a complement of $a$. Since complements are unique, $a = b$. This contradicts the assumption that $b$ is an immediate predecessor of $a$. Thus the only join-irreducible elements of $L$ are its atoms.

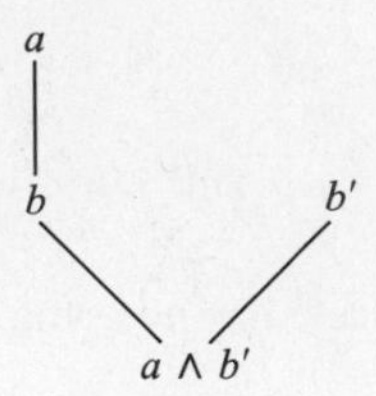


**Fig. 7-22**

# Supplementary Problems

## ORDERED SETS AND SUBSETS

**7.41.** Let $A = \{1, 2, 3, 4, 5, 6\}$ be ordered as in Fig. 7-23($a$).

($a$) Find all minimal and maximal elements of $A$.

($b$) Does $A$ have a first or last element?

($c$) Find all linearly ordered subsets of $A$, each of which contains at least three elements.

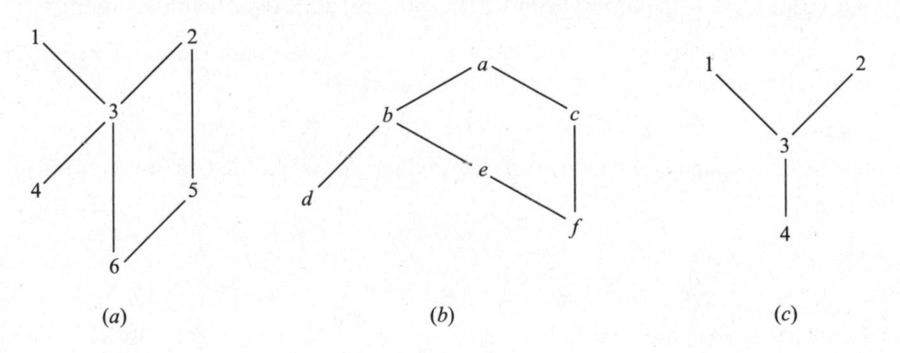


**Fig. 7-23**

**7.42.** Let $B = \{a, b, c, d, e, f\}$ be ordered as in Fig. 7-23($b$).

($a$) Find all minimal and maximal elements of $B$.

($b$) Does $B$ have a first or last element?

($c$) List two and find the number of consistent enumerations of $B$ into the set $\{1, 2, 3, 4, 5, 6\}$.

**7.43.** Let $C = \{1, 2, 3, 4\}$ be ordered as in Fig. 7-23($c$). Let $L(C)$ denote the collection of all nonempty chains in $C$ ordered by set inclusion. Draw a diagram of $L(C)$.

**7.44.** Draw the diagrams of the partitions of $m$ (see Example 7.4) where: ($a$) $m = 4$; ($b$) $m = 6$.

**7.45.** Let $\mathbf{D}_m$ denote the positive divisors of $m$ ordered by divisibility. Draw the Hasse diagrams of:

($a$) $\mathbf{D}_{12}$; ($b$) $\mathbf{D}_{15}$; ($c$) $\mathbf{D}_{16}$; ($d$) $\mathbf{D}_{17}$.

**7.46.** Let $S = \{a, b, c, d, e, f\}$ be an ordered set. Suppose, under the relation $<<$ (immediately precedes), there are exactly six pairs of elements as follows:

$$f << a, \quad f << d, \quad e << b, \quad c << f, \quad e << c, \quad b << f$$

(*a*) Find all minimal and maximal elements of $S$.

(*b*) Does $S$ have any first or last element?

(*c*) Find all pairs of elements, if any, which are noncomparable.

**7.47.** State whether each of the following is true or false and, if it is false, give a counterexample.

(*a*) If an ordered set $S$ has only one maximal element $a$, then $a$ is a last element.

(*b*) If a finite ordered set $S$ has only one maximal element $a$, then $a$ is a last element.

(*c*) If a linearly ordered set $S$ has only one maximal element $a$, then $a$ is a last element.

**7.48.** Let $S = \{a, b, c, d, e\}$ be ordered as in Fig. 7-24(*a*).

(*a*) Find all minimal and maximal elements of $S$.

(*b*) Does $S$ have any first or last element?

(*c*) Find all subsets of $S$ in which $c$ is a minimal element.

(*d*) Find all subsets of $S$ in which $c$ is a first element.

(*e*) List all linearly ordered subsets with three or more elements.

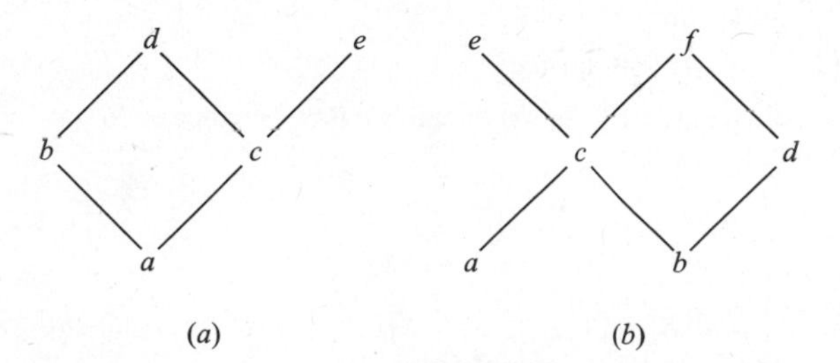


**Fig. 7-24**

**7.49.** Let $S = \{a, b, c, d, e, f\}$ be ordered as in Fig. 7-24(*b*)

(*a*) Find all minimal and maximal elements of $S$.

(*b*) Does $S$ have any first or last element?

(*c*) List all chains (linearly ordered subsets) with three or more elements.

**7.50.** Let $S = \{a, b, c, d, e, f, g\}$ be ordered as in Fig. 7-15. Find the number $n$ of chains in $S$ with:

(*a*) four elements; (*b*) five elements.

**7.51.** Let $S = \{1, 2, \ldots, 7, 8\}$ be ordered as in Fig. 7-16. Find the number $n$ of chains in $S$ with:

(*a*) five elements; (*b*) six elements.

**7.52.** Give an example of an ordered set with one minimal element but no first element.

## CONSISTENT ENUMERATIONS

**7.53.** Let $S = \{a, b, c, d, e\}$ be ordered as in Fig. 7.24(*a*). List all consistent enumerations of $S$ into $\{1, 2, 3, 4, 5\}$.

**7.54.** Let $S = \{a, b, c, d, e, f\}$ be ordered as in Fig. 7-24(*b*). Find the number $n$ of consistent enumerations of $S$ into $\{1, 2, 3, 4, 5, 6\}$.

**7.55.** Suppose the following are three consistent enumerations of an ordered set $A = \{a, b, c, d\}$:

$$[(a, 1), (b, 2), (c, 3), (d, 4)], \quad [(a, 1), (b, 3), (c, 2), (d, 4)], \quad [(a, 1), (b, 4), (c, 2), (d, 3)]$$

Assuming the Hasse diagram $D$ of $A$ is connected (any two points are connected by a path), draw $D$.

## SET CONSTRUCTIONS AND ORDER

**7.56.** Let $M = \{2, 3, 4, \ldots\}$ and let $M^2 = M \times M$ be ordered as follows:

$$(a, b) \leq (c, d) \quad \text{if } a|c \text{ and } b \leq d$$

Find all minimal and maximal elements of $M \times M$.

**7.57.** Consider the English alphabet $\mathbf{A} = \{\text{a}, \text{b}, \text{c}, \ldots, \text{y}, \text{z}\}$ with the usual (alphabetical) order. Recall that the Kleene closure $\mathbf{A}^*$ consists of all words in $\mathbf{A}$. Let $L$ consist of the following elements in $\mathbf{A}^*$:

gone, or, arm, go, an, about, gate, one, at, occur

(*a*) Sort $L$ according to the short-lex order, i.e., first by length and then alphabetically.

(*b*) Sort $L$ alphabetically.

**7.58.** Consider the ordered sets $A$ and $B$ appearing in Fig. 7-23(*a*) and (*b*), respectively. Suppose $S = A \times B$ is given the product order, i.e.,

$$(a, b) \leq (a', b') \quad \text{if } a \leq a' \text{ and } b \leq b'$$

Insert the correct symbol, $<$, $>$, or $\|$, between each pair of elements of $S$:

(*a*) $(4, b)$ ____ $(2, e)$ (*c*) $(5, d)$ ____ $(1, a)$

(*b*) $(3, a)$ ____ $(6, f)$ (*d*) $(6, e)$ ____ $(2, b)$

**7.59.** Suppose $\mathbf{P} = \{1, 2, 3, \ldots\}$ and $\mathbf{A} = \{\text{a}, \text{b}, \text{c}, \ldots, \text{y}, \text{z}\}$ are given the usual orders, and $S = \mathbf{P} \times \mathbf{A}$ is ordered lexicographically. Sort the following elements of $S$:

$$(2, z), (1, c), (2, c), (1, y), (4, b), (4, z), (3, b), (2, a)$$

**7.60.** Consider the set $\mathbf{P}$ of positive integers, the English alphabet $\mathbf{A}$, and the set $B$ of negative integers with the usual orders:

$$\mathbf{P} = \{1, 2, 3, \ldots\}, \qquad \mathbf{A} = \{\text{a}, \text{b}, \text{c}, \ldots, \text{y}, \text{z}\}, \qquad B = \{\ldots, -3, -2, -1\}$$

Suppose $S = \mathbf{P} \cup \mathbf{A} \cup B$, $T = \mathbf{P} \cup B \cup \mathbf{A}$, $U = B \cup \mathbf{A} \cup \mathbf{P}$, $V = B \cup \mathbf{P} \cup \mathbf{A}$ are each given the concatenation order. (Here the sets $\mathbf{P}, \mathbf{A}, B$ in $S$, $T$, $U$, $V$ are ordered as shown in the union.)

(*a*) Which of the sets $S$, $T$, $U$, $V$ has a minimal element?

(*b*) Which of the sets $S$, $T$, $U$, $V$ has a maximal element?

(*c*) Which element or elements in the sets $S$, $T$, $U$, $V$ have no immediate predecessor?

(*d*) Which element or elements in the sets $S$, $T$, $U$, $V$ have no immediate successor?

## UPPER AND LOWER BOUNDS, SUPREMUM AND INFIMUM

**7.61.** Let $S = \{a, b, c, d, e, f, g\}$ be ordered as in Fig. 7-15. Consider the subset $A = \{a, c, d\}$ of $S$.

(*a*) Find the set of upper bounds of $A$. (*c*) Does $\sup(A)$ exist?

(*b*) Find the set of lower bounds of $A$. (*d*) Does $\inf(A)$ exist?

**7.62.** Repeat Problem 7.61 for subset $B = \{b, c, e\}$ of $S$.

**7.63.** Let $S = \{1, 2, \ldots, 7, 8\}$ be ordered as in Fig. 7-16. Consider the subset $A = \{3, 6, 7\}$ of $S$.

(*a*) Find the set of upper bounds of $A$. (*c*) Does $\sup(A)$ exist?

(*b*) Find the set of lower bounds of $A$. (*d*) Does $\inf(A)$ exist?

**7.64.** Repeat Problem 7.63 for the subset $B = \{1, 2, 4, 7\}$ of $S$.

**7.65.** Consider the set $\mathbf{Q}$ of rational numbers with the usual order $\leq$. Let $A = \{x \in \mathbf{Q} : 5 < x^3 < 27\}$.

(*a*) Is $A$ bounded above or below? (*b*) Do $\sup(A)$ and $\inf(A)$ exist?

**7.66.** Consider the set $\mathbf{R}$ of real numbers with the usual order $\leq$. Let $B = \{x \in \mathbf{R} : x \in \mathbf{Q} \text{ and } 5 < x^3 < 27\}$.

(*a*) Is $B$ bounded above or below? (*b*) Do $\sup(B)$ and $\inf(B)$ exist?

## ORDER-ISOMORPHIC SETS, SIMILARITY MAPPINGS

**7.67.** Let $S$ be the ordered set in Fig. 7-24(*a*). Suppose $A = \{1, 2, 3, 4, 5\}$ is order-isomorphic to $S$ and the following is a similarity mapping from $S$ onto $A$:

$$f = \{(a, 1), (b, 4), (c, 5), (d, 2), (e, 3)\}$$

Draw the Hasse diagram of $A$.

**7.68.** Find the number of nonisomorphic ordered sets with three elements $a, b, c$, and draw their diagrams.

**7.69.** Find the number of connected nonisomorphic ordered sets with four elements $a, b, c, d$, and draw their diagrams.

**7.70.** Find the number of similarity mappings $f: S \to S$ if $S$ is the ordered set in:

(*a*) Fig. 7-23(*a*); (*b*) Fig. 7-23(*b*); (*c*) Fig. 7-23(*c*).

**7.71.** Suppose $\mathbf{P} = \{1, 2, 3, \ldots\}$ and $\mathbf{A} = \{\text{a}, \text{b}, \text{c}, \ldots, \text{z}\}$ are given the usual orders, and each of $S = \mathbf{P} \cup \mathbf{A}$ and $T = \mathbf{A} \cup \mathbf{P}$ is given the concatenation order. Which of the sets $\mathbf{P}$, $\mathbf{A}$, $S$, $T$ are order-isomorphic?

**7.72.** Which of the sets $S$, $T$, $U$, $V$ in Problem 7.60 are order-isomorphic?

**7.73.** Determine whether or not $\zeta = \zeta^*$ where $\zeta$ is the order type of each of the following sets (with the usual order):

(*a*) $\mathbf{R}$; (*b*) $A = \{\ldots, -3, -2, -1\}$; (*c*) $B = \{\ldots, -4, -2, 0, 2, 4, \ldots\}$.

**7.74.** Determine which of the sets in Problem 7.73 have the same order type as: (*a*) $\mathbf{P}$; (*b*) $\mathbf{Z}$, (*c*) $\mathbf{Q}$.

**7.75.** Let $C$ be the ordered set in Fig. 7-23(*c*). (*a*) Draw the Hasse diagram of the collection $p(C)$ of predecessor sets ordered by set inclusion. (*b*) Is $C$ order-isomorphic to $p(C)$?

**LATTICES**

**7.76.** Consider the lattice $L$ in Fig. 7-25($a$). ($a$) Find all sublattices with five elements. ($b$) Find all join-irreducible elements, and atoms. ($c$) Find complements of $a$ and $b$, if they exist. ($d$) Is $L$ distributive? Complemented?

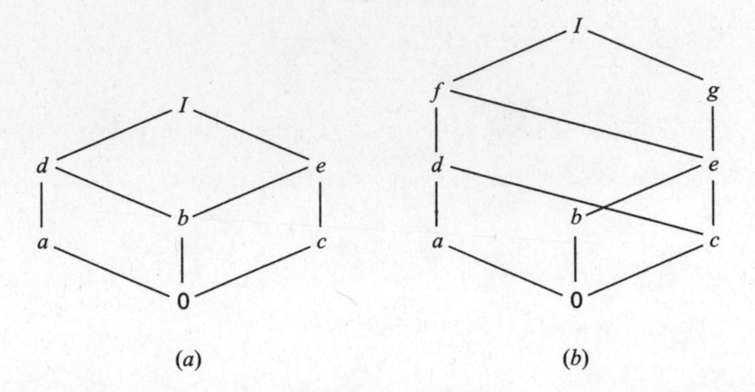


**Fig. 7-25**

**7.77.** Consider the lattice $M$ in Fig. 7-25($b$). ($a$) Find join-irreducible elements. ($b$) Find the atoms. ($c$) Find complements of $a$ and $b$, if they exist. ($d$) Express each $x$ in $M$ as the join of irredundant join-irreducible elements. ($e$) Is $M$ distributive? Complemented?

**7.78.** Consider the bounded lattice $L$ in Fig. 7-26($a$).

($a$) Find the complements, if they exist, of $e$ and $f$.

($b$) Express $I$ in an irredundant join-irreducible decomposition in as many ways as possible.

($c$) Is $L$ distributive?

($d$) Describe the isomorphisms of $L$ with itself.

**7.79.** Consider the bounded lattice $L$ in Fig. 7-26($b$).

($a$) Find the complements, if they exist, of $a$ and $c$.

($b$) Express $I$ in an irredundant join-irreducible decomposition in as many ways as possible.

($c$) Is $L$ distributive?

($d$) Describe the isomorphisms of $L$ with itself.

**7.80.** Redo Problem 7.79 for the bounded lattice $L$ in Fig. 7-26($c$).

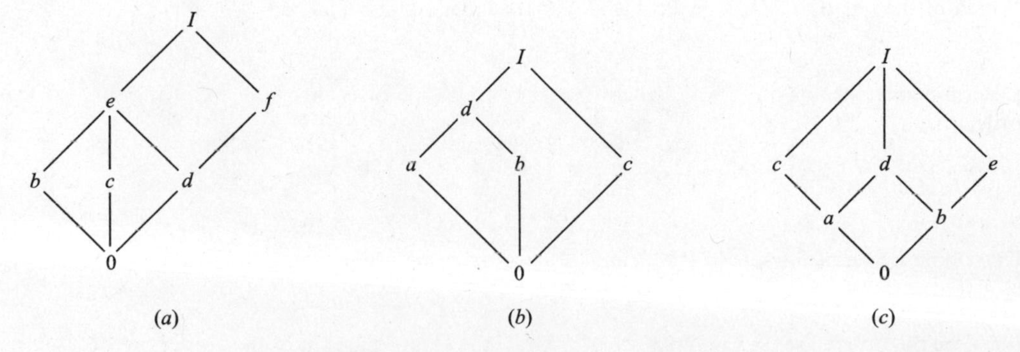


**Fig. 7-26**

**7.81.** Let $\mathbf{D}_{60} = \{1, 2, 3, 4, 5, 6, 10, 12, 15, 20, 30, 60\}$, the divisors of 60, be ordered by divisibility.

(*a*) Draw the Hasse diagram of $\mathbf{D}_{60}$.

(*b*) Which elements are join-irreducible? Atoms?

(*c*) Find the complements of 2 and 10, if they exist.

(*d*) Express each number x as the join of a minimum number of irredundant join-irreducible elements.

**7.82.** Consider the lattice **P** of positive integers ordered by divisibility. (*a*) Which elements are join-irreducible? (*b*) Which elements are atoms?

**7.83.** Show that the following "weak" distributive laws hold for any lattice:

(*a*) $a \vee (b \wedge c) \leq (a \vee b) \wedge (a \wedge c)$

(*b*) $a \wedge (b \vee c) \leq (a \wedge b) \vee (a \wedge c)$

**7.84.** Let $S = \{1, 2, 3, 4\}$. Three partitions of $S$ follow:

$$P_1 = [12, 3, 4], \qquad P_2 = [12, 34], \qquad P_3 = [13, 2, 4]$$

(Here [12, 3, 4] is short for $[\{1, 2\}, \{3\}, \{4, \}]$.)

(*a*) Find the other nine partitions of $S$.

(*b*) Let $L$ be the collection of the twelve partitions of $S$ ordered by *refinement*, that is, $P_i \leq P_j$ if each cell of $P_i$ is a subset of a cell of $P_j$. For example, $P_1 \leq P_2$, but $P_2$ and $P_3$ are noncomparable. Show that $L$ is a bounded lattice and draw its Hasse diagram.

**7.85.** An element $a$ in a lattice $L$ is said to be *meet-irreducible* if $a = x \wedge y$ implies $a = x$ or $a = y$. Find all meet-irreducible elements in: (*a*) Fig. 7-25(*a*); (*b*) Fig. 7-25(*b*); (*c*) $\mathbf{D}_{60}$ (see Problem 7.81).

**7.86.** A lattice $M$ is said to be *modular* if whenever $a \leq c$ we have the law

$$a \vee (b \wedge c) = (a \vee b) \wedge c$$

(*a*) Prove that every distributive lattice is modular.

(*b*) Verify that the nondistributive lattice in Fig. 7-7(*b*) is modular; hence the converse of (*a*) is not true.

(*c*) Show that the nondistributive lattice in Fig. 7-7(*a*) is nonmodular. [In fact, one can prove that every nonmodular lattice contains a sublattice isomorphic to Fig. 7-7(*a*).]

## Answers to Supplementary Problems

**7.41.** (*a*) Minimal: 4 and 6; Maximal: 1 and 2. (*b*) First: none; Last: none. (*c*) $\{1, 3, 4\}$, $\{1, 3, 6\}$, $\{2, 3, 4\}$, $\{2, 3, 6\}$, $\{2, 5, 6\}$.

**7.42.** (*a*) Minimal: *d* and *f*; Maximal: *a*. (*b*) First: none; Last: *a*. (*c*) There are eleven: *dfebca*, *dfecba*, *dfceba*, *fdebca*, *fdecba*, *fdceba*, *fedbca*, *fedcba*, *fcdeba*, *fecdba*, *fcedba*.

**7.43.** See Fig. 7-27.

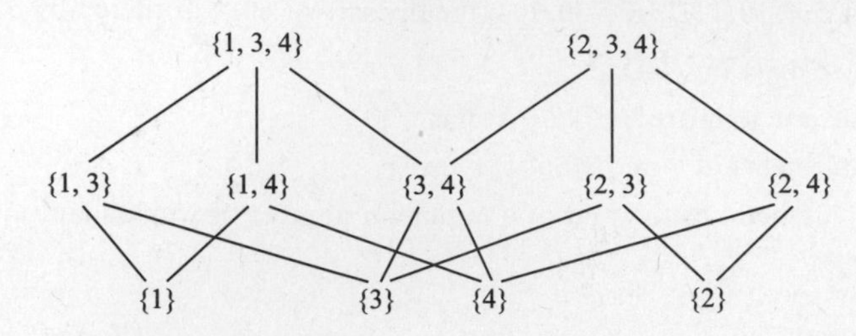


**Fig. 7-27**

**7.44.** See Fig. 7-28.

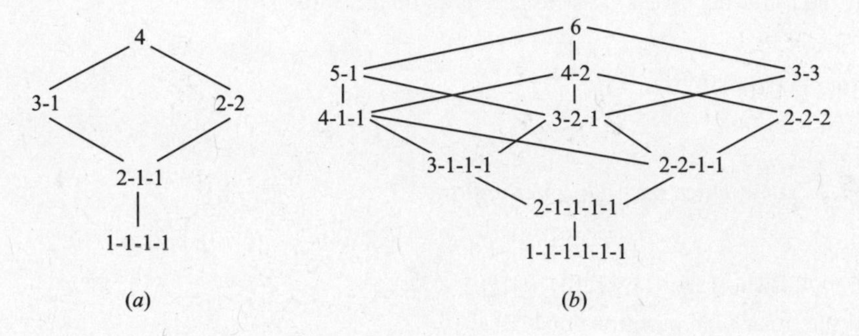


**Fig. 7-28**

**7.45.** See Fig. 7-29.

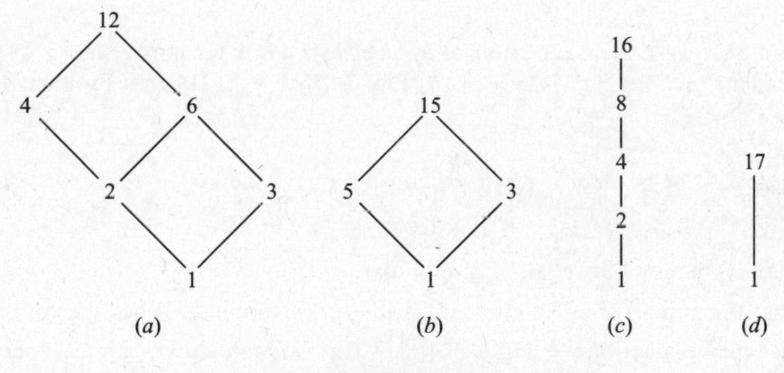


**Fig. 7-29**

**7.46.** Hint: Draw diagram of $S$. (*a*) Minimal: $e$; Maximal: $a$, $d$. (*b*) First: $e$; Last: none. (*c*) $\{a, d\}$, $\{b, c\}$.

**7.47.** (*a*) False. Example: $\mathbf{P} \cup \{a\}$ where $1 << a$, and $\mathbf{P}$ ordered by $\leq$. (*b*) True. (*c*) True.

**7.48.** (*a*) Minimal: $a$; Maximal: $d$ and $e$. (*b*) First: $a$; Last: none. (*c*) Any subset which contains $c$ and omits $a$; that is, $c$, $cb$, $cd$, $ce$, $cbd$, $cbe$, $cde$, $cbde$. (*d*) $c$, $cd$, $ce$, $cde$. (*e*) $abd, acd, ace$.

**7.49.** (*a*) Minimal: $a$ and $b$; Maximal: $e$ and $f$. (*b*) First: none; Last: none. (*c*) $ace, acf, bce, bcf, bdf$.

**7.50.** (*a*) Four; (*b*) none

**7.51.** (*a*) Six; (*b*) none

**7.52.** $S = \{a\} \cup A$ where $A = \{\ldots, -3, -2, -1, 0\}$ has the usual order and where $a << 0$.

**7.53.** $abcde, abced, acbde, acbed, acebd$

**7.54.** Eleven

**7.55.** $a << b, \quad a << c, \quad c << d$

**7.56.** Minimal: $(p, 2)$ where $p$ is a prime. Maximal: none.

**7.57.** (*a*) an, at, go, or, arm, one, gate, gone, about, occur
(*b*) an, about, arm, at, gate, go, gone, occur, one, or

**7.58.** (*a*) $\|$; (*b*) $>$; (*c*) $\|$; (*d*) $<$

**7.59.** $1c, 1y, 2a, 2c, 2z, 3b, 4b, 4z$

**7.60.** (*a*) $S$ and $T$; (*b*) $T$ and $V$; (*c*) $1, a \in S$, $1 \in T$, $a \in V$; (*d*) $-1, z \in S$, $z \in T$, $z \in V$

**7.61.** (*a*) $e, f, g$; (*b*) $a$; (*c*) $\sup(A) = e$; (*d*) $\inf(A) = a$

**7.62.** (*a*) $e, f, g$; (*b*) none; (*c*) $\sup(B) = e$; (*d*) none

**7.63.** (*a*) 1, 2, 3; (*b*) 8; (*c*) $\sup(A) = 3$; (*d*) $\inf(A) = 8$

**7.64.** (*a*) None; (*b*) 8; (*c*) none; (*d*) $\inf(B) = 8$

**7.65.** (*a*) Both; (*b*) $\sup(A) = 3$, $\inf(A)$ does not exist.

**7.66.** (*a*) Both; (*b*) $\sup(A) = 3$, $\inf(A) = \sqrt[3]{5}$.

**7.67.** See Fig. 7-30.

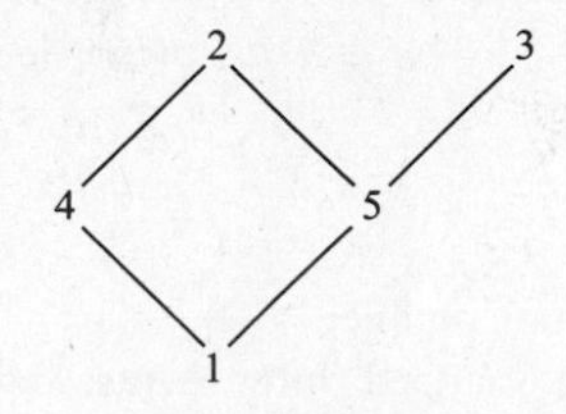


**Fig. 7-30**

**7.68.** Four: (1) $a, b, c$; (2) $a, b << c$; (3) $a << b$, $a << c$; (4) $a << b << c$.

**7.69.** Four: See Fig. 7-31.

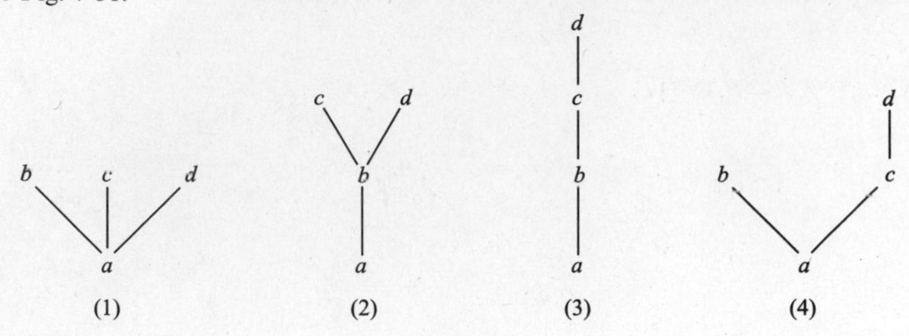


**Fig. 7-31**

**7.70.** (*a*) One: identity mapping; (*b*) one; (*c*) two

**7.71.** **P** and $T$

**7.72.** None

**7.73.** (*a*) Yes; (*b*) no; (*c*) yes

**7.74.** (*a*) None; (*b*) $B$; (*c*) none

**7.75.** (*a*) See Fig. 7-32; (*b*) yes (always)

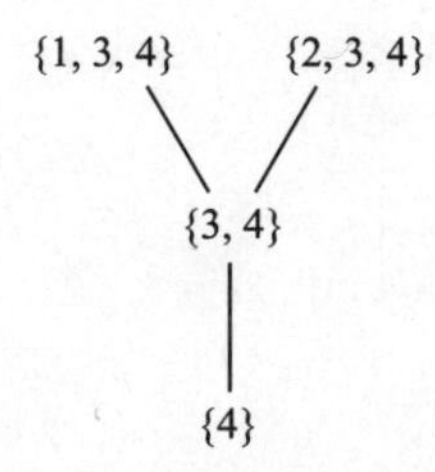


**Fig. 7-32**

**7.76.** (*a*) Six: $0abdI, 0acdI, 0adeI, 0bceI, 0aceI, 0cdeI$
(*b*) (i) $a, b, e, 0$; (iii) $a, b, c$
(*c*) $c$ and $e$ are complements of $a$; $b$ has no complement.
(*d*) No; no

**7.77.** (*a*) $a, b, c, g, 0$; (*b*) $a, b, c$; (*c*) $g$ is the complement of $a$; $b$ has no complement.
(*d*) $I = a \vee g, f = a \vee b = a \vee c, e = b \vee c, d = a \vee c$; other elements are join irreducible. (*e*) No; no

**7.78.** (*a*) $e$ has none; $f$ has $b$ and $c$.
(*b*) $I = c \vee d \vee f = b \vee c \vee f = b \vee d \vee f$
(*c*) No, since decompositions are not unique.
(*d*) Two: 0, $d$, $e$, $f$, $I$ must be mapped into themselves. Then $F = 1_L$, identity map on $L$, or $F = \{(b, c), (c, b)\}$.

**7.79.** (*a*) $a$ has $c$; $c$ has $a$ and $b$. (*b*) $I = a \vee c = b \vee c$.
(*c*) No. (*d*) Two: 0, $c$, $d$, $I$ must be mapped into themselves. Then $f = 1_L$ or $f = \{(a, b), (b, a)\}$.

**7.80.** (*a*) $a$ has $e$, $c$ has $b$ and $e$. (*b*) $I = a \vee e = b \vee c = c \vee e$. (*c*) No.
(*d*) Two: 0, $d$, $I$ are mapped into themselves. Then $f = 1_L$ or $f = \{(a, b), (b, a), (c, d), (d, c)\}$.

**7.81.** (*a*) See Fig. 7-33. (*b*) 1, 2, 3, 4, 5; the atoms are 2, 3, and 5. (*c*) 2 has none, 10 has 3.

(*d*) $60 = 4 \vee 3 \vee 5$ $\quad 30 = 2 \vee 3 \vee 5$ $\quad 20 = 4 \vee 5$

$15 = 3 \vee 5$ $\quad 12 = 3 \vee 4$ $\quad 10 = 2 \vee 5$ $\quad 6 = 2 \vee 3$

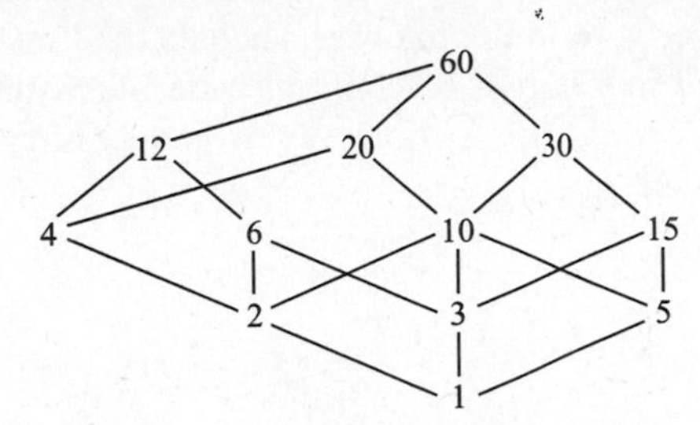


Fig. 7-33

**7.82.** (*a*) Powers of primes and 1; (*b*) primes

**7.84.** (*a*) [1, 2, 3, 4], [14, 2, 3], [13, 24], [14, 23], [123, 4], [124, 3], [134, 2], [234, 1], [1234]
(*b*) See Fig. 7-34.

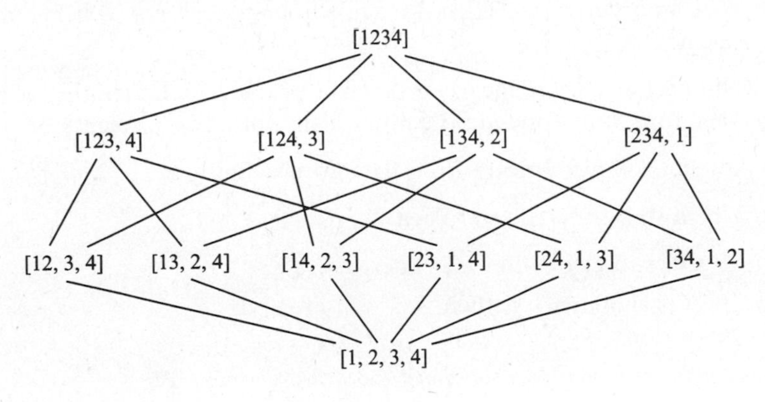


Fig. 7-34

**7.85.** Geometrically, an element $a \neq I$ is meet-irreducible if and only if $a$ has only one immediate successor:
(*a*) $a, c, d, e, I$; (*b*) $a, b, d, f, g, I$; (*c*) 4, 6, 10, 12, 15, 60.

**7.86.** (*a*) If $a \leq c$ then $a \vee c = c$. Hence

$$a \vee (b \wedge c) = (a \vee b) \wedge (a \vee c) = (a \vee b) \wedge c$$

(*c*) Here $a \leq c$. But $a \vee (b \wedge c) = a \vee 0 = a$ and $(a \vee b) \wedge c = I \wedge c = c$; hence

$$a \vee (b \wedge c) \neq (a \vee b) \wedge c$$

# Chapter 8

# Ordinal Numbers

## 8.1 INTRODUCTION

Numbers are usually used for two different things. One is to measure quantity, such as the number of students in a class, and the other is to indicate order, such as the first student, the second student, and so on. Cardinal numbers, covered in Chapter 6, essentially measure quantity, whereas ordinal numbers, covered in this chapter, indicate order. First, however, it is necessary to discuss a special kind of an ordered set, called a well-ordered set.

## 8.2 WELL-ORDERED SETS

Not every ordered set, even if it is linearly ordered, need have a first element. For example, $\mathbf{Z}$ is linearly ordered but it does not have a first element. On the other hand, one of the fundamental properties of the set

$$\mathbf{P} = \{1, 2, 3, \ldots\}$$

of counting numbers (positive integers) is that $\mathbf{P}$ and every subset of $\mathbf{P}$ has a first element. Such an ordered set is said to be well-ordered. Namely:

**Definition 8.1**: Let $A$ be an ordered set. Then $A$ is said to be *well-ordered* if every subset of $A$ contains a first element.

Note that any well-ordered set $A$ is linearly ordered. For if $a, b \in A$, then the subset $\{a, b\}$ of $A$ has a first element which, therefore, must precede the other; hence any two elements of $A$ are comparable.

The following theorem follows directly from the above definition.

**Theorem 8.1**: Let $A$ be a well-ordered set. Then:

(i) Every subset of $A$ is well-ordered.

(ii) If $B$ is similar to $A$, then $B$ is well-ordered.

**EXAMPLE 8.1** Consider the following two subsets of the well-ordered set $\mathbf{P}$:

$$A_1 = \{1, 3, 5, \ldots\} \qquad \text{and} \qquad A_2 = \{2, 4, 6, \ldots\}$$

Then $A_1$ and $A_2$ are also well-ordered. Suppose the union

$$S = A_1 \cup A_2 = \{1, 3, 5, \ldots; 2, 4, 6, \ldots\}$$

is ordered from left to right, as shown. Then $S$ is also well-ordered. This shows that a set, such as $\mathbf{P} = A_1 \cup A_2$, can be well-ordered in more than one way.

Suppose $\{A_i : i \in I\}$ is a linearly ordered collection of disjoint linearly ordered sets, that is, $I$ is linearly ordered and each $A_i$ is linearly ordered. Then the union $S = \bigcup_i A_i$ will be linearly ordered as follows:

$$a < b \qquad \text{if} \begin{cases} a \in A_i,\ b \in A_j,\ i < j \\ a, b \in A_i,\ a < b \text{ in } A_i \end{cases}$$

This ordering will be called the *usual ordering* on the union $S$. (It is also called the concatenation or sum ordering on $S$.) The ordering is somewhat analogous to a lexicographical ordering in the sense that the index ordering has the first priority. This ordering is sometimes pictured by listing the elements of $A_i$ before the elements of $A_j$ when $i < j$. Example 8.1 is an instance of such an ordering and its picture.

The following theorem applies.

**Theorem 8.2**: Suppose $\{A_i : i \in I\}$ is a well-ordered family of disjoint well-ordered sets, that is, $I$ is well-ordered and each $A_i$ is well-ordered. Then the union $S = \bigcup_i A_i$, with the usual ordering, is well-ordered.

**EXAMPLE 8.2** Let $V = \{a_1, a_2, \ldots, a_n\}$ be any finite linearly ordered set. Then $V$ may be written in the form

$$V = \{a_{i_1}, a_{i_2}, \ldots, a_{i_n}\}$$

where the elements are ordered as shown. Notice that $V$ is well-ordered. Furthermore, notice that any other linearly ordered set $W$ with $n$ elements, say

$$W = \{b_{i_1}, b_{i_2}, \ldots, b_{i_n}\}$$

is similar to $V$.

We formally state the comment in Example 8.3.

**Theorem 8.3**: All finite linearly ordered sets with the same number $n$ of elements are well-ordered and are similar to each other.

## 8.3 TRANSFINITE INDUCTION

The reader is familiar with the principle of mathematical induction. Namely:

**Principle of Mathematical Induction**: Let $S$ be a subset of the set $\mathbf{P}$ of counting numbers with the following two properties:

(1) $1 \in S$.
(2) If $n \in S$, then $n + 1 \in S$.

Then $S$ is the set of all counting numbers, that is, $S = \mathbf{P}$.

The above principle is one of Peano's axioms for the counting numbers $\mathbf{P}$. The principle can be shown to be a consequence of the fact that $\mathbf{P}$ is well-ordered. In fact, there is a somewhat similar statement which is true for any well-ordered set (proved in Problem 8.1).

**Theorem 8.4 (Principle of Transfinite Induction)**: Let $S$ be a subset of a well-ordered set $A$ with the following two properties:

(1) $a_0 \in S$.
(2) If $s(a) \subseteq S$, then $a \in S$.

Then $S$ is the entire set $A$, that is, $S = A$.

Here $a_0$ is the first element of $A$ and $s(a)$, called the *initial segment* of $a$, is defined to be the set of all elements in $A$ which strictly precedes $a$.

Initial segments will be discussed below, and Chapter 9 will discuss transfinite induction in much more detail.

## 8.4 LIMIT ELEMENTS

Let $A$ be an ordered set, and let $a, b$ belong to $A$. Recall that $a$ is called an *immediate predecessor* of $b$, and that $b$ is called an *immediate successor* of $a$, written

$$a << b$$

if $a < b$ but no element in $A$ lies between $a$ and $b$, that is, there does not exist an element $c$ in $A$ such that $a < c < b$.

**EXAMPLE 8.3**

(*a*) Let $A = \{a, b, c, d, e\}$ be ordered as in Fig. 8-1. Then $e$ is an immediate predecessor of $b$ and $c$, and $b$ is an immediate successor of $d$ and $e$.

(*b*) Consider the set **Q** of rational numbers with the usual order. Even though **Q** is linearly ordered, no element in **Q** has an immediate predecessor or an immediate successor. For if $a, b \in \mathbf{Q}$, say $a < b$, then $(a+b)/2$ belongs to **Q** and

$$a < (a+b)/2 < b$$

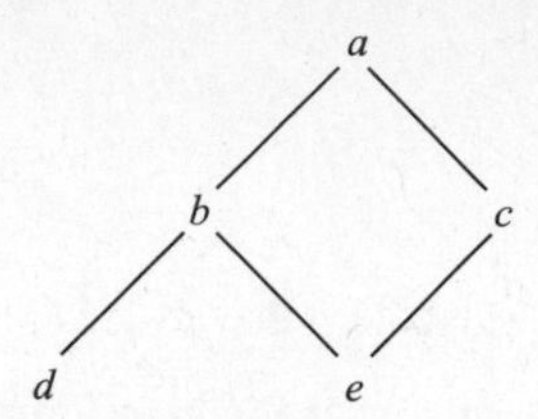


**Fig. 8-1**

Example 8.3 shows that linearly ordered sets need not have any immediate predecessors or any immediate successors. This is not true in the case of well-ordered sets. That is:

**Theorem 8.5**: Every element in a well-ordered set $A$ has a unique immediate successor except the last element.

*Proof*: Let $a \in A$, and let $M(a)$ denote the set of elements of $A$ which strictly succeeds $a$. If $a$ is not the last element, then $M(a) \neq 0$. Since $A$ is well-ordered, $M(a)$ has a first element, say $b$. We claim $b$ is an immediate successor of $a$. Otherwise, there is an element $c \in A$ such that $a < c < b$. Then $c \in M(a)$ and this contradicts the fact that $b$ is the first element of $M(a)$. We claim $b$ is the only immediate successor of $a$. Otherwise, there is another immediate successor of $a$, say $d$. Then $d \in M(a)$ and, since $b$ is a first element of $M(a)$, we have $a < b < d$. This contradicts the assumption that $d$ is an immediate successor of $a$. Thus the first element $b$ of $M(a)$ is the unique immediate successor of $a$.

There is no analogous statement to Theorem 8.4 about immediate predecessors, that is, there do exist elements in well-ordered sets, besides the first element, which do not have immediate predecessors. For example, the set

$$S = A_1 \cup A_2 = \{1, 3, 5, \ldots; 2, 4, 6, \ldots\}$$

in Example 8.1 is well-ordered, and both 1 and 2 do not have immediate predecessors.

In view of the above comment and example, we introduce the following definition.

**Definition 8.2**: An element $a$ in a well-ordered set $A$ is called a *limit element* if it does not have an immediate predecessor and if it is not the first element.

According to this definition, the element 2 in the above set $S = A_1 \cup A_2$ is a limit element.

## 8.5 INITIAL SEGMENTS

Let $A$ be a well-ordered set. The *initial segment* $s(a)$ of an element $a \in A$ consists of all elements in $A$ which strictly precede $a$. In other words,

$$s(a) = \{x : x \in A, x < a\}$$

Notice that $s(a)$ is a subset of $A$.

**EXAMPLE 8.4** Consider again the well-ordered $S$ in Example 8.1, that is,

$$S = A_1 \cup A_2 = \{1, 3, 5, \ldots; 2, 4, 6, \ldots\}$$

Then $s(1) = \varnothing$, $s(5) = \{1, 3\}$, $s(2) = \{1, 3, 5, \ldots\}$, and $s(8) = \{1, 3, 5, \ldots; 2, 4, 6\}$.

One basic property of initial segments is contained in the next theorem (proved in Problem 8.2).

**Theorem 8.6**: Let $S(A)$ denote the collection of all initial segments of elements in a well-ordered set $A$, and let $S(A)$ be ordered by set inclusion. Then $A$ is similar (order-isomorphic) to $S(A)$ and, in particular, the function $f\colon A \to S(A)$ defined by $f(x) = s(x)$ is a similarity mapping between $A$ and $S(A)$.

## 8.6 SIMILARITY BETWEEN A WELL-ORDERED SET AND ITS SUBSETS

Consider the set $\mathbf{P}$ of counting numbers, and the subset $E = \{2, 4, 6, \ldots\}$ of $\mathbf{P}$. The function $f\colon \mathbf{P} \to E$ defined by $f(x) = 2x$ is a similarity mapping of $\mathbf{P}$ onto its subset $E$. Notice that, for every $x \in \mathbf{P}$,

$$x \leq f(x)$$

This property, which is true in general, is the content of the next theorem (proved in Problem 8.3). Namely:

**Theorem 8.7**: Let $A$ be a well-ordered set, let $B$ be a subset of $A$, and let the function $f\colon A \to B$ be a similarity mapping of $A$ onto $B$. Then, for every $a \in A$,

$$a \leq f(a)$$

The following important properties of well-ordered sets (proved in Problems 8.4 and 8.5) are consequences of the preceding theorem.

**Theorem 8.8**: Let $A$ and $B$ be similar well-ordered sets. Then there exists only one similarity mapping of $A$ onto $B$.

**Theorem 8.9**: A well-ordered set cannot be similar to one of its initial segments.

## 8.7 COMPARISON OF WELL-ORDERED SETS

The next theorem (proved in Problem 8.12) gives an important relationship between any two well-ordered sets.

**Theorem 8.10**: Let $A$ and $B$ be well-ordered sets. Then $A$ and $B$ are similar, or one of them is similar to an initial segment of the other.

Suppose $A$ and $B$ are well-ordered sets, and suppose $A$ is similar to an initial segment of $B$. Then $A$ is said to be *shorter* than $B$, and $B$ is said to be *longer* than $A$. With these definitions, Theorem 8.10 can be restated as follows:

**Theorem 8.10′**: Let $A$ and $B$ be well-ordered sets. Then $A$ is shorter than $B$, $A$ is similar to $B$, or $A$ is longer than $B$.

The preceding theorem can be strengthened as follows:

**Theorem 8.11**: Let $\mathscr{A}$ be a collection of pairwise nonsimilar well-ordered sets. Then there exists a set $A$ in $\mathscr{A}$ such that $A$ is shorter than every other set in $\mathscr{A}$.

**EXAMPLE 8.5**

(*a*) Consider two finite well-ordered sets

$$A = \{a_1, a_2, \ldots, a_m\} \quad \text{and} \quad B = \{b_1, b_2, \ldots, b_n\}$$

Suppose $m < n$. Then $A$ is similar to the initial segment $\{b_1, b_2, \ldots, b_m\}$ of $B$, and hence $A$ would be shorter than $B$. Similarly, if $m > n$ then $A$ would be longer than $B$.

(*b*) The set $\mathbf{P} = \{1, 2, 3, \ldots\}$ is shorter than the well-ordered set

$$S = \{1, 3, 5, \ldots; 2, 4, 6, \ldots\}$$

since $\mathbf{P}$ is similar to the initial segment $\{1, 3, 5, \ldots\}$ of $S$.

## 8.8 ORDINAL NUMBERS

Consider a collection $\mathscr{S}$ of well-ordered sets. Each well-ordered set $A$ in $\mathscr{S}$ is assigned a symbol in such a way that any two well-ordered sets $A$ and $B$ are assigned the same symbol if and only if $A$ and $B$ are similar (order-isomorphic). This symbol is called the *ordinal number* of $A$. We will write

$$\lambda = \operatorname{ord}(A)$$

to indicate that $\lambda$ is the ordinal number of $A$.

Recall (Theorem 7.5) that the relation of similarity (order-isomorphism), denoted by

$$A \simeq B$$

is an equivalence relation in any collection of ordered sets. Thus by the fundamental theorem on equivalence relations, all ordered sets, and in particular all well-ordered sets, are partitioned into disjoint classes of similar sets. One may view an ordinal number as the equivalence class of all similar well-ordered sets.

Recall (Section 7.9) that every linearly ordered set $S$ is assigned an order type. Thus an ordinal number may also be viewed as the order type of a well-ordered set.

**Definition 8.3**: The ordinal number of each of the well-ordered sets

$$\varnothing, \{1\}, \{1, 2\}, \{1, 2, 3\}, \ldots$$

is denoted by $0, 1, 2, 3, \ldots$ respectively, and is called a *finite* ordinal number. All other ordinals are called *transfinite* numbers.

**Definition 8.4**: The ordinal number of the set $\mathbf{P}$ of counting numbers is denoted by

$$\omega = \operatorname{ord}(\mathbf{P})$$

Although the symbols $0, 1, 2, 3, \ldots$ are used to denote natural numbers (nonnegative integers), cardinal numbers and, now, ordinal numbers, the context in which the symbols appear determines their particular meaning. Furthermore, since any two finite well-ordered sets with the same number of elements are similar, $0, 1, 2, 3, \ldots$ are the only finite ordinal numbers.

## 8.9 INEQUALITIES AND ORDINAL NUMBERS

An inequality relation is defined for the ordinal numbers as follows:

**Definition 8.5**: Let $\lambda$ and $\mu$ be ordinal numbers and let $A$ and $B$ be two well-ordered sets such that $\lambda = \operatorname{ord}(A)$ and $\mu = \operatorname{ord}(B)$. Then

$$\lambda < \mu$$

if $A$ is similar to an initial segment of $B$.

Accordingly, for $\lambda = \text{ord}(A)$ and $\mu = \text{ord}(B)$, we have the following:

$$\begin{array}{ll} \lambda < \mu & \text{if } A \text{ is shorter than } B, \\ \lambda = \mu & \text{if } A \text{ is similar to } B, \\ \lambda > \mu & \text{if } A \text{ is longer than } B, \\ \lambda \leq \mu & \text{if } \lambda < \mu \text{ or } \lambda = \mu, \\ \lambda \geq \mu & \text{if } \lambda > \mu \text{ or } \lambda = \mu. \end{array}$$

**EXAMPLE 8.6**

(*a*) Consider two finite well-ordered sets $A$ and $B$, say

$$A = \{a_1, a_2, \ldots, a_m\} \quad \text{and} \quad B = \{b_1, b_2, \ldots, b_n\}$$

Say $m < n$. Then $A$ is similar to the initial segment $\{b_1, b_2, \ldots, b_m\}$ of $B$. Hence $\text{ord}(A) < \text{ord}(B)$.

In other words, $m < n$ as ordinal numbers if and only if $m < n$ as nonnegative integers. Thus the inequality relation for ordinal numbers is an extension of the inequality relation in the set $\mathbf{N}$ of natural numbers.

(*b*) Let $\lambda = \text{ord}(S) = \text{ord}(\{1, 3, 5, \ldots; 2, 4, 6, \ldots\})$. Since the set $\mathbf{P} = \{1, 2, 3, \ldots\}$ is similar to the initial segment $\{1, 3, 5, \ldots\}$ of $S$, we have

$$\omega < \lambda$$

### Properties of the Inequality Relation on Ordinal Numbers

Theorem 8.10 tells us that any two well-ordered sets $A$ and $B$ are similar or one of them is similar to an initial segment of the other. Accordingly, the next theorem is a direct consequence of Theorem 8.10 and the above definition.

**Theorem 8.12**: Any set of ordinal numbers is linearly ordered by the relation $\lambda \leq \mu$

In view of Theorem 8.10, the preceding theorem can be strengthened as follows:

**Theorem 8.13**: Any set of ordinal numbers is well-ordered by the relation $\lambda \leq \mu$.

Now let $\lambda$ be any ordinal number and let $s(\lambda)$ denote the set of ordinal numbers less than $\lambda$. By the preceding theorem, $s(\lambda)$ is a well-ordered set and, therefore, $\text{ord}(s(\lambda))$ exists.

| Question: What is the relationship between $\lambda$ and $\text{ord}(s(\lambda))$? |
|---|

The answer is given in the next theorem (proved in Problem 8.16).

**Theorem 8.14**: Let $s(\lambda)$ be the set of ordinals less than the ordinal $\lambda$. Then $\lambda = \text{ord}(s(\lambda))$.

Since the ordinal numbers are themselves well-ordered, every ordinal has an immediate successor. Some nonzero ordinals, for example $\omega$, do not have immediate predecessors; these are called *limit ordinal numbers* or, simply, *limit numbers*.

## 8.10 ORDINAL ADDITION

An operation of *addition* is defined for ordinal numbers as follows:

**Definition 8.6**: Let $\lambda$ and $\mu$ be ordinal numbers, and let $A$ and $B$ be disjoint sets such that $\lambda = \text{ord}(A)$ and $\mu = \text{ord}(B)$. Then

$$\lambda + \mu = \text{ord}(\{A; B\})$$

Recall that $\{A; B\}$ is given the usual order where every element of $A$ precedes every element of $B$.

**EXAMPLE 8.7** Recall $\omega = \text{ord}(P) = \text{ord}(\{1, 2, \ldots\})$ and $n = \text{ord}(\{a_1, a_2, \ldots, a_n\})$. Then

$$n + \omega = \text{ord}(\{a_1, a_2, \ldots, a_n; 1, 2, \ldots\}) = \omega$$

But

$$\omega + n = \text{ord}(\{1, 2, \ldots; a_1, a_2, \ldots, a_n\}) > \omega$$

since **P** is similar to $s(a_1)$, the initial segment of $a_1$.

Example 8.7 tells us that the operation of addition of ordinal numbers is not commutative. However, the following conditions do hold.

**Theorem 8.15**: (1) Addition of ordinal numbers satisfies the associative law, i.e.,

$$(\lambda + \mu) + \eta = \lambda + (\mu + \eta)$$

(2) The ordinal 0 is an additive identity element, i.e.,

$$0 + \lambda = \lambda + 0 = \lambda$$

**EXAMPLE 8.8** (Addition of Finite Ordinals) Here we will denote the finite ordinals by

$$0^*, 1^*, 2^*, \ldots$$

Consider, now, two finite well-ordered disjoint sets

$$A = \{a_1, a_2, \ldots, a_m\} \quad \text{and} \quad B = \{b_1, b_2, \ldots, b_n\}$$

Then $m^* = \text{ord}(A)$ and $n^* = \text{ord}(B)$. Therefore,

$$m^* + n^* = \text{ord}(A) + \text{ord}(B) = \text{ord}(\{A; B\}) = (m + n)^*$$

Thus the operation of addition for finite ordinal numbers corresponds to the operation of addition for the set **N** of natural numbers (nonnegative integers).

Note once again that the set of ordinal numbers is itself a well-ordered set; hence every ordinal has an immediate successor. For the finite ordinals, i.e., the natural numbers, it is easily seen that $n + 1$ is the immediate successor to $n$. The next theorem (proved in Problem 8.17) states that this property is true in general.

**Theorem 8.16**: Let $\lambda$ be any ordinal number. Then $\lambda + 1$ is the immediate successor of $A$.

### General Addition of Ordinal Numbers

Addition of real numbers, which include the natural numbers, is a binary operation and can be extended by induction to any finite sum

$$a_1 + a_2 + \ldots + a_n$$

The sum of an infinite number of real numbers, such as

$$1 + 2 + 3 + 4 + \cdots \quad \text{or} \quad 1 + \tfrac{1}{2} + \tfrac{1}{4} + \cdots$$

has no meaning (unless one introduces the concepts of limits). On the other hand, it is possible to define the sum of an infinite number of ordinal numbers as follows.

Let $\{\lambda_i : i \in I\}$ be any well-ordered collection, finite or infinite, of ordinal numbers. In other words, $I$ is a well-ordered set and to each $i \in I$ there corresponds an ordinal number $\lambda_i$. For each $i \in I$, let $A_i$ be a set such that

$$\lambda_i = \text{ord}(A_i)$$

Then the collection of sets $\{A_i \times \{i\} : i \in I\}$ is a well-ordered collection of pairwise disjoint well-ordered sets. By Theorem 8.2,

$$S = \bigcup\{A_i \times \{i\} : i \in I\}$$

is a well-ordered set. Thus the following definition is meaningful.

**Definition 8.7**: Let $\{\lambda_i : i \in I\}$ be a well-ordered collection of ordinal numbers such that $\lambda_i = \text{ord}(A_i)$. Then

$$\sum_{i \in I} \lambda_i = \text{ord}(\bigcup \{A_i \times \{i\} : i \in I\})$$

According to the above definition, we have

$$1 + 1 + 1 + \cdots = \omega$$

In fact, if each $\lambda_i$ is finite (and not 0), then

$$\lambda_1 + \lambda_2 + \lambda_3 + \cdots = \sum_{i \in \mathbf{P}} \lambda_i = \omega$$

## 8.11 ORDINAL MULTIPLICATION

An operation of multiplication is defined for ordinal numbers as follows:

**Definition 8.8**: Let $\lambda$ and $\mu$ be ordinal numbers and let $A$ and $B$ be well-ordered sets such that $\lambda = \text{ord}(A)$ and $\mu = \text{ord}(B)$. Then

$$\lambda\mu = \text{ord}(A \times B)$$

where $A \times B$ is ordered *reverse lexicographically*.

The product set $A \times B$ is ordered reverse lexicographically means that

$$(a, a') < (b, b') \qquad \text{if} \quad \begin{cases} a' < b' \\ \text{or } a' = b' \text{ but } a < b \end{cases}$$

Unless otherwise stated, the product set $A \times B$ of two well-ordered sets $A$ and $B$ is to be ordered reverse lexicographically.

**EXAMPLE 8.9** Note first that $2 = \text{ord}(\{a, b\})$ and $\omega = \text{ord}(\{1, 2, 3, \ldots\})$. Then

$$2\omega = \text{ord}(\{(a, 1), (b, 1), (a, 2), (b, 2), \ldots\}) = \omega$$

But

$$\omega 2 = \text{ord}(\{(1, a), (2, a), \ldots; (1, b), (2, b), \ldots\}) > \omega$$

since $\mathbf{P} = \{1, 2, 3, \ldots\}$ is similar to the initial segment $\{(1, a), (2, a), \ldots\}$.

The above Example 8.9 tells us that the operation of multiplication of ordinal numbers is not commutative. However, the following conditions do hold.

**Theorem 8.17**: (1) The associative law for multiplication holds, i.e.,

$$(\lambda\mu)\eta = \lambda(\mu\eta)$$

(2) The left distributive law of multiplication over addition holds; i.e.,

$$\lambda(\mu + \eta) = \lambda\mu + \lambda\eta$$

(3) The ordinal 1 is a multiplicative identity element, i.e.,

$$1\lambda = \lambda 1 = \lambda$$

## 8.12 STRUCTURE OF ORDINAL NUMBERS

We now write down many of the ordinal numbers according to their order. First come the finite ordinals

$$0,\ 1,\ 2,\ 3,\ \ldots$$

and then comes the first limit ordinal $\omega$ and its successors:

$$\omega,\ \omega+1,\ \omega+2,\ \ldots$$

By Example 8.9, $\text{ord}(\{0,1,2,\ldots;\ \omega,\ \omega+1,\omega+2,\ldots\}) = \omega 2$. Hence next comes the second limit ordinal $\omega 2$ and its successors:

$$\omega 2\ ,\omega 2+1,\ \omega 2+2,\ \omega 2+3,\ \ldots$$

The next limit number is $\omega 3$. We proceed as follows:

$$\omega 3,\ \omega 3+1,\ \ldots,\ \omega 4,\ \ldots,\ \omega 5,\ \ldots,\ \ldots,\ \omega\omega = \omega^2$$

Here $\omega\omega = \omega^2$ is the limit number following the limit numbers $\omega n$ where $n \in \mathbf{P}$. We continue:

$$\omega^2,\ \omega^2+1,\ \ldots,\ \omega^2+\omega,\ \omega^2+\omega+1,\ \ldots,\ \omega^2+\omega 2,\ \ldots,\ \omega^2+\omega 3,\ \ldots,\ \ldots,\ \omega^2+\omega^2 = \omega^2 2$$

Then

$$\omega^2 2,\ \ldots,\ \omega^2 3,\ \ldots,\ \omega^2 4,\ \ldots,\ \omega^2\omega = \omega^3$$

Then we have the powers of $\omega$:

$$\omega^3,\ \omega^3+1,\ \ldots,\ \omega^4,\ \ldots,\ \omega^5,\ \ldots,\ \ldots,\ \omega^\omega$$

Here $\omega^\omega$ is the limit number after the limit numbers $\omega^n$ where $n \in \mathbf{P}$. We proceed:

$$\omega^\omega,\ \ldots,\ (\omega^\omega)^\omega,\ \ldots,\ ((\omega^\omega)^\omega)^\omega,\ \ldots,\ \ldots$$

After all these ordinals we have the ordinal $\epsilon_0$. We can continue:

$$\epsilon_0,\ \epsilon_0+1,\ \ldots$$

We note that each of the ordinal numbers we have enumerated is still the ordinal number of a countable set.

## 8.13 AUXILIARY CONSTRUCTION OF ORDINAL NUMBERS

Recall again the following theorem.

**Theorem 8.14**: Let $s(\lambda)$ be the set of ordinal numbers which precede $\lambda$. Then

$$\lambda = \text{ord}(s(\lambda))$$

Some authors use the above property of ordinal numbers to actually define the ordinal numbers. Roughly speaking, an ordinal number is defined to be the set of ordinal numbers which precede it. Specifically:

**Definition:**

$$\begin{aligned} 0 &\equiv \varnothing \\ 1 &\equiv \{0\} \\ 2 &\equiv \{0,1\} \\ 3 &\equiv \{0,1,2\} \\ &\vdots \\ \omega &\equiv \{0,1,2,\ldots\} \\ \omega+1 &\equiv \{0,1,2,\ldots,\omega\} \end{aligned} \qquad \begin{aligned} \omega+2 &\equiv \{0,1,2,\ldots,\omega,\omega+1\} \\ &\vdots \\ \omega 2 &\equiv \{0,1,\ldots,\omega,\omega+1,\ldots\} \\ \omega 2+1 &\equiv \{0,1,\ldots,\omega,\omega+1,\ldots,\omega 2\} \\ &\vdots \end{aligned}$$

One main reason the ordinal numbers are developed as above is in order to avoid certain inherent contradictions which appear in the preceding development of the ordinal numbers (which are discussed in Chapter 9).

## Solved Problems

### WELL-ORDERED SETS

**8.1.** Prove Theorem 8.4 (Principle of Transfinite Induction): Let $S$ be a subset of a well-ordered set $A$ with the following properties: (1) $a_0 \in S$, (2) $s(a) \subseteq S$ implies $a \in S$. Then $S = A$.

Suppose $S \neq A$, i.e., suppose $A \backslash S = T$ is not empty. Since $A$ is well-ordered, $T$ has a first element $t_0$. Each element $x \in s(t_0)$ precedes $t_0$ and, therefore, cannot belong to $T$, i.e., belongs to $S$; hence $s(t_0) \subseteq S$. By (2), $t_0 \in S$. This contradicts the fact that $t_0 \in A \backslash S$. Hence the original assumption that $S \neq A$ is not true; in other words, $S = A$. (Note that (1) is in fact a consequence of (2) since $\varnothing = s(a_0)$ is a subset of $S$ and, therefore, implies $a_0 \in S$.)

**8.2.** Prove Theorem 8.6: Let $S(A)$ denote the collection of all initial segments of elements in a well-ordered set $A$, and let $S(A)$ be ordered by set inclusion. Then $A$ is similar to $S(A)$ and, in particular, the function $f\colon A \to S(A)$ defined by $f(x) = s(x)$ is a similarity mapping between $A$ and $S(A)$.

By definition $f$ is onto. We show that $f$ is one-to-one. Suppose $x \neq y$. Then one of them, say $x$, strictly precedes the other; hence $x \in s(y)$. But, by definition of initial segment, $x \notin s(x)$. Thus $s(x) \neq s(y)$, and hence $f$ is one-to-one.

We show that $f$ preserves order, that is,

$$x \leq y \quad \text{if and only if} \quad s(x) \subseteq s(y)$$

Let $x \leq y$. Suppose $a \in s(x)$. Then $a \leq x$ and hence $a \leq y$; thus $a \in s(y)$. Since $a \in s(x)$ implies $a \in s(y)$, $s(x)$ is a subset of $s(y)$. Now suppose $x \not\leq y$, that is, $x > y$. Then $y \in s(x)$. But, by definition of initial segment, $y \notin s(y)$; hence $s(x) \not\subseteq s(y)$. In other words, $x \leq y$ if and only if $s(x) \subseteq s(y)$.

**8.3.** Prove Theorem 8.7: Let $A$ be a well-ordered set, let $B$ be a subset of $A$, and let $f\colon A \to B$ be a similarity mapping of $A$ onto $B$. Then, for every $a \in A$, $a \leq f(a)$.

Let $D = \{x : f(x) < x\}$. If $D$ is empty the theorem is true. Suppose $D \neq \varnothing$. Then, since $A$ is well-ordered, $D$ has a first element $d_0$. Note $d_0 \in D$ means $f(d_0) < d_0$. Since $f$ is a similarity mapping,

$$f(d_0) < d_0 \qquad \text{implies} \qquad f(f(d_0)) < f(d_0)$$

Consequently, $f(d_0)$ also belongs to $D$. But $f(d_0) < d_0$ and $f(d_0) \in D$ contradicts the fact that $d_0$ is the first element of $D$. Hence the original assumption that $D \neq \varnothing$ leads to a contradiction. Therefore $D$ is empty and the theorem is true.

**8.4.** Prove Theorem 8.8: Let $A$ and $B$ be similar well-ordered sets. Then there exists only one similarity mapping of $A$ into $B$.

Let $f\colon A \to B$ and $g\colon A \to B$ be similarity mappings. Suppose $f \neq g$. Then there exists an element $x \in A$ such that $f(x) \neq g(x)$. Consequently, either $f(x) < g(x)$ or $g(x) < f(x)$. Say $f(x) < g(x)$. Since $g\colon A \to B$ is a similarity mapping, $g^{-1}\colon B \to A$ is also a similarity mapping. Furthermore, $g^{-1} \circ f\colon A \to A$, the composition of two similarity mappings, is also a similarity mapping. But

$$f(x) < g(x) \qquad \text{implies} \qquad (g^{-1} \circ f)(x) < (g^{-1} \circ g)(x) = x$$

We have $g^{-1} \circ f$ is a similarity mapping and $(g^{-1} \circ f)(x) < x$. This contradicts Theorem 8.7. Hence the assumption that $f \neq g$ leads to a contradiction. Accordingly, there can be only one similarity mapping of $A$ into $B$.

**8.5.** Prove Theorem 8.9: A well-ordered set cannot be similar to one of its initial segments.

Let $A$ be a well-ordered set and let $f\colon A \to s(a)$ be a similarity mapping of $A$ onto one of its initial segments. Then $f(a) \in s(a)$. Therefore

$$f(a) < a$$

This last fact contradicts Theorem 8.7. Therefore $A$ cannot be similar to one of its initial segments.

**8.6.** Prove: Let $A$ be a well-ordered set and let $S$ be a subset of $A$ with the following property:

If $a \leq b$ and $b \in S$, then $a \in S$.

Then $S = A$ or $S$ is an initial segment of $A$.

Suppose $S \neq A$. Then $A \backslash S$ has a first element $a_0$ where $a_0 \notin S$. We show that $S = s(a_0)$. Suppose $x \in s(a)$. Then $x < a_0$ and hence $x \notin A \backslash S$. Therefore $x \in S$. Thus $s(a) \subseteq S$.

Now suppose $y \notin s(a_0)$, that is, suppose $a_0 < y$. But $y \in S$ and $a_0 < y$ implies $a_0 \in S$, which contradicts the fact that $a_0 \notin S$. Hence $y \notin S$. We have shown that $y \notin s(a_0)$ implies $y \notin S$, which means that $S \subseteq s(a_0)$.

Both inclusions imply $S = s(a_0)$.

**8.7.** Prove: Two different initial segments of a well-ordered set cannot be similar.

Let $s(a)$ and $s(b)$ be two different initial segments, that is, $a \neq b$. Either $a < b$ or $b < a$; say $a < b$. Then $s(a)$ is an initial segment of the well-ordered set $s(b)$. Hence, by Theorem 8.9, $s(b)$ is not similar to $s(a)$.

**8.8.** Prove: Let $A$ and $B$ be well-ordered sets, and let an initial segment $s(a)$ of $A$ be similar to an initial segment of $B$. Then $s(a)$ is similar to a unique initial segment $s(b)$ of $B$.

Let $s(a) \simeq s(b)$ and $s(a) \simeq s(b')$ where $b, b' \in B$. Then $s(b) \simeq s(b')$. By Problem 8.7, $s(b) = s(b')$. Therefore, $b = b'$.

**8.9.** Prove: Let $A$ and $B$ be well-ordered sets such that an initial segment $s(a)$ of $A$ is similar to an initial segment $s(b)$ of $B$. Then each initial segment of $s(a)$ is similar to an initial segment of $s(b)$, that is,

$$a' \leq a \quad \text{implies} \quad s(a') \simeq s(b') \quad \text{where} \quad b' \leq b$$

Furthermore, if $f$: $s(a) \to s(b)$ is the similarity mapping of $s(a)$ onto $s(b)$, then $f$ restricted to $s(a')$ is the similarity mapping of $s(a')$ onto $s(b') = f(s(a'))$.

Let $f(a') = b'$. Note that $f$ restricted to $s(a')$ is one-to-one and preserves order; hence $s(a') \simeq f(s(a'))$. Furthermore, since $f$ is a similarity mapping,

$$a^* < a' \quad \text{if and only if} \quad f(a^*) < b'$$

Then $f(s(a')) = s(b')$, and therefore $s(a') \simeq s(b')$.

**8.10.** Prove: Let $A$ and $B$ be well-ordered sets and let

$$S = \{x : x \in A,\ s(x) \simeq s(y) \text{ where } y \in B\}$$

(In other words, each element $x \in S$ has the property that its initial segment $s(x)$ is similar to an initial segment $s(y)$ of $B$.) Then $S = A$ or $S$ is an initial segment of $A$.

Let $x \in S$ and $y < x$. By Problem 8.9, $s(y)$ is similar to an initial segment of $B$; hence $y \in S$. In other words,

$$y < x \text{ and } x \in S \quad \text{implies} \quad y \in S$$

By Problem 8.6, $S = A$ or $S$ is an initial segment of $A$.

**8.11.** Prove: Let $A$ and $B$ be well-ordered sets and let

$$S = \{x : x \in A,\ s(x) \simeq s(y) \text{ where } y \in B\}$$
$$T = \{y : y \in B,\ s(y) \simeq s(x) \text{ where } x \in A\}$$

Then $S$ is similar to $T$.

Let $x \in S$. Then, by Problem 8.8, $s(x)$ is similar to a unique segment $s(y)$ of $B$. Thus to each $x \in S$ there corresponds a unique $y \in Y$ such that $s(x) \simeq s(y)$, and vice versa. Hence the function $f$: $S \to T$ defined by

$$f(x) = y \quad \text{if} \quad s(x) \simeq s(y)$$

is one-to-one and onto.

Now let $x', x \in S, f(x) = y, f(x') = y'$ and $x' < x$. The theorem is proven if we can show that $y' < y$, that is, that $f$ preserves order.

Let $g$: $s(x) \to s(y)$ be the similarity mapping of $s(x)$ into $s(f(x)) = s(y)$. By Problem 8.9, $g$ restricted to $s(x')$ is a similarity mapping of $s(x')$ into the initial segment $s(g(x'))$ of $B$. But, by Problem 8.8, there exists only one similarity mapping of $s(x')$ into $B$. Consequently, $g(x') = f(x') = y'$. Since $g(x') \in s(y)$,

$$g(x') = y' < y$$

Since we have shown that $y' < y, f$ preserves order. Therefore, $S$ is similar to $T$.

**8.12.** Prove Theorem 8.10: Let $A$ and $B$ be well-ordered sets. Then $A$ is shorter than $B$, $A$ is similar to $B$, or $A$ is longer than $B$.

Let $S$ and $T$ be defined as in the preceding problem. Note $S \simeq T$. By Problem 8.10, there are four possibilities:

*Case I*: $S = A$ and $T = B$. Then $A$ is similar to $B$.
*Case II*: $S = A$ and $T = s(b)$, an initial segment of $B$. Then $A$ is shorter than $B$.
*Case III*: $T = B$ and $S = s(a)$, an initial segment of $A$. Then $A$ is longer than $B$.
*Case IV*: $S = s(a)$ and $T = s(b)$. Then $a \in S$ since its initial segment $s(a)$ is similar to an initial segment $s(b)$ of $B$. But $a$ cannot belong to its own initial segment; hence this case is impossible.

Thus the theorem is true.

**8.13.** Prove: Let $\mathscr{A}$ be a collection of initial segments of a well-ordered set $A$. Then there is an initial segment $s(a) \in \mathscr{A}$ such that $s(a) \subseteq s(x)$ for any other initial segment $s(x)$ in $\mathscr{A}$; that is, there is an initial segment $s(a) \in \mathscr{A}$ which is shorter than every other initial segment in $\mathscr{A}$.

By Theorem 8.6, $A$ is similar to $S(A)$, the family of all initial segments of elements in $A$, ordered by set inclusion. Since $A$ is well-ordered, $S(A)$ is also well-ordered. Since $\mathscr{A}$ is a subset of $S(A)$, it has a first element $s(a)$. Therefore $s(a) \subseteq s(x)$ for any other initial segment $s(x) \in \mathscr{A}$.

**8.14.** Prove Theorem 8.11: Let $\mathscr{A}$ be a collection of pairwise nonsimilar well-ordered sets. Then there exists a set $A_0$ in $\mathscr{A}$ such that $A_0$ is shorter than every other set in $\mathscr{A}$.

Let $B$ be any set in $\mathscr{A}$. Define

$$\mathscr{B} = \{X : X \in \mathscr{A}, X \text{ is shorter than } B\}$$

If $\mathscr{B}$ is empty, then $B$ satisfies the requirements of the theorem. Suppose $\mathscr{B} \neq \varnothing$. If we show that $\mathscr{B}$ has a shortest set $A_0$ then, considering the way $\mathscr{B}$ was defined, $A_0$ will also be the shortest set in $\mathscr{A}$.

Now, by Theorem 8.10, every set $A \in \mathscr{B}$ is similar to an initial segment $s(a)$ of $B$. Let $\mathscr{B}'$ be the collection of those initial segments of $B$ each of which is similar to a set in $\mathscr{B}$. By Problem 8.13, $\mathscr{B}'$ contains an initial segment $s(a_0)$ which is shorter than every other initial segment in $\mathscr{B}'$. Consequently, the set $A_0 \in \mathscr{B}$, which is similar to $s(a_0)$, is shorter than every other set in $\mathscr{B}$.

Therefore, $A_0$ satisfies the requirements of the theorem.

## ORDINAL NUMBERS

**8.15.** Prove: Let $\lambda = \text{ord}(A)$ and let $\mu < \lambda$. Then there is a unique initial segment $s(a)$ of $A$ such that $\mu = \text{ord}(s(a))$.

Let $\mu = \text{ord}(B)$. Since $\mu < \lambda$, $B$ is shorter than $A$, that is, $B$ is similar to an initial segment $s(a)$ of $A$. Therefore, $\mu = \text{ord}(s(a))$. Furthermore, $s(a)$ is the only initial segment whose ordinal number is $\mu$ since, by Problem 8.7, two different initial segments of $A$ cannot be similar.

**8.16.** Prove Theorem 8.14: Let $s(\lambda)$ be the set of ordinals less than the ordinal $\lambda$. Then $\lambda = \text{ord}(s(\lambda))$.

Let $\lambda = \text{ord}(A)$, and let $S(A)$ denote the collection of all initial segments of $A$ ordered by set inclusion. By Theorem 8.5, $A \simeq S(A)$; hence $\lambda = \text{ord}(S(A))$. If we show that $s(\lambda)$ is similar to $S(A)$, the theorem will follow.

Let $\mu \in s(\lambda)$; then $\mu < \lambda$. By Problem 8.15, there is a unique initial segment $s(a)$ of $A$ such that $\mu = \text{ord}(s(a))$. Hence the function $f: s(\lambda) \to S(A)$ defined by

$$f(\mu) = s(a) \quad \text{if} \quad \mu = \text{ord}(s(a))$$

is one-to-one. Furthermore, we show that $f$ is onto. Suppose $s(b) \in S(A)$. Then $s(b)$ is shorter than $A$ and therefore $\text{ord}(s(b)) = \nu < \text{ord}(A) = \lambda$. This means $\nu \in s(\lambda)$. Hence $f(\nu) = s(b)$, and so $f$ is onto.

To complete the proof of the theorem, it is only necessary to show that $f$ preserves order; then $f$ is a similarity mapping and $s(\lambda) \simeq S(A)$. Let $\mu < \nu$, where $\mu, \nu \in s(\lambda)$. Then $\mu = \text{ord}(s(a))$ and $\nu = \text{ord}(s(b))$, that is, $f(\mu) = s(a)$ and $f(\nu) = s(b)$. Since $\mu < \nu$, $s(a)$ is an initial segment of $s(b)$; hence $s(a)$ is a proper subset of $s(b)$. In other words, under the ordering of $S(A)$, $s(a) < s(b)$. Thus $f$ preserves order.

**8.17.** Prove Theorem 8.16: Let $\lambda$ be any ordinal number. Then $\lambda + 1$ is the immediate successor of $\lambda$.

Let $\mu$ be the immediate successor of $\lambda$. Then, by definition of $s(\mu)$,

$$s(\mu) = s(\lambda) \cup \{\lambda\}$$

Hence

$$\text{ord}(s(\mu)) = \text{ord}(s(\lambda)) + \text{ord}(\{\lambda\})$$

That is, $\mu = \lambda + 1$.

**8.18.** Prove, by giving a counterexample, that the right distributive law of multiplication over addition (for the ordinal numbers) is not true in general. In other words, exhibit three ordinal numbers $\lambda$, $\mu$, $\nu$ such that

$$(\lambda + \mu)\nu \neq \lambda\nu + \mu\nu$$

By Example 8.9, $(1 + 1)\omega = 2\omega = \omega$. On the other hand, using the left distributive law,

$$1\omega + 1\omega = \omega + \omega = \omega 1 + \omega 1 = \omega(1 + 1) = \omega 2 > \omega$$

Therefore, $(1 + 1)\omega \neq 1\omega + 1\omega$.

**8.19.** Let $\{A_i : i \in I\}$ be a well-ordered collection of pairwise disjoint well-ordered sets. Suppose $\text{ord}(I) = \omega$ and $\text{ord}(A_i) = \omega$ for every $i \in I$. Find $\text{ord}(\bigcup_i A_i)$.

$$\text{ord}(\textstyle\bigcup_i A_i) = \omega\omega + \omega + \cdots = \omega(1 + 1 + 1 + \cdots) = \omega\omega = \omega^2$$

**8.20.** Prove: $\omega + \omega = \omega 2$.

**Method 1:** Using the left distributive law we get

$$\omega + \omega = \omega 1 + \omega 1 = \omega(1 + 1) = \omega 2$$

**Method 2:** Consider the well-ordered sets

$$A = \{a_1, a_2, \ldots\}, \qquad B = \{b_1, b_2, \ldots\}, \qquad C = \{c_1, c_2, \ldots\}, \qquad D = \{r, s\}$$

Note that

$$\omega = \text{ord}(A) = \text{ord}(B) = \text{ord}(C) \quad \text{and} \quad 2 = \text{ord}(D)$$

Then

$$\omega + \omega = \text{ord}(\{A; B\}) = \text{ord}(\{a_1, a_2, \ldots; b_1, b_2, \ldots\})$$
$$\omega 2 = \text{ord}(C \times D) = \text{ord}(\{(c_1, r),\ (c_2, r), \ldots; (c_1, s),\ (c_2, s), \ldots\})$$

But the function $f$: $\{A; B\} \to \{C \times D\}$ defined by

$$f(x) = \begin{cases} (c_i, r) & \text{if } x = a_i \\ (c_i, s) & \text{if } x = b_i \end{cases}$$

is a similarity mapping of $\{A; B\}$ onto $C \times D$. Hence

$$\omega + \omega = \text{ord}(\{A; B\}) = \text{ord}(\{C \times D\}) = \omega 2$$

## Supplementary Problems

**8.21.** Prove Theorem 8.1: Let $A$ be a well-ordered set. Then: (i) Every subset of $A$ is well-ordered. (ii) If $B$ is similar to $A$, then $B$ is well-ordered

**8.22.** Prove Theorem 8.2: Let $\{A_i : i \in I\}$ be a well-ordered family of pairwise disjoint well-ordered sets. Then the union $S = \bigcup_i A_i$ (with the usual ordering) is well-ordered.

**8.23.** Assume that the set **P** of counting numbers with the usual order is well-ordered. Prove the Principle of Mathematical Induction: Let $S$ be a subset of **P** with the properties:

$$(1)\ 1 \in S \text{ and } (2)\ n \in S \text{ implies } n+1 \in S;$$

then $S = \mathbf{P}$.

**8.24.** Prove that 0 is the identity element for addition of ordinal numbers, that is, for any ordinal $\lambda$, we have $0 + \lambda = \lambda + 0 = \lambda$.

**8.25.** Prove that 1 is the identity element for multiplication of ordinal numbers, that is, for any ordinal $\lambda$, we have $1\lambda = \lambda 1 = \lambda$.

**8.26.** Prove: If each $\lambda_i$, $i \in \mathbf{P}$, is a finite ordinal, then $\lambda_1 + \lambda_2 + \cdots = \sum_i \lambda_i = \omega$.

**8.27.** Prove: Let $\lambda$ be any infinite ordinal number. Then $\lambda = \mu + n$, where $\mu$ is a limit number and $n$ is a finite ordinal.

**8.28.** State whether each of the following statements about ordinals is true or false; if it is true prove it, and if it is false give a counterexample: (*a*) If $\lambda \neq 0$, then $\mu < \lambda + \mu$. (*b*) If $\lambda \neq 0$, then $\mu < \mu + \lambda$.

**8.29.** State whether each of the following statements concerning ordinals is true or false; if it is true prove it, and if it is false give a counterexample:

(*a*) If $\lambda \neq 0$ and $\mu < \nu$, then $\lambda + \mu < \lambda + \nu$.

(*b*) If $\lambda \neq 0$ and $\mu < \nu$, then $\mu + \lambda < \nu + \lambda$.

**8.30.** Prove: The left distributive law of multiplication over addition holds for ordinal numbers, that is,

$$\lambda(\mu + \nu) = \lambda\mu + \lambda\nu$$

## Answers to Supplementary Problems

**8.27.** *Hint*: Note that a well-ordered set cannot contain an ordered subset $A = \{\cdots < a_3 < a_2 < a_1\}$, since $A$ is not well-ordered.

**8.28.** (*a*) False. (*b*) True

**8.29.** (*a*) True. (*b*) False

# Chapter 9

# Axiom of Choice, Zorn's Lemma, Well-Ordering Theorem

## 9.1 INTRODUCTION

Many properties of well-ordered sets were investigated in the preceding Chapter 8. We have not said much about the existence of such sets. Central to the theory of set theory is the fact that any set can be well-ordered! This was proved by E. Zermelo in 1904. Specifically, this "well-ordering theorem" can be shown to be equivalent to the axiom of choice and Zorn's lemma. This equivalence and some of its consequences will be treated in this chapter. We will end the chapter with some paradoxes in set theory.

## 9.2 CARTESIAN PRODUCTS AND CHOICE FUNCTIONS

The following theorem applies.

**Definition 9.1**: Let $\{A_i : i \in I\}$ be a nonempty family of nonempty sets. Then the cartesian product of $\{A_i : i \in I\}$, denoted by

$$\prod\{A_i : i \in I\} \text{ or } \prod_i A_i$$

is the set of all choice functions defined on $\{A_i : i \in I\}$.

Recall that a function $f$: $\{A_i : i \in I\} \to X$, where each $A_i$ is a subset of $X$, is called a choice function if $f(A_i) = a_i$ belongs to $A_i$, for every $i \in I$. In other words, $f$ "chooses" a point $a_i \in A_i$ for each set $A_i$.

**EXAMPLE 9.1** Let $\{A_1, A_2, \ldots, A_n\}$ be a finite family of sets. Recall (Chapter 2) that the cartesian product of the $n$ sets,

$$A_1 \times A_2 \times \cdots \times A_n = \prod_{i=1}^{n} A_i$$

is defined to be the set of $n$-tuples

$$(a_1, a_2, \ldots, a_n)$$

where $a_i \in A_i$ for $i = 1, 2, \ldots, n$. On the other hand, each choice function $f$ defined on $\{A_1, A_2, \ldots, A_n\}$ corresponds to the unique $n$-tuple

$$(f(A_1), f(A_2), \ldots, f(A_n))$$

and vice versa. Accordingly, in the finite case, Definition 9.1 agrees with the previous definition of the cartesian product.

The main reason for introducing Definition 9.1 is that it applies to any family of sets: finite, denumerable, or even nondenumerable. The previous definition, which used the concept of $n$-tuples, applied only to a finite family of sets.

**Remark**: Although a choice function is defined for a family of subsets, any family of sets $\{A_i : i \in I\}$ can be considered to be a family of subsets of their union $\bigcup_i A_i$.

## 9.3 AXIOM OF CHOICE

The axiom of choice lies at the foundations of mathematics and, in particular, the theory of sets. This "innocent looking" axiom, which follows, has as a consequence some of the most powerful and important results in mathematics.

**Axiom of Choice:** The cartesian product of a nonempty family of nonempty sets is nonempty.

Using Definition 9.1, the axiom of choice can be stated as follows:

**Axiom of Choice:** There exists a choice function for any nonempty family of nonempty sets.

The axiom of choice is equivalent to the following postulate:

**Zermelo's Postulate:** Let $\{A_i : i \in I\}$ be any nonempty family of disjoint nonempty sets. Then there exists a subset $B$ of the union $\bigcup_i A_i$ such that the intersection of $B$ and each set $A_i$ consists of exactly one element.

Observe that in Zermelo's postulate the sets are disjoint whereas in the axiom of choice they may not be disjoint.

## 9.4 WELL-ORDERING THEOREM, ZORN'S LEMMA

The following theorem is attributed to Zermelo, who proved the theorem directly from the axiom of choice.

**Well-Ordering Theorem:** Every set can be well-ordered.

Zorn's lemma, which follows, is one of the most important tools in mathematics; it establishes the existence of certain types of elements although no constructive process is given to find these elements.

**Zorn's Lemma:** Let $X$ be a nonempty partially ordered set in which every chain (linearly ordered subset) has an upper bound in $X$. Then $X$ contains at least one maximal element.

We formally state and prove (Problem 9.4) the following basic result of set theory:

**Theorem 9.1:** The following are equivalent:

(i) Axiom of choice;
(ii) Well-ordering theorem;
(iii) Zorn's lemma.

## 9.5 CARDINAL AND ORDINAL NUMBERS

Let $\lambda = \text{ord}(A)$ be an ordinal number. Then we can associate with $\lambda$ the unique cardinal number $\alpha = |A|$. We call $\alpha$ the cardinal number of $\lambda$ and denote it by

$$\alpha = \overline{\lambda}$$

This function from the ordinal numbers to the cardinal numbers is not one-to-one, that is, there are different ordinal numbers with the same cardinal number. For example,

$$\omega = \text{ord}(\{1, 2, 3, \ldots\}) \qquad \text{and} \qquad \omega 2 = \text{ord}(\{a_1, a_2, \ldots; b_1, b_2, \ldots\})$$

are both ordinal numbers of denumerable sets with the same cardinal number $\aleph_0$. In other words,

$$\overline{\omega} = \aleph_0 = \overline{\omega 2}$$

The well-ordering theorem implies that the above function from the ordinal numbers to the cardinal numbers is onto. For, suppose $\alpha = |A|$ is any cardinal number. By the well-ordering theorem, $A$ can be well-ordered; say $\lambda = \text{ord}(A)$. Then $\alpha = \overline{\lambda}$. Hence $\alpha$ is the cardinal number of at least one ordinal number $\lambda$. (Here, $A$ is used both as the original set and then as the well-ordered set.)

**Correspondence between Ordinal and Cardinal Numbers**

The following correspondence between the ordinal and cardinal numbers is easily established.

**Theorem 9.2:** Let $\alpha = \overline{\lambda}$ and $\beta = \overline{\mu}$ be cardinal numbers. Then:

(1) If $\alpha < \beta$, then $\lambda < \mu$.

(2) If $\lambda < \mu$, then $\alpha \leq \beta$.

The next result, mentioned previously, is a direct consequence of the well-ordering theorem.

**Theorem 6.12 (Law of Trichotomy):** Let $\alpha$ and $\beta$ be any cardinal numbers. Then one of the following holds:

$$\alpha < \beta, \qquad \alpha = \beta, \qquad \alpha > \beta$$

That is, the cardinal numbers are linearly ordered by the inequality relation defined for the cardinal numbers. Since the ordinal numbers are themselves well-ordered, we can make an even stronger statement.

**Theorem 9.3:** Any set of cardinal numbers is well-ordered by the relation $\alpha \leq \beta$.

## 9.6 ALEPHS

Recall that the cardinal number of denumerable sets is denoted by

$$\aleph_0$$

(Here aleph, $\aleph$, is the first letter of the Hebrew alphabet.) Since the cardinal numbers are well-ordered, the following system of notation is used to denote cardinal numbers. The immediate successor of $\aleph_0$ is denoted by $\aleph_1$, and its immediate successor by $\aleph_2$, and so on. The cardinal number which succeeds all the $\aleph_n$ is denoted by $\aleph_\omega$. In fact every infinite cardinal can be uniquely denoted by an $\aleph$ with an ordinal number as a subscript as follows:

**Notation:** Let $\alpha$ be any infinite cardinal number. Let $s(\alpha)$ be the set of infinite cardinal numbers less than $\alpha$. Note that $s(\alpha)$ is well-ordered; say $\lambda = \text{ord}(s(\alpha))$. Then

$$\aleph_\lambda$$

denotes the cardinal number $\alpha$.

The continuum hypothesis can now be reformulated as follows:

**Continuum Hypothesis:** $\aleph_1 = \mathbf{c}$.

## 9.7 PARADOXES IN SET THEORY

The theory of sets was first studied as a mathematical discipline by Cantor (1845–1918) in the latter part of the nineteenth century. Today, the theory of sets lies at the foundations of mathematics and has revolutionized almost every branch of mathematics. At about the same time that set theory began to influence other branches of mathematics, various contradictions, called paradoxes, were discovered, the first by Burali-Forti in 1897. In this section, some of these paradoxes are presented. Although it is possible to eliminate these known contradictions by a strict axiomatic development of set theory, there are still many questions which are unanswered.

**Set of All Sets (Cantor's Paradox)**

Let $\mathscr{C}$ be the set of all sets. Then every subset of $\mathscr{C}$ is also a member of $\mathscr{C}$; hence the power set $\mathscr{P}(\mathscr{C})$ of $\mathscr{C}$ is a subset of $\mathscr{C}$, that is,

$$\mathscr{P}(\mathscr{C}) \subseteq \mathscr{C}$$

But $\mathscr{P}(\mathscr{C}) \subseteq \mathscr{C}$ implies that

$$|\mathscr{P}(\mathscr{C})| \leq |\mathscr{C}|$$

However, according to Cantor's theorem,

$$|\mathscr{C}| < |\mathscr{P}(\mathscr{C})|$$

Thus the concept of the set of all sets leads to a contradiction.

**Russell's Paradox**

Let $Z$ be the collection of all sets which do not contain themselves as members, that is,

$$Z = \{X : X \notin X\}$$

| Question: Does $Z$ belong to itself or not? |
|---|

If $Z$ does not belong to $Z$ then, by definition of $Z$, the set $Z$ does belong to itself. On the other hand, if $Z$ does belong to $Z$ then, by definition of $Z$, the set $Z$ does not belong to itself. In either case we are led to a contradiction.

The above paradox is somewhat analogous to the following popular paradox: In a certain town, there is a barber who shaves only and all those men who do not shave themselves. Question: Who shaves the barber?

**Set of All Ordinal Numbers (Burali-Forti Paradox)**

Let $\Delta$ be the set of all ordinal numbers. By a previous theorem $\Delta$ is a well-ordered set, say $\alpha = \operatorname{ord}(\Delta)$. Now consider $s(\alpha)$, the set of all ordinal numbers less than $\alpha$. Note:

(1) Since $s(\alpha)$ consists of all elements in $\Delta$ which precede $\alpha$, $s(\alpha)$ is an initial segment of $\Delta$.
(2) By a previous theorem $\alpha = \operatorname{ord}(s(\alpha))$; hence $\operatorname{ord}(s(\alpha)) = \alpha = \operatorname{ord}(\Delta)$.

Therefore $\Delta$ is similar to one of its initial segments, which is not possible. Thus the concept of the set of all ordinal numbers leads to a contradiction of Theorem 8.9.

**Set of All Cardinal Numbers**

Let $\mathscr{A}$ be the set of all cardinal numbers. Then for each cardinal $\alpha \in \mathscr{A}$ there is a set $A_\alpha$ such that $\alpha = |A_\alpha|$. Let

$$A = \bigcup(A_\alpha : \alpha \in \mathscr{A})$$

Consider the power set $\mathscr{P}(A)$ of $A$. Note $\mathscr{P}(A) \approx A_{|\mathscr{P}(A)|}$, which is a subset of $A$. Hence

$$|\mathscr{P}(A)| \leq |A|$$

But by Cantor's theorem,

$$|A| < |\mathscr{P}(A)|$$

Thus the concept of the set of all cardinal numbers leads to a contradiction.

**Class of All Sets Equipotent to a Set**

Let $A = \{a, b, \ldots\}$ be any set (not necessarily countable) and let $\mathscr{A} = \{i, j, \ldots\}$ be any other set. Consider the sets

$$A_i = \{(a, i), (b, i), \ldots\}$$
$$A_j = \{(a, j)(b, j), \ldots\}$$
$$\cdots\cdots\cdots\cdots\cdots\cdots$$
$$\cdots\cdots\cdots\cdots\cdots\cdots$$
$$\cdots\cdots\cdots\cdots\cdots\cdots$$

that is, the class of sets $\{A_i : i \in \mathscr{A}\}$. Note that

$$|\{A_i : i \in \mathscr{A}\}| = |\mathscr{A}|$$

and $A_i \approx A$ for every $i \in \mathscr{A}$.

Now let $\alpha$ be the class of all sets equipotent to $A$. Consider the power set $\mathscr{P}(\alpha)$ of $\alpha$, and define the class of sets $\{A_i : i \in \mathscr{P}(\alpha)\}$ as above. Since each $A_i \approx A$, we have

$$\{A_i : i \in \mathscr{P}(\alpha)\} \subseteq \alpha$$

Hence

$$|\mathscr{P}(\alpha)| = |\{A_i : i \in \mathscr{P}(\alpha)\}| \leq |\alpha|$$

But by Cantor's theorem, $|\alpha| < |\mathscr{P}(\alpha)|$. Thus the concept of the class of all sets equipotent to a set leads to a contradiction.

**Class of All Sets Similar to a Well-Ordered Set**

Let $A$ be any well-ordered set. Then the set $A_i$, defined as above and ordered by

$$(a, i) \leq (b, i) \qquad \text{if} \qquad a \leq b$$

is well-ordered and is similar to $A$, that is, $A_i \simeq A$.

Now let $\lambda$ be the class of all sets similar to the well-ordered set $A$. Consider the power set $\mathscr{P}(\lambda)$ of $\lambda$ and define the class of sets $\{A_i : i \in \mathscr{P}(\lambda)\}$ as above. Since each set $A_i$ is similar to $A$, we have

$$\{A_i : i \in \mathscr{P}(\lambda)\} \subseteq \lambda$$

Hence

$$|\mathscr{P}(\lambda)| = |\{A_i : i \in \mathscr{P}(\lambda)\}| \leq |\lambda|$$

But by Cantor's theorem, $|\lambda| < |\mathscr{P}(\lambda)|$. Thus the concept of the class of all sets similar to a well-ordered set leads to a contradiction.

# Solved Problems

## AXIOM OF CHOICE

**9.1.** Show that the axiom of choice is equivalent to Zermelo's postulate.

Let $\{A_i : i \in I\}$ be a nonempty family of disjoint nonempty sets and let $f$ be a choice function on $\{A_i : i \in I\}$. Set $B = \{f(A_i) : i \in I\}$. Then

$$A_i \cap B = \{f(A_i)\}$$

consists of exactly one element since the $A_i$ are disjoint and $f$ is a choice function. Accordingly, the axiom of choice implies Zermelo's postulate.

Now let $\{A_i : i \in I\}$ be any nonempty family of nonempty sets which may or may not be disjoint. Set

$$A_i^* = \{A_i\} \times \{i\} \qquad \text{for every } i \in I$$

Then certainly $\{A_i^*\}$ is a disjoint family of sets since $i \neq j$ implies $A_i \times \{i\} \neq A_j \times \{j\}$, even if $A_i = A_j$. By Zermelo's postulate, there exists a subset $B$ of $\bigcup(A_i^* : i \in I\}$ such that

$$B \cap A_i^* = \{(a_i, i)\}$$

consists of exactly one element. Then $a_i \in A_i$, and so the function $f$ on $\{A_i : i \in I\}$ defined by $f(A_i) = a_i$ is a choice function. Accordingly, Zermelo's postulate implies the axiom of choice.

**9.2.** Prove the well-ordering theorem (Zermelo): Every nonempty set $X$ can be well-ordered.

Let $f$ be a choice function on the collection $\mathscr{P}(X)$ of all subsets of $X$, that is,

$$f\colon \mathscr{P}(X) \to X \qquad \text{with} \qquad f(A) \in A, \qquad \text{for} \qquad \text{every } A \subseteq X$$

A subset $A$ of $X$ will be called *normal* if it has a well-ordering with the additional property that, for every $a \in A$,

$$f(X - s_A(a)) = a \qquad \text{where} \qquad s_A(a) = \{x \in A : x < a\}$$

i.e., $s_A(a)$ is the initial segment of $a$ in the ordering of $A$. We show that normal sets exist. Set

$$x_0 = f(X), \qquad x_1 = f(X \backslash \{x_0\}), \qquad x_2 = f(X \backslash \{x_0, x_1\})$$

Then $A = \{x_0, x_1, x_2\}$ is normal. We claim that if $A$ and $B$ are normal subsets of $X$, then either $A = B$ or one is an initial segment of the other. Since $A$ and $B$ are well-ordered, one of them, say $A$, is similar to $B$ or to an initial segment of $B$ (Theorem 8.10). Thus there exists a similarity mapping $\alpha : A \to B$. Set

$$A^* = \{x \in A : \alpha(x) \neq x\}$$

If $A^*$ is empty, then $A = B$ or $A$ is an initial segment of $B$. Suppose $A^* \neq \varnothing$, and let $a_0$ be the first element of $A^*$. Then $s_A(a_0) = s_B(\alpha(a_0))$. But $A$ and $B$ are normal, and so

$$a_0 = f(X) \backslash s_A(a_0)) = f(X \backslash s_B(\alpha(a_0))) = \alpha(a_0)$$

But this contradicts the definition of $A^*$, and so $A = B$ or $A$ is an initial segment of $B$. In particular, if $a \in A$ and $b \in B$ then either $a, b \in A$ or $a, b \in B$. Furthermore, if $a, b \in A$ and $a, b \in B$ then $a \leq b$ as elements of $A$ if and only if $a \leq b$ as elements of $B$.

Now let $Y$ consist of all those elements in $X$ which belong to at least one normal set. If $a, b \in Y$, then $a \in A$ and $b \in B$ where $A$ and $B$ are normal and so, as noted above, $a, b \in A$ or $a, b \in B$. We define an order in $Y$ as follows: $a \leq b$ as elements of $Y$ iff $a \leq b$ as elements of $A$ or as elements of $B$. This order is well-defined, i.e., independent of the particular choice of $A$ and $B$, and, furthermore, it is a linear order. Now let $Z$ be any nonempty subset of $Y$ and let $a$ be any arbitrary element in $Z$. Then $a$ belongs to a normal set $A$. Hence $A \cap Z$ is a nonempty subset of the well-ordered set $A$ and so contains a first element $a_0$. Furthermore, $a_0$ is a first element of $Z$ (Problem 9.13); thus $Y$ is, in fact, well-ordered.

We next show that $Y$ is normal. If $a \in Y$, then $a$ belongs to a normal set $A$. Furthermore, $s_A(a) = s_y(a)$ (Problem 9.13), and so

$$f(X \setminus s_y(a)) = f(X \setminus s_A(a)) = a$$

that is, $Y$ is normal. Lastly, we claim that $Y = X$. Suppose not, i.e., suppose $X \setminus Y \neq \emptyset$ and, say, $a = f(X \setminus Y)$. Set $Y^* = Y \cup \{a\}$ and let $Y^*$ be ordered by the order in $Y$ together with $a$ dominating every element in $Y$. Then $f(X \setminus s_Y(a)) = f(X \setminus Y) = a$ and so $Y^*$ is normal. Thus $a \in Y$. But this contradicts the fact that $f$ is a choice function, i.e., $f(X \setminus Y) = a \in X \setminus Y$ which is disjoint from $Y$. Hence $Y = X$, and so $X$ is well-ordered.

**9.3.** Prove (using the well-ordering theorem): Let $X$ be a partially ordered set. Then $X$ contains a maximal chain (linearly ordered subset), i.e., a chain which is not a proper subset of any other chain.

The result clearly holds if $X$ is empty (or even finite); hence we can assume that $X$ is not empty and that $X$ can be well-ordered with, say, first element $x_0$. (Observe that $X$ now has both a partial ordering and a well-ordering; the terms initial segment of $X$ and first element of a subset of $X$ will only be used with respect to the well-ordering, and the term comparable will only be used with respect to the partial ordering.)

Let $A$ be an initial segment of $X$ (where we allow $A = X$). A function $f$: $A \to A$ will be called *special* if

$$f(x) = \begin{cases} x, & \text{if } x \text{ is comparable to every element of } f[s(x)] \\ x_0, & \text{otherwise.} \end{cases}$$

Here $s(x)$ denotes the initial segment of $x$. We claim that if a special function exists then it is unique. If not, then there exist special functions $f$ and $f'$ on $A$ and a first element $a_0$ for which $f(a_0) \neq f'(a_0)$; hence $f$ and $f'$ agree on $s(a_0)$, which implies $f(a_0) = f'(a_0)$, a contradiction.

**Remark**: If $A$ and $A'$ are initial segments with special functions $f$ and $f'$ respectively and if $A \subseteq A'$, then the uniqueness of $f$ on $A$ implies that $f'$ restricted to $A$ equals $f$, i.e., $f'(a) = f(a)$ for every $a \in A$.

Now let $B$ be the union of those $A_i$ which admit a special function $f_i$. Since the $A_i$ are initial segments, so is $B$. Furthermore, $B$ admits the special function $g : B \to B$ defined by $g(b) = f_i(b)$ where $b \in A_i$. By the above remark, $g$ is well-defined. We next show that $B = X$. Let $y \in X$ be the first element for which $y \notin B$. Then $C = B \cup \{y\}$ is an initial segment. Moreover, $C$ admits the special function $h : C \to C$ defined as follows: $h(c) = g(c)$ if $c \in B$, and $h(y) = y$ or $x_0$ according as $y$ is or is not comparable to every element in $h[B]$. It now follows that $y \in B$, a contradiction. Thus no such $y$ exists and so $B = X$.

Lastly, we claim that $g[B]$, i.e., $g[X]$, is a maximal chain (linearly ordered subset) of $X$. If not, then there exists an element $z \in X$ such that $z \notin g[X]$ but $z$ is comparable to every element of $g[X]$. Thus, in particular, $z$ is comparable to every element of $g[s(z)]$. By definition of a special function, $g(z) = z$ which implies $z \in g[X]$, a contradiction. Thus $g[X]$ is a maximal chain of $X$, and the theorem is proved.

**9.4.** Prove Theorem 9.1: The following are equivalent: (i) axiom of choice, (ii) well-ordering theorem, (iii) Zorn's lemma.

By Problem 9.2, (i) implies (ii). We use Problem 9.3 to prove that (ii) implies (iii). Let $X$ be a partially ordered set in which every chain (linearly ordered subset) has an upper bound. We need to show that $X$ has a maximal element. By Problem 9.3, $X$ has a maximal chain, say $Y$. By hypothesis, $Y$ has an upper bound $m$ in $X$. We claim that $m$ is a maximal element of $X$. If not, then there exists $z \in X$ such that $z$ dominates $m$. It follows that $z \notin Y$ since $m$ is an upper bound for $Y$, and that $Y \cup \{z\}$ is linearly ordered. This contradicts the maximality of $Y$. Thus $m$ is a maximal element of $X$ and, consequently, (ii) implies (iii).

It remains to show that (iii) implies (i). By Problem 9.1, it suffices to prove that (iii) implies Zermelo's postulate. Let $\{A_i\}$ be a nonempty family of disjoint nonempty sets. Let $\mathscr{B}$ be the class of all subsets of $\bigcup_i A_i$ which intersect each $A_i$ in at most one element. We partially order $\mathscr{B}$ by set inclusion. Let $\mathscr{C} = \{B_j\}$ be a chain of $\mathscr{B}$. We claim that $B = \bigcup_j B_j$ belongs to $\mathscr{B}$. If not, then $B$ intersects some $A_{i_0}$ in more than one element; say $a, b \in B \cap A_{i_0}$ where $a \neq b$. Since $a, b \in B$, there exist $B_{j_1}$ and $B_{j_2}$ such that $a \in B_{j_1}$ and $b \in B_{j_2}$. But $\mathscr{C} = \{B_j\}$ is linearly ordered by set inclusion; hence $a$ and $b$ belong to either $B_{j_1}$ or $B_{j_2}$. This implies that $B_{j_1}$ or $B_{j_2}$ intersects $A_{i_0}$ in more than one element, a contradiction. Accordingly, $B$ belongs to $\mathscr{B}$, and so $B$ is an upper bound for the chain $\mathscr{C}$.

We have shown that every chain in $\mathscr{B}$ has an upper bound. By Zorn's lemma, $\mathscr{B}$ has a maximal element $M$. If $M$ does not intersect each $A_i$ in exactly one point, then $M$ and some $A_{i_k}$ are disjoint. Say $c \in A_{i_k}$. Then $M \cup \{c\}$ belongs to $\mathscr{B}$, which contradicts the maximality of $M$. Thus $M$ intersects each $A_i$ in exactly one point, and therefore (iii) implies Zermelo's postulate.

Thus the theorem is proved.

## APPLICATIONS OF ZORN'S LEMMA

**9.5.** Let $R$ be a relation from $A$ to $B$, that is, let $R$ be a subset of $A \times B$. Suppose the domain of $R$ is $A$. Prove that there exists a subset $f^*$ of $R$ such that $f^*$ is a function from $A$ into $B$.

Let $\mathscr{A}$ be the family of subsets of $R$ in which each $f \in \mathscr{A}$ is a function from a subset of $A$ into $B$. Partially order $\mathscr{A}$ by set inclusion. Note that if $f\colon A_1 \to B$ is a subset of $g\colon A_2 \to B$ then $A_1 \subseteq A_2$.

Now suppose $\mathscr{C} = \{f_i\colon A_i \to B\}$ is a chain (linearly ordered subset) of $\mathscr{A}$. Then (Problem 9.14) $f = \bigcup_i f_i$ is a function from $\bigcup_i A_i$ into $B$ and, therefore, $f$ is an upper bound of $\mathscr{C}$. By Zorn's lemma, $\mathscr{A}$ has a maximal element $f^*\colon A^* \to B$. If we show that $A^* = A$, then the theorem is proved.

Suppose $A^* \neq A$. Then there exists an element $a \in A$ such that $a \notin A^*$. Furthermore, since the domain of $R$ is $A$, there exists an ordered pair $(a, b) \in R$. Then $f^* \cup \{(a, b)\}$ is a function from $A^* \cup \{a\}$ into $B$. But this contradicts the fact that $f^*$, which would be a proper subset of $f^* \cup \{(a, b)\}$, is a maximal element of $\mathscr{A}$. Therefore $A^* = A$, and the theorem is proved.

**9.6.** (Application to Linear Algebra.) Prove that every vector space $V$ has a basis.

If $V$ consists of the zero vector alone then, by definition, the empty set is a basis for $V$; hence we assume $V$ contains a nonzero vector $a$. Let $\mathscr{B}$ be the family of independent sets of vectors in $V$. In other words, each element $B \in \mathscr{B}$ is an independent set of vectors. Note that $\mathscr{B}$ is nonempty since, e.g., $\{a\}$ belongs to $\mathscr{B}$. Partial order $\mathscr{B}$ by set inclusion.

Now suppose $\mathscr{C} = \{B_i\}$ is a chain of $\mathscr{B}$. If we show that $A = \bigcup_i B_i$ belongs to $\mathscr{B}$, i.e., $A$ is an independent set of vectors, then $A$ would be an upper bound of $\mathscr{C}$. Assume that $A$ is dependent. Then there exist vectors $a_1, a_2, \ldots, a_n$ in $A$ and scalars $c_1, c_2, \ldots, c_n$, not all zero, such that

$$c_1 a_1 + c_2 a_2 + \cdots + c_n a_n = 0 \qquad (1)$$

Since each $a_j \in A$, there exists $B_{j'}$ in $\mathscr{C}$ such that $a_j \in B_{j'}$. Since $\mathscr{C} = \{B_i\}$ is linearly ordered, one of the sets $B_{1'}, B_{2'}, \ldots B_{n'}$, say $B_{1'}$, is a superset of the others; hence $a_1, a_2, \ldots, a_n$ all belong to $B_{1'}$. In view of (1), $B_{1'}$ would be dependent, which is a contradiction. Thus $A$ is independent, $A$ belongs to $\mathscr{B}$, and $A$ is an upper bound of $\mathscr{C}$.

By Zorn's lemma, $\mathscr{B}$ has an upper bound $B^*$. $B^*$ can then be shown to be a basis for $V$.

**Remark**: The main part of the proof consists in showing that $A = \bigcup_i B_i$ does belong to $\mathscr{B}$. This is a typical example of how Zorn's lemma is used.

**9.7.** (Application to Algebra.) Let $R$ be a ring with unity 1. Prove that every proper ideal $J$ of $R$ is contained in a maximal ideal.

Recall that an ideal $J$ is proper if $J \neq R$, and an ideal $M$ is maximal if no ideal $K$ properly lies between $M$ and $R$, that is, if $M \subseteq K \subseteq R$, then $M = K$ or $K = R$. Also, when $R$ has a unity element 1, an ideal $J$ is proper if and only if $1 \notin J$.

Let $J$ be any proper ideal of $R$. Let $\mathscr{A}$ be the collection of all proper ideals of $R$ which contain $J$. $\mathscr{A}$ is not empty since $J \in \mathscr{A}$. Partially order $\mathscr{A}$ by set inclusion. Suppose $\mathscr{C}$ is a chain in $\mathscr{A}$. Let $M$ be the union of the ideals in $\mathscr{C}$. Now $M$ is an ideal since the union of an ascending chain of ideals is an ideal. Since 1 is not in any ideal of $\mathscr{C}$, $1 \notin M$ and hence $M$ is a proper ideal. Thus $M \in \mathscr{A}$. Clearly, $M$ is an upper bound for $\mathscr{C}$. By Zorn's lemma, $\mathscr{A}$ has a maximal element $J^*$. Then $J^*$ is a maximal ideal containing $J$.

## Supplementary Problems

**9.8.** State whether each of the following statements about cardinal numbers is true or false and give reasons for your answer:

(*a*) $\aleph_0 + \aleph_\lambda = \aleph_\lambda$; (*b*) $\aleph_\lambda + \aleph_\mu = \aleph_{\lambda+\mu}$.

**9.9.** Prove Theorem 9.2: Let $\alpha = \overline{\lambda}$ and $\beta = \overline{\mu}$ be cardinal numbers. Then:

(i) $\alpha < \beta$ implies $\lambda < \mu$; (ii) $\lambda < \mu$ implies $\alpha \leq \beta$.

**9.10.** Prove Theorem 6.12 (Law of Trichotomy). For any cardinal numbers $\alpha$ and $\beta$, exactly one of the following holds:

$$\alpha < \beta, \ \alpha = \beta, \ \alpha > \beta.$$

**9.11.** Prove Theorem 9.3: Any set of cardinal numbers is well-ordered by the relation $\alpha \leq \beta$.

**9.12.** Consider the proof of the following statement:

There exists a finite set of natural numbers which is not a proper subset of another finite set of natural numbers.

*Proof*: Let $\mathscr{B}$ be the family of all finite sets of natural numbers. Partially order $\mathscr{B}$ by set inclusion. Now let $\mathscr{C} = \{B_i\}$ be a chain of $\mathscr{B}$. Let $A = \bigcup_i B_i$. Note that each $B_i \subseteq A$. Hence $A$ is an upper bound of $\mathscr{C} = \{B_i\}$. Thus every chain of $\mathscr{B}$ has an upper bound. By Zorn's lemma, $\mathscr{B}$ has a maximal element, a finite set which is not a proper subset of another finite set.

*Question*: Since the statement is obviously false, which step in the proof is incorrect?

**9.13.** Prove the following two statements which were assumed in the proof in Problem 9.2:

(i) The first element $a_0$ of the set $A \cap Z$ is a first element of the set $Z$.

(ii) $s_A(a) = s_Y(a)$.

**9.14.** Prove the following statement which was assumed in the proof in Problem 9.5: Let $\{f_i : A_i \to B\}$ be a collection of functions which is linearly ordered by set inclusion. Then $\bigcup_i f_i$ is a function from $\bigcup_i A_i$ into $B$.

## Answers to Supplementary Problems

**9.8.** (*a*) True. For $\aleph_0$ is the cardinal number of a denumerable set and, as proven previously, the union of a denumerable set and an infinite set does not change the cardinality of the infinite set.

(*b*) False. If not, since the addition of cardinals is commutative, we would have

$$\aleph_{\lambda+\mu} = \aleph_\lambda + \aleph_\mu = \aleph_\mu + \aleph_\lambda = \aleph_{\mu+\lambda}$$

This would imply that the addition of ordinal numbers is commutative, which is not true.

**9.12.** $A$ does not belong to $\mathscr{C} = \{B_i\}$.

# PART III: **Related Topics**

# Chapter 10

## Logic and Propositional Calculus

### 10.1 INTRODUCTION

Many proofs in mathematics and many algorithms in computer science use logical expressions such as

"IF $p$ THEN $q$" or "IF $p_1$ AND $p_2$, THEN $q_1$ OR $q_2$"

It is therefore necessary to know the cases in which these expressions are either TRUE or FALSE: what we refer to as the truth values of such expressions. We discuss these issues in this chapter.

We also investigate the truth value of quantified statements, which are statements which use the logical quantifiers "for every" and "there exists".

### 10.2 PROPOSITIONS AND COMPOUND PROPOSITIONS

A *proposition* (or *statement*) is a declarative sentence which is true or false, but not both. Consider, for example, the following eight sentences:

(i) Paris is in France.
(ii) $1+1=2$.
(iii) $2+2=3$.
(iv) London is in Denmark.
(v) $9<6$.
(vi) $x=2$ is a solution of $x^2=4$.
(vii) Where are you going?
(viii) Do your homework.

All of them are propositions except (vii) and (viii). Moreover, (i), (ii), and (vi) are true, whereas, (iii), (iv), and (v) are false.

#### Compound Propositions

Many propositions are *composite*, that is, composed of *subpropositions* and various connectives discussed subsequently. Such composite propositions are called *compound propositions*. A proposition is said to be *primitive* if it cannot be broken down into simpler propositions, that is, if it is not composite.

**EXAMPLE 10.1**

(*a*) "Roses are red and violets are blue" is a compound proposition with subpropositions "Roses are red" and "Violets are blue".

(*b*) "John is intelligent or studies every night" is a compound proposition with subpropositions "John is intelligent" and "John studies every night".

(*c*) The above propositions (i) through (vi) are all primitive propositions; they cannot be broken down into simpler propositions.

The fundamental property of a compound proposition is that its truth value is completely determined by the truth values of its subpropositions together with the way in which they are connected to form the compound propositions.

The next section studies some of these connectives.

## 10.3 BASIC LOGICAL OPERATIONS

This section discusses the three basic logical operations of conjunction, disjunction, and negation which correspond, respectively, to the English words "and", "or", and "not".

### Conjunction $p \wedge q$

Any two propositions can be combined by the word "and" to form a compound proposition called the *conjunction* of the original propositions. Symbolically,

$$p \wedge q$$

read "$p$ and $q$", denotes the conjunction of $p$ and $q$. Since $p \wedge q$ is a proposition it has a truth value, and this truth value depends only on the truth values of $p$ and $q$. Specifically:

**Definition 10.1**: If $p$ and $q$ are true, then $p \wedge q$ is true; otherwise $p \wedge q$ is false.

The truth value of $p \wedge q$ may be defined equivalently by the table in Fig. 10-1(*a*). Here, the first line is a short way of saying that if $p$ is true and $q$ is true, then $p \wedge q$ is true. The second line says that if $p$ is true and $q$ is false, then $p \wedge q$ is false. And so on. Observe that there are four lines corresponding to the four possible combinations of T and F for the two subpropositions $p$ and $q$. Note that $q \wedge q$ is true only when both $p$ and $q$ are true.

| $p$ | $q$ | $p \wedge q$ |
|---|---|---|
| T | T | T |
| T | F | F |
| F | T | F |
| F | F | F |

(*a*) "$p$ and $q$"

| $p$ | $q$ | $p \vee q$ |
|---|---|---|
| T | T | T |
| T | F | T |
| F | T | T |
| F | F | F |

(*b*) "$p$ or $q$"

| $p$ | $\neg p$ |
|---|---|
| T | F |
| F | T |

(*c*) "not $p$"

**Fig. 10-1**

**EXAMPLE 10.2** Consider the following four statements:

(i) Paris is in France and $2 + 2 = 4$.
(ii) Paris is in France and $2 + 2 = 5$.
(iii) Paris is in England and $2 + 2 = 4$.
(iv) Paris is in England and $2 + 2 = 5$.

Only the first statement is true. Each of the other statements is false since at least one of its substatements is false.

### Disjunction, $p \vee q$

Any two propositions can be combined by the word "or" to form a compound proposition called the *disjunction* of the original propositions. Symbolically,

$$p \vee q$$

read "$p$ or $q$", denotes the disjunction of $p$ and $q$. The truth value of $p \vee q$ depends only on the truth values of $p$ and $q$ as follows.

**Definition 10.2**: If $p$ and $q$ are false, then $p \vee q$ is false; otherwise $p \vee q$ is true.

The truth value of $p \vee q$ may be defined equivalently by the table in Fig. 10-1($b$). Observe that $p \vee q$ is false only in the fourth case when both $p$ and $q$ are false.

**EXAMPLE 10.3** Consider the following four statements:

(i) Paris is in France or $2+2=4$. (iii) Paris is in England or $2+2=4$.
(ii) Paris is in France or $2+2=5$. (iv) Paris is in England or $2+2=5$.

Only the last statement (iv) is false. Each of the other statements is true since at least one of its substatements is true.

**Remark**: The English word "or" is commonly used in two distinct ways. Sometimes it is used in the sense of "$p$ or $q$ or both", i.e., at least one of the two alternatives occurs, as above, and sometimes it is used in the sense of "$p$ or $q$ but not both", i.e., exactly one of the two alternatives occurs. For example, the sentence "He will go to Harvard or to Yale" uses "or" in the latter sense, called the *exclusive disjunction*. Unless otherwise stated, "or" shall be used in the former sense. This discussion points out the precision we gain from our symbolic language: $p \vee q$ is defined by its truth table and *always* means "$p$ and/or $q$".

### Negation, $\neg p$

Given any proposition $p$, another proposition, called the *negation* of $p$, can be formed by writing "It is not the case that . . ." or "It is false that . . ." before $p$ or, if possible, by inserting in $p$ the word "not". Symbolically,

$$\neg p$$

read "not $p$", denotes the negation of $p$. The truth value of $\neg p$ depends on the truth value of $p$ as follows.

**Definition 10.3**: If $p$ is true, then $\neg p$ is false; and if $p$ is false, then $\neg p$ is true.

The truth value of $\neg p$ may be defined equivalently by the table in Fig. 10-3($c$). Thus the truth value of the negation of $p$ is always the opposite of the truth value of $p$.

**EXAMPLE 10.4** Consider the following six statements.

($a_1$) Paris is in France. ($b_1$) $2+2=5$.

($a_2$) It is not the case that Paris is in France. ($b_2$) It is not the case that $2+2=5$.

($a_3$) Paris is not in France. ($b_3$) $2+2 \neq 5$.

Then ($a_2$) and ($a_3$) are each the negation of ($a_1$); and ($b_2$) and ($b_3$) are each the negation of ($b_1$). Since ($a_1$) is true, ($a_2$) and ($a_3$) are false; and since ($b_1$) is false, ($b_2$) and ($b_3$) are true.

**Remark**: The logical notation for the connectives "and", "or", and "not" are not completely standard. For example, some texts use:

| | |
|---|---|
| $p\ \&\ q$, $p \cdot q$ or $pq$ | for $p \wedge q$ |
| $p+q$ | for $p \vee q$ |
| $p'$, $\bar{p}$ or $\sim p$ | for $\neg p$ |

## 10.4 PROPOSITIONS AND TRUTH TABLES

Let $P(p, q, \ldots)$ denote an expression constructed from logical variables $p, q, \ldots$, which take on the value TRUE (T) or FALSE (F), and the logical connectives $\wedge$, $\vee$, and $\neg$ (and others discussed subsequently). Such an expression $P(p, q, \ldots)$ will be called a *proposition*.

The main property of a proposition $P(p, q, \ldots)$ is that its truth value depends exclusively upon the truth values of its variables, that is, the truth value of a proposition is known once the truth value of each of its variables is known. A simple concise way to show this relationship is through a *truth table*. We describe a way to obtain such a truth table below.

Consider, for example, the proposition $\neg(p \wedge \neg q)$. Figure 10-2(*a*) indicates how the truth table of $\neg(p \wedge \neg q)$ is constructed. Observe that the first columns of the table are for the variables $p, q, \ldots$ and and that there are enough rows in the table to allow for all possible combinations of T and F for these variables. (For 2 variables, 4 rows are necessary; for 3 variables, 8 rows are necessary; and, in general, for $n$ variables $2^n$ rows are required.) There is then a column for each "elementary" stage of the construction of the proposition, the truth table at each step being determined from the previous stages by the definitions of the connectives $\wedge$, $\vee$, $\neg$. Finally we obtain the truth value of the proposition, which appears in the last column.

The actual truth table of the proposition $\neg(p \wedge \neg q)$ is shown in Fig. 10-2(*b*). It consists precisely of the columns in Fig. 10-2(*a*) which appear under the variables and under the proposition; the other columns were merely used in the construction of the truth table.

| $p$ | $q$ | $\neg q$ | $p \wedge \neg q$ | $\neg(p \wedge \neg q)$ |
|---|---|---|---|---|
| T | T | F | F | T |
| T | F | T | T | F |
| F | T | F | F | T |
| F | F | T | F | T |

(*a*)

| $p$ | $q$ | $\neg(p \wedge \neg q)$ |
|---|---|---|
| T | T | T |
| T | F | F |
| F | T | T |
| F | F | T |

(*b*)

**Fig. 10-2**

**Remark**: In order to avoid an excessive number of parentheses, we sometimes adopt an order of precedence for the logical connectives. Specifically:

$\neg$ has precedence over $\wedge$ which has precedence over $\vee$.

For example, $\neg p \wedge q$ means $(\neg p) \wedge q$ and not $\neg(p \wedge q)$.

### Alternative Method for Constructing a Truth Table

Another way to construct the truth table for $\neg(p \wedge \neg q)$ follows:

(*a*) First we construct the truth table shown in Fig. 10-3. That is, first we list all the variables and the combinations of their truth values. Then the proposition is written on the top row to the right of its variables with sufficient space so that there is a column under each variable and each connective in the proposition. Also there is a final row labeled "Step".

| $p$ | $q$ | $\neg$ | $(p$ | $\wedge$ | $\neg$ | $q)$ |
|---|---|---|---|---|---|---|
| T | T | | | | | |
| T | F | | | | | |
| F | T | | | | | |
| F | F | | | | | |
| Step | | | | | | |

**Fig. 10-3**

(*b*) Next, additional truth values are entered into the truth table in various steps as shown in Fig. 10-4. That is, first the truth values of the variables are entered under the variables in the proposition, and then there is a column of truth values entered under each logcial operation. We also indicate the step in which each column of truth values is entered in the table.

The truth table of the proposition then consists of the original columns under the variables and the last step, that is, the last column entered into the table.

| $p$ | $q$ | $\neg$ | $(p$ | $\wedge$ | $\neg$ | $q)$ |
|---|---|---|---|---|---|---|
| T | T | | T | | | T |
| T | F | | T | | | F |
| F | T | | F | | | T |
| F | F | | F | | | F |
| Step | | | 1 | | | 1 |

(*a*)

| $p$ | $q$ | $\neg$ | $(p$ | $\wedge$ | $\neg$ | $q)$ |
|---|---|---|---|---|---|---|
| T | T | | T | | F | T |
| T | F | | T | | T | F |
| F | T | | F | | F | T |
| F | F | | F | | T | F |
| Step | | | 1 | | 2 | 1 |

(*b*)

| $p$ | $q$ | $\neg$ | $(p$ | $\wedge$ | $\neg$ | $q)$ |
|---|---|---|---|---|---|---|
| T | T | | T | F | F | T |
| T | F | | T | T | T | F |
| F | T | | F | F | F | T |
| F | F | | F | F | T | F |
| Step | | | 1 | 3 | 2 | 1 |

(*c*)

| $p$ | $q$ | $\neg$ | $(p$ | $\wedge$ | $\neg$ | $q)$ |
|---|---|---|---|---|---|---|
| T | T | T | T | F | F | T |
| F | T | F | T | T | T | F |
| F | F | T | F | F | F | T |
| F | F | T | F | F | T | F |
| Step | | 4 | 1 | 3 | 2 | 1 |

(*d*)

**Fig. 10-4**

## 10.5 TAUTOLOGIES AND CONTRADICTIONS

Some propositions $P(p, q, \ldots)$ contain only T in the last column of their truth tables or, in other words, they are true for any truth values of their variables. Such propositions are called *tautologies*. Analogously, a proposition $P(p, q, \ldots)$ is called a *contradiction* if it contains only F in the last column of its truth table or, in other words, if it is false for any truth values of its variables. For example, the proposition "$p$ or not $p$", that is, $p \vee \neg p$, is a tautology, and the proposition "$p$ and not $p$", that is, $p \wedge \neg p$, is a contradiction. This is verified by looking at their truth tables in Fig. 10-5. (The truth tables have only two rows since each proposition has only the one variable $p$.)

| $p$ | $\neg p$ | $p \vee \neg p$ |
|---|---|---|
| T | F | T |
| F | T | T |

(*a*) $p \vee \neg p$

| $p$ | $\neg p$ | $p \wedge \neg p$ |
|---|---|---|
| T | F | F |
| F | T | F |

(*b*) $p \wedge \neg p$

**Fig. 10-5**

Note that the negation of a tautology is a contradiction since it is always false, and the negation of a contradiction is a tautology since it is always true.

Now let $P(p, q, \ldots)$ be a tautology, and let $P_1(p, q, \ldots)$, $P_2(p, q, \ldots) \ldots$ be any propositions. Since $P(p, q, \ldots)$ does not depend upon the particular truth values of its variables $p, q, \ldots$, we can substitute $P_1$ for $p$, $P_2$ for $q, \ldots$ in the tautology $P(p, q, \ldots)$ and still have a tautology. We state this result formally.

**Theorem 10.1** **(Principle of Substitution)**: If $P(p, q, \ldots)$ is a tautology, then $P(P_1, P_2, \ldots)$ is a tautology for any propositions $P_1, P_2, \ldots$.

## 10.6 LOGICAL EQUIVALENCE

Two propositions $P(p, q, \ldots)$ and $Q(p, q, \ldots)$ are said to be *logically equivalent*, or simply *equivalent* or *equal*, denoted by

$$P(p, q, \ldots) \equiv Q(p, q, \ldots)$$

if they have identical truth tables. Consider, for example, the truth tables of $\neg(p \wedge q)$ and $\neg p \vee \neg q$ appearing in Fig. 10-6. Observe that both truth tables are the same, that is, both propositions are false in the first case and true in the other three cases. Accordingly, we can write

$$\neg(p \wedge q) \equiv \neg p \vee \neg q$$

In other words, the propositions are logically equivalent.

| $p$ | $q$ | $p \wedge q$ | $\neg(p \wedge q)$ |
|---|---|---|---|
| T | T | T | F |
| T | F | F | T |
| F | T | F | T |
| F | F | F | T |

(*a*) $\neg(p \wedge q)$

| $p$ | $q$ | $\neg p$ | $\neg q$ | $\neg p \vee \neg q$ |
|---|---|---|---|---|
| T | T | F | F | F |
| T | F | F | T | T |
| F | T | T | F | T |
| F | F | T | T | T |

(*b*) $\neg p \vee \neg q$

**Fig. 10-6**

**Remark**: Consider the statement

"It is not the case that roses are red and violets are blue"

This statement can be written in the form $\neg(p \wedge q)$, where

$p$ is "roses are red" and $q$ is "violets are blue"

However, as noted above, $\neg(p \wedge q) \equiv \neg p \vee \neg q$. Thus the statement

"Roses are not red, or violets are not blue"

has the same meaning as the given statement.

## 10.7 ALGEBRA OF PROPOSITIONS

Propositions satisfy various laws which are listed in Table 10-1. (In this table, T and F are restricted to the truth values "true" and "false" respectively.) We state this result formally.

**Theorem 10.2**: Propositions satisfy the laws of Table 10-1.

**Table 10-1 Laws of the Algebra of Propositions**

| | |
|---|---|
| **Idempotent laws** | |
| (1*a*) $p \vee p \equiv p$ | (1*b*) $p \wedge p \equiv p$ |
| **Associative laws** | |
| (2*a*) $(p \vee q) \vee r \equiv p \vee (q \vee r)$ | (2*b*) $(p \wedge q) \wedge r \equiv p \wedge (q \wedge r)$ |
| **Commutative laws** | |
| (3*a*) $p \vee q \equiv q \vee p$ | (3*b*) $p \wedge q \equiv q \wedge p$ |
| **Distributive laws** | |
| (4*a*) $p \vee (q \wedge r) \equiv (p \vee q) \wedge (p \vee r)$ | (4*b*) $p \wedge (q \vee r) \equiv (p \wedge q) \vee (p \wedge r)$ |
| **Identity laws** | |
| (5*a*) $p \vee \mathrm{T} \equiv p$ | (5*b*) $p \wedge \mathrm{F} \equiv p$ |
| (6*a*) $p \vee \mathrm{T} \equiv \mathrm{T}$ | (6*b*) $p \wedge \mathrm{F} \equiv \mathrm{F}$ |
| **Complement laws** | |
| (7*a*) $p \vee \neg p \equiv \mathrm{T}$ | (8*a*) $\neg \mathrm{T} \equiv \mathrm{F}$ |
| (7*b*) $p \wedge \neg p \equiv \mathrm{F}$ | (8*b*) $\neg \mathrm{F} \equiv \mathrm{T}$ |
| **Involution law** | |
| (9) $\neg\neg p \equiv p$ | |
| **DeMorgan's laws** | |
| (10*a*) $\neg(p \vee q) \equiv \neg p \wedge \neg q$ | (10*b*) $\neg(p \wedge q) \equiv \neg p \vee \neg q$ |

## 10.8 CONDITIONAL AND BICONDITIONAL STATEMENTS

Many statements, particularly in mathematics, are of the form "If $p$ then $q$". Such statements are called *conditional* statements, and are denoted by

$$p \rightarrow q$$

The conditional $p \rightarrow q$ is frequently read "$p$ implies $q$" or "$p$ only if $q$".

Another common statement is of the form "$p$ if and only if $q$". Such statements are called *biconditional* statements, and are denoted by

$$p \leftrightarrow q$$

The truth values of $p \rightarrow q$ and $p \leftrightarrow q$ are defined by the tables in Fig. 10-7. Observe that:

(*a*) The conditional $p \rightarrow q$ is false only when the first part $p$ is true and the second part $q$ is false. Accordingly, when $p$ is false, the conditional $p \rightarrow q$ is true regardless of the truth value of $q$.

(*b*) The biconditional $p \leftrightarrow q$ is true whenever $p$ and $q$ have the same truth values and false otherwise.

| $p$ | $q$ | $p \rightarrow q$ |
|---|---|---|
| T | T | T |
| T | F | F |
| F | T | T |
| F | F | T |

(*a*) $p \rightarrow q$

| $p$ | $q$ | $p \leftrightarrow q$ |
|---|---|---|
| T | T | T |
| T | F | F |
| F | T | F |
| F | F | T |

(*b*) $p \leftrightarrow q$

**Fig. 10-7**

The truth table of the proposition $\neg p \vee q$ appears in Fig. 10-8. Observe that the truth tables of $\neg p \vee q$ and $p \rightarrow q$ are identical, that is, they are both false only in the second case. Accordingly, $p \rightarrow q$ is logically equivalent to $\neg p \vee q$; that is,

$$p \rightarrow q \equiv \neg p \vee q$$

In other words, the conditional statement "If $p$ then $q$" is logically equivalent to the statement "Not $p$ or $q$" which only involves the connectives $\vee$ and $\neg$ and thus was already a part of our language. We may regard $p \rightarrow q$ as an abbreviation for an oft-recurring statement.

| $p$ | $q$ | $\neg p$ | $\neg p \vee q$ |
|---|---|---|---|
| T | T | F | T |
| T | F | F | F |
| F | T | T | T |
| F | F | T | T |

$\neg p \vee q$

Fig. 10-8

## 10.9 ARGUMENTS

An *argument* is an assertion that a given set of propositions $P_1, P_2, \ldots, P_n$, called *premises*, yields (has as a consequence) another proposition $Q$, called the *conclusion*. Such an argument is denoted by

$$P_1, P_2, \ldots, P_n \vdash Q$$

The notion of a "logical argument" or "valid argument" is formalized as follows.

**Definition 10.4**: An argument $P_1, P_2, \ldots, P_n \vdash Q$ is said to be *valid* if $Q$ is true whenever all the premises $P_1, P_2, \ldots, P_n$ are true. An argument which is not valid is called a *fallacy*.

**EXAMPLE 10.5**

(*a*) The following argument is valid:

$$p, p \rightarrow q \vdash q \text{ (Law of Detachment)}$$

The proof of this rule follows from the truth table in Fig. 10-9. Specifically, $p$ and $p \rightarrow q$ are true simultaneously only in Case (row) 1, and in this case $q$ is true.

| $p$ | $q$ | $p \rightarrow q$ |
|---|---|---|
| T | T | T |
| T | F | F |
| F | T | T |
| F | F | T |

Fig. 10-9

(*b*) The following argument is a fallacy:

$$p \rightarrow q,\ q \vdash p$$

For $p \rightarrow q$ and $q$ are both true in Case (row) 3 in the truth table in Fig. 10-9, but in this case $p$ is false.

Now the propositions $P_1, P_2, \ldots, P_n$ are true simultaneously if and only if the proposition $P_1 \wedge P_2 \wedge \cdots \wedge P_n$ is true. Thus the argument $P_1, P_2, \ldots, P_n \vdash Q$ is valid if and only if $Q$ is true whenever $P_1 \wedge P_2 \wedge \cdots \wedge P_n$ is true or, equivalently, if the proposition $(P_1 \wedge P_2 \wedge \cdots \wedge P_n) \rightarrow Q$ is a tautology. We state this result formally.

**Theorem 10.3**: The argument $P_1, P_2, \ldots, P_n \vdash Q$ is valid if and only if the proposition $(P_1 \wedge P_2 \wedge \cdots \wedge P_n) \rightarrow Q$ is a tautology.

We apply this theorem in the next example.

**EXAMPLE 10.6** A fundamental principle of logical reasoning states:

"If $p$ implies $q$ and $q$ implies $r$, then $p$ implies $r$"

That is, the following argument is valid:

$$p \rightarrow q, \; q \rightarrow r \; \vdash \; p \rightarrow r \quad \textit{(Law of Syllogism)}$$

This fact is verified by the truth table in Fig. 10-10, which shows that the following proposition is a tautology:

$$[(p \rightarrow q) \wedge (q \rightarrow r)] \rightarrow (p \rightarrow r)$$

Equivalently, the argument is valid since the premises $p \rightarrow q$ and $q \rightarrow r$ are true simultaneously only in Cases (rows) 1, 5, 7, 8 and in these cases the conclusion $p \rightarrow r$ is also true. (Observe that the truth table required $2^3 = 8$ lines since there are three variables, $p, q, r$.)

| $p$ | $q$ | $r$ | $([p$ | $\rightarrow$ | $q)$ | $\wedge$ | $(q$ | $\rightarrow$ | $r)]$ | $\rightarrow$ | $(p$ | $\rightarrow$ | $r)$ |
|---|---|---|---|---|---|---|---|---|---|---|---|---|---|
| T | T | T | T | T | T | T | T | T | T | T | T | T | T |
| T | T | F | T | T | T | F | T | F | F | T | T | F | F |
| T | F | T | T | F | F | F | F | T | T | T | T | T | T |
| T | F | F | T | F | F | F | F | T | F | T | T | F | F |
| F | T | T | F | T | T | T | T | T | T | T | F | T | T |
| F | T | F | F | T | T | F | T | F | F | T | F | T | F |
| F | F | T | F | T | F | T | F | T | T | T | F | T | T |
| F | F | F | F | T | F | T | F | T | F | T | F | T | F |
| Step | | | 1 | 2 | 1 | 3 | 1 | 2 | 1 | 4 | 1 | 2 | 1 |

**Fig. 10-10**

We now apply the above theory to arguments involving specific statements. We emphasize that the validity of an argument does not depend upon the truth values nor the content of the statements appearing in the argument, but upon the particular form of the argument. This is illustrated in the following example.

**EXAMPLE 10.7** Consider the following argument:

$S_1$: If a man is a bachelor, he is unhappy.
$S_2$: If a man is unhappy, he dies young.

.................................................................

$S$: Bachelors die young.

Here the statement $S$ below the line denotes the conclusion of the argument, and the statements $S_1$ and $S_2$ above the line denote the premises. We claim that the argument $S_1, S_2 \vdash S$ is valid. For the argument is of the form

$$p \rightarrow q, \; q \rightarrow r \; \vdash \; p \rightarrow r$$

where $p$ is "He is a bachelor", $q$ is "He is unhappy" and $r$ is "He dies young"; and by Example 10.6 this argument (law of syllogism) is valid.

## 10.10 LOGICAL IMPLICATION

A proposition $P(p, q, \ldots)$ is said to *logically imply* a proposition $Q(p, q, \ldots)$, written

$$P(p, q, \ldots) \Rightarrow Q(p, q, \ldots)$$

if $Q(p, q, \ldots)$ is true whenever $P(p, q, \ldots)$ is true.

**EXAMPLE 10.8** We claim that $p$ logically implies $p \vee q$. For consider the truth table in Fig. 10-11. Observe that $p$ is true in Cases (rows) 1 and 2, and in these cases $p \vee q$ is also true. Thus $p \Rightarrow p \vee q$.

| $p$ | $q$ | $p \vee q$ |
|---|---|---|
| T | T | T |
| T | F | T |
| F | T | T |
| F | F | F |

**Fig. 10-11**

Now if $Q(p, q, \ldots)$ is true whenever $P(p, q, \ldots)$ is true, then the argument

$$P(p, q, \ldots) \vdash Q(p, q, \ldots)$$

is valid; and conversely. Furthermore, the argument $P \vdash Q$ is valid if and only if the conditional statement $P \rightarrow Q$ is always true, i.e., a tautology. We state this result formally.

**Theorem 10.4**: For any propositions $P(p, q, \ldots)$ and $Q(p, q, \ldots)$ the following three statements are equivalent:

(i) $P(p, q, \ldots)$ logically implies $Q(p, q, \ldots)$.

(ii) The argument $P(p, q, \ldots) \vdash Q(p, q, \ldots)$ is valid.

(iii) The proposition $P(p, q, \ldots) \rightarrow Q(p, q, \ldots)$ is a tautology.

We note that some logicians and many texts use the word "implies" in the same sense as we use "logically implies", and so they distinguish between "implies" and "if . . . then". These two distinct concepts are, of course, intimately related as seen in the above theorem.

## 10.11 PROPOSITIONAL FUNCTIONS, QUANTIFIERS

Let $A$ be a given set. A *propositional function* (or an *open sentence* or *condition*) defined on $A$ is an expression

$$p(x)$$

which has the property that $p(a)$ is true or false for each $a \in A$. That is, $p(x)$ becomes a statement (with a truth value) whenever any element $a \in A$ is substituted for the variable $x$. The set $A$ is called the *domain* of $p(x)$, and the set $T_p$ of all elements of $A$ for which $p(a)$ is true is called the *truth set* of $p(x)$. In other words,

$$T_p = \{x : x \in A,\ p(x) \text{ is true}\} \quad \text{or} \quad T_p = \{x : p(x)\}$$

Frequently, when $A$ is some set of numbers, the condition $p(x)$ has the form of an equation or inequality involving the variable $x$.

**EXAMPLE 10.9** Find the truth set $T_p$ of each propositional function $p(x)$ defined on the set $\mathbf{P} = \{1, 2, 3, \ldots\}$.

(*a*) Let $p(x)$ be "$x + 2 > 7$". Then

$$T_p = \{x : x \in \mathbf{P},\ x + 2 > 7\} = \{6, 7, 8, \ldots\}$$

consisting of all integers greater than 5.

(*b*) Let $p(x)$ be "$x + 5 < 3$". Then

$$T_p = \{x : x \in \mathbf{P},\ x + 5 < 3\} = \varnothing$$

the empty set. In other words, $p(x)$ is not true for any positive integer in **P**.

(*c*) Let $p(x)$ be "$x + 5 > 1$". Then

$$T_p = \{x : x \in \mathbf{P},\ x + 5 > 1\} = \mathbf{P}$$

Thus $p(x)$ is true for every element in **P**.

**Remark**: The above example shows that if $p(x)$ is a propositional function defined on a set $A$ then $p(x)$ could be true for all $x \in A$, for some $x \in A$, or for no $x \in A$. The next two subsections discusses quantifiers related to such propositional functions.

### Universal Quantifier

Let $p(x)$ be a propositional function defined on a set $A$. Consider the expression

$$(\forall x \in A)p(x) \qquad \text{or} \qquad \forall x\, p(x) \tag{10.1}$$

which reads "For every $x$ in $A$, $p(x)$ is a true statement" or, simply, "For all $x$, $p(x)$". The symbol

$$\forall$$

which reads "for all" or "for every" is called the *universal quantifier*. The statement (*10.1*) is equivalent to the statement

$$T_p = \{x : x \in A,\ p(x)\} = A \tag{10.2}$$

that is, that the truth set of $p(x)$ is the entire set $A$.

The expression $p(x)$ by itself is an open sentence or condition and therefore has no truth value. However, $\forall x, p(x)$ that is, $p(x)$ preceded by the quantifier $\forall$, does have a truth value which follows from the equivalence of (*10.1*) and (*10.2*). Specifically:

| $Q_1$: If $\{x : x \in A,\ p(x)\} = A$ then $\forall x, p(x)$ is true; otherwise, $\forall x,\ p(x)$ is false. |
|---|

**EXAMPLE 10.10**

(*a*) The proposition $(\forall n \in \mathbf{P})\ (n + 4 > 3)$ is true since

$$\{n : n + 4 > 3\} = \{1, 2, 3, \ldots\} = \mathbf{P}$$

(*b*) The proposition $(\forall n \in \mathbf{P})\ (n + 2 > 8)$ is false since

$$\{n : n + 2 > 8\} = \{7, 8, \ldots\} \neq \mathbf{P}$$

(*c*) The symbol $\forall$ can be used to define the intersection of an indexed collection $\{A_i : i \in I\}$ of sets $A_i$ as follows:

$$\bigcap(A_i : i \in I) = \{x : \forall i \in I,\ x \in A_i\}$$

**Existential Quantifier**

Let $p(x)$ be a propositional function defined on a set $A$. Consider the expression

$$(\exists x \in A)p(x) \quad \text{or} \quad \exists x,\ p(x) \tag{10.3}$$

which reads "There exists an $x$ in $A$ such that $p(x)$ is a true statement" or, simply, "For some $x$, $p(x)$". The symbol

$$\exists$$

which reads "there exists" or "for some" or "for at least one" is called the *existential quantifier*. Statement (*10.3*) is equivalent to the statement

$$T_p = \{x : x \in A,\ p(x)\} \neq \varnothing \tag{10.4}$$

i.e., that the truth set of $p(x)$ is not empty. Accordingly, $\exists x, p(x)$, that is, $p(x)$ preceded by the existential quantifier $\exists$ does have a truth value. Specifically,

> $Q_2$: If $\{x : p(x)\} \neq \varnothing$ then $\exists x, p(x)$ is true; otherwise, $\exists x, p(x)$ is false.

**EXAMPLE 10.11**

(*a*) The proposition $(\exists n \in \mathbf{P})\ (n + 4 < 7)$ is true since

$$\{n : n + 4 < 7\} = \{1, 2\} \neq \varnothing$$

(*b*) The proposition $(\exists n \in \mathbf{P})\ (n + 6 < 4)$ is false since

$$\{n : n + 6 < 4\} = \varnothing$$

(*c*) The symbol $\exists$ can be used to define the union of an indexed collection $\{A_i : i \in I\}$ of sets $A_i$ as follows:

$$\bigcup(A_i : i \in I) = \{x : \exists i \in I,\ x \in A_i\}$$

**Notation**

Let $A = \{2, 3, 5\}$ and let $p(x)$ be the sentence "$x$ is a prime number" or, simply "$x$ is prime". Then the proposition

"Two is prime and three is prime and five is prime" (*)

can be denoted by

$$p(2) \wedge p(3) \wedge p(5) \quad \text{or} \quad \wedge\,(a \in A,\ p(a))$$

which is equivalent to the statement

"Every number in $A$ is prime" or $\forall a \in A,\ p(a)$ (**)

Similarly, the proposition

"Two is prime or three is prime or five is prime"

can be denoted by

$$p(2) \vee p(3) \vee p(5) \quad \text{or} \quad \vee\,(a \in A,\ p(a))$$

which is equivalent to the statement

"At least one number in $A$ is prime" or $\exists a \in A,\ p(a)$

Alternatively, we can write

$$\wedge(a \in A,\ p(a)) \equiv \forall a \in A,\ p(a) \quad \text{and} \quad \vee\,(a \in A,\ p(a)) \equiv \exists a \in a,\ p(a)$$

where the symbols $\wedge$ and $\vee$ are used instead of $\forall$ and $\exists$.

**Remark**: If $A$ were an infinite set, then a statement of the form $(*)$ could be made since the sentence would not end; but a statement of the form $(**)$ can always be made, even when $A$ is infinite.

## 10.12 NEGATION OF QUANTIFIED STATEMENTS

Consider the statement: "All math majors are male". Its negation is either of the following equivalent statements:

"It is not the case that all math majors are male"

"There exists at least one math major who is a female (not male)"

Symbolically, using $M$ to denoted the set of math majors, the above can be written as

$$\neg(\forall x \in M)\ (x \text{ is male}) \equiv (\exists x \in M)\ (x \text{ is not male})$$

or, when $p(x)$ denotes "$x$ is male",

$$\neg(\forall x \in M)p(x) \equiv (\exists x \in M)\neg p(x) \qquad \text{or} \qquad \neg\forall x, p(x) \equiv \exists x \neg p(x)$$

The above is true for any proposition $p(x)$. That is:

**Theorem 10.5 (DeMorgan)**: $\neg(\forall x \in A)p(x) \equiv (\exists x \in A)\neg p(x)$.

In other words, the following two statements are equivalent:

(1) It is not true that, for all $a \in A$, $p(a)$ is true.

(2) There exists an $a \in A$ such that $p(a)$ is false.

There is an analogous theorem for the negation of a proposition which contains the existential quantifier.

**Theorem 10.6 (DeMorgan)**: $\neg(\exists x \in A)p(x) \equiv (\forall x \in A)\neg p(x)$.

That is, the following two statements are equivalent:

(1) It is not true that for some $a \in A$, $p(a)$ is true.

(2) For all $a \in A$, $p(a)$ is false.

**EXAMPLE 10.12**

(*a*) The following statements are negatives of each other:

"For all positive integers $n$ we have $n + 2 > 8$"

"There exists a positive integer $n$ such that $n + 2 \not> 8$"

(*b*) The following statements are also negatives of each other:

"There exists a college student who is 60 years old"

"Every college student is not 60 years old"

**Remark**: The expression $\neg p(x)$ has the obvious meaning; that is:

"The statement $\neg p(a)$ is true when $p(a)$ is false, and vice versa"

Previously, $\neg$ was used as an operation on statements; here $\neg$ is used as an operation on propositional functions. Similarly, $p(x) \wedge q(x)$, read "$p(x)$ and $q(x)$", is defined by:

"The statement $p(a) \wedge q(a)$ is true when $p(a)$ and $q(a)$ are true"

Similarly, $p(x) \vee q(x)$, read "$p(x)$ or $q(x)$", is defined by:

"The statement $p(a) \vee q(a)$ is true when $p(a)$ or $q(a)$ is true"

Thus in terms of truth sets:

(i) $\neg p(x)$ is the complement of $p(x)$.

(ii) $p(x) \wedge q(x)$ is the intersection of $p(x)$ and $q(x)$.

(iii) $p(x) \vee q(x)$ is the union of $p(x)$ and $q(x)$.

One can also show that the laws for propositions also hold for propositional functions. For example, we have DeMorgan's laws:

$$\neg(p(x) \wedge q(x)) \equiv \neg p(x) \vee \neg q(x) \quad \text{and} \quad \neg(p(x) \vee q(x)) \equiv \neg p(x) \wedge \neg q(x)$$

**Counterexample**

Theorem 10.6 tells us that to show that a statement $\forall x, p(x)$ is false, it is equivalent to show that $\exists x \neg p(x)$ is true or, in other words, that there is an element $x_0$ with the property that $p(x_0)$ is false. Such an element $x_0$ is called a *counterexample* to the statement $\forall x, p(x)$.

**EXAMPLE 10.13**

(*a*) Consider the statement $\forall x \in \mathbf{R}, |x| \neq 0$. The statement is false since 0 is a counterexample, that is, $|0| \neq 0$ is not true.

(*b*) Consider the statement $\forall x \in \mathbf{R}, x^2 \geq x$. The statement is not true since, for example, 1/2 is a counterexample. Specifically, $(1/2)^2 \geq 1/2$ is not true, that is, $(1/2)^2 < 1/2$.

(*c*) Consider the statement $\forall x \in \mathbf{P}, x^2 \geq x$. This statement is true where **P** is the set of positive integers. In other words, there does not exist a positive integer $n$ for which $n^2 < n$.

**Propositional Functions with More than One Variable**

A propositional function (of $n$ variables) defined over a product set $A = A_1 \times \cdots \times A_n$ is an expression

$$p(x_1, x_2, \ldots, x_n)$$

which has the property that $p(a_1, a_2, \ldots, a_n)$ is true or false for any $n$-tuple $(a_1, \ldots, a_n)$ in $A$. For example,

$$x + 2y + 3z < 18$$

is a propositional function on $\mathbf{P}^3 = \mathbf{P} \times \mathbf{P} \times \mathbf{P}$. Such a propositional function has no truth value. However, we do have the following:

**Basic Principle:** A propositional function preceded by a quantifier for each variable, for example,

$$\forall x \, \exists y, \; p(x, y) \quad \text{or} \quad \exists x \, \forall y \, \exists z, \; p(x, y, z)$$

denotes a statement and has a truth value.

**EXAMPLE 10.14** Let $B = \{1, 2, 3, \ldots, 9\}$ and let $p(x, y)$ denote "$x + y = 10$". Then $p(x, y)$ is a propositional function on $A = B^2 = B \times B$.

(*a*) The following is a statement since there is a quantifier for each variable:

$$\forall x \, \exists y, \; p(x, y) \quad \text{that is,} \quad \text{"For every } x\text{, there exists a } y \text{ such that } x + y = 10\text{"}$$

This statement is true. For example, if $x = 1$, let $y = 9$; if $x = 2$, let $y = 8$, and so on.

(*b*) The following is also a statement:

$\exists y\, \forall x,\ p(x, y)$, that is, "There exists a $y$ such that, for every $x$, we have $x + y = 10$"

No such $y$ exists; hence this statement is false.

> **Warning!** Observe that the only difference between (*a*) and (*b*) in the above Example 10.14 is the order of the quantifiers. Thus a different ordering of the quantifiers may yield a different statement.

We note that, when translating quantified statements into English, the expression "such that" frequently follows "there exists".

### Negating Quantified Statements with More than One Variable

Quantified statements with more than one variable may be negated by successively applying Theorems 10.5 and 10.6. Thus each $\forall$ is changed to $\exists$, and each $\exists$ is changed to $\forall$ as the negation symbol $\neg$ passes through the statement from left to right. For example

$$\begin{aligned}\neg[\forall x \exists y \exists z,\ p(x, y, z)] &\equiv \exists x \neg[\exists y \exists z,\ p(x, y, z)] \equiv \exists x \forall y[\neg \exists z,\ p(x, y, z)\\ &\equiv \exists x \forall y \forall z,\ \neg p(x, y, z)\end{aligned}$$

Naturally, we do not put in all the steps when negating such quantified statements.

**EXAMPLE 10.15**

(*a*) Consider the quantified statement:

"Every student has at least one course where the lecturer is a teaching assistant"

Its negation is the statement:

"There is a student such that in every course the lecturer is not a teaching assistant"

(*b*) The formal definition that $L$ is the limit of a sequence $a_1, a_2, \ldots$ follows:

$$\forall \varepsilon > 0,\ \exists n_0 \in \mathbf{P},\ \forall n > n_0,\ |a_n - L| < \varepsilon$$

Thus $L$ is not the limit of the sequence $a_1,\ a_2, \ldots$ when

$$\exists \varepsilon > 0,\ \forall n_0 \in \mathbf{P},\ \exists n > n_0,\ |a_n - L| \geq \varepsilon$$

## Solved Problems

### PROPOSITIONS AND LOGICAL OPERATIONS

**10.1.** Let $p$ be "It is cold" and let $q$ be "It is raining". Give a simple verbal sentence which describes each of the following statements: (*a*) $\neg p$; (*b*) $p \wedge q$; (*c*) $p \vee q$; (*d*) $q \vee \neg p$.

In each case, translate $\wedge$, $\vee$ and $\sim$ to read "and", "or", and "It is false that" or "not", respectively, and then simplify the English sentence.

(*a*) It is not cold. (*c*) It is cold or it is raining.

(*b*) It is cold and raining. (*d*) It is raining or it is not cold.

**10.2.** Let $p$ be "Erik reads *Newsweek*", let $q$ be "Erik reads *The New Yorker*", and let $r$ be "Erik reads *Time*". Write each of the following in symbolic form:

(*a*) Erik reads *Newsweek* or *The New Yorker*, but not *Time*.

(*b*) Erik reads *Newsweek* and *The New Yorker*, or he does not read *Newsweek* and *Time*.

(*c*) It is not true that Erik reads *Newsweek* but not *Time*.

(*d*) It is not true that Erik reads *Time* or *The New Yorker* but not *Newsweek*.

Use $\vee$ for "or", $\wedge$ for "and" (or, its logical equivalent, "but"), and $\neg$ for "not" (negation).

(*a*) $(p \vee q) \wedge \neg r$; (*b*) $(p \wedge q) \vee \neg (p \wedge r)$; (*c*) $\neg (p \wedge \neg r)$; (*d*) $\neg [(r \vee q) \wedge \neg p]$.

## TRUTH VALUES AND TRUTH TABLES

**10.3.** Determine the truth value of each of the following statements:

(*a*) $4 + 2 = 5$ and $6 + 3 = 9$.

(*b*) $3 + 2 = 5$ and $6 + 1 = 7$.

(*c*) $4 + 5 = 9$ and $1 + 2 = 4$.

(*d*) $3 + 2 = 5$ and $4 + 7 = 11$.

The statement "$p$ and $q$" is true only when both substatements are true. Thus:

(*a*) false, (*b*) true; (*c*) false; (*d*) true.

**10.4.** Find the truth table of $\neg p \wedge q$.

See Fig. 10.12, which gives both methods for constructing the truth table.

| $p$ | $q$ | $\neg p$ | $\neg p \wedge q$ |
|---|---|---|---|
| T | T | F | F |
| T | F | F | F |
| F | T | T | T |
| F | F | T | F |

(*a*) Method 1

| $p$ | $q$ | $\neg$ | $p$ | $\wedge$ | $q$ |
|---|---|---|---|---|---|
| T | T | F | T | F | T |
| T | F | F | T | F | F |
| F | T | T | F | T | T |
| F | F | T | F | F | F |
| | Step | 2 | 1 | 3 | 1 |

(*b*) Method 2

**Fig. 10-12**

**10.5.** Verify that the proposition $p \vee \neg (p \wedge q)$ is a tautology.

Construct the truth table of $p \vee \neg (p \wedge q)$ as shown in Fig. 10.13. Since the truth value of $p \vee \neg (p \wedge q)$ is T for all values of $p$ and $q$, the proposition is a tautology.

| $p$ | $q$ | $p \wedge q$ | $\neg (p \wedge q)$ | $p \vee \neg (p \wedge q)$ |
|---|---|---|---|---|
| T | T | T | F | T |
| T | F | F | T | T |
| F | T | F | T | T |
| F | F | F | T | T |

**Fig. 10-13**

**10.6.** Show that the propositions $\neg(p \wedge q)$ and $\neg p \vee \neg q$ are logically equivalent.

Construct the truth tables for $\neg(p \wedge q)$ and $\neg p \vee \neg q$ as in Fig. 10.14. Since the truth tables are the same (both propositions are false in the first case and true in the other three cases), the propositions $\neg(p \wedge q)$ and $\neg p \vee \neg q$ are logically equivalent and we can write

$$\neg(p \wedge q) \equiv \neg p \vee \neg q$$

| $p$ | $q$ | $p \wedge q$ | $\neg(p \wedge q)$ |
|---|---|---|---|
| T | T | T | F |
| T | F | F | T |
| F | T | F | T |
| F | F | F | T |

(*a*) $\neg(p \wedge q)$

| $p$ | $q$ | $\neg p$ | $\neg q$ | $\neg p \vee \neg q$ |
|---|---|---|---|---|
| T | T | F | F | F |
| T | F | F | T | T |
| F | T | T | F | T |
| F | F | T | T | T |

(*b*) $\neg p \vee \neg q$

**Fig. 10-14**

**10.7.** Using the laws in Table 10-1 to show that $\neg(p \vee q) \vee (\neg p \wedge q) \equiv \neg p$.

| **Statement** | **Reason** |
|---|---|
| (1) $\neg(p \vee q) \vee (\neg p \wedge q) \equiv (\neg p \wedge \neg q) \vee (\neg p \wedge q)$ | DeMorgan's law |
| (2) $\equiv \neg p \wedge (\neg q \vee q)$ | Distributive law |
| (3) $\equiv \neg p \wedge t$ | Complement law |
| (4) $\equiv \neg p$ | Identity law |

## CONDITIONAL STATEMENTS

**10.8.** Rewrite the following statements without using the conditional:

(*a*) If it is cold, he wears a hat.

(*b*) If productivity increases, then wages rise.

Recall that "If $p$ then $q$" is equivalent to "Not $p$ or $q$"; that is, $p \rightarrow q \equiv \neg p \vee q$. Hence,

(*a*) It is not cold or he wears a hat.

(*b*) Productivity does not increase or wages rise.

**10.9.** Determine the contrapositive of each statement:

(*a*) If John is a poet, then he is poor.

(*b*) Only if Marc studies will he pass the test.

(*a*) The contrapositive of $p \rightarrow q$ is $\neg q \rightarrow \neg p$. Hence the contrapositive of the given statement is

"If John is not poor, then he is not a poet"

(*b*) The given statement is equivalent to "If Marc passes the test, then he studied". Hence its contrapositive is

"If Marc does not study, then he will not pass the test"

**10.10.** Write the negation of each statement as simply as possible.

(*a*) If she works, she will earn money.
(*b*) He swims if and only if the water is warm.
(*c*) If it snows, then they do not drive the car.

(*a*) Note that $\neg(p \rightarrow q) \equiv p \wedge \neg q$; hence the negation of the statement follows:

"She works or she will not earn money"

(*b*) Note that $\neg(p \leftrightarrow q) \equiv p \leftrightarrow \neg q \equiv \neg p \leftrightarrow q$; hence the negation of the statement is either of the following:

"He swims if and only if the water is not warm"

"He does not swim if and only if the water is warm"

(*c*) Note that $\neg(p \rightarrow \neg q) \equiv p \wedge \neg\neg q \equiv p \wedge q$. Hence the negation of the statement follows:

"It snows and they drive the car"

## ARGUMENTS

**10.11.** Show that the following argument is a fallacy: $p \rightarrow q, \neg p \vdash \neg q$.

Construct the truth table for $[(p \rightarrow q) \wedge \neg p] \rightarrow \neg q$ as in Fig. 10.15. Since the proposition $[(p \rightarrow q)(\wedge \neg p] \rightarrow \neg q$ is not a tautology, the argument is a fallacy. Equivalently, the argument is a fallacy since in third line of the truth table $p \rightarrow q$ and $\neg p$ are true but $\neg q$ is false.

| $p$ | $q$ | $p \rightarrow q$ | $\neg p$ | $(p \rightarrow q) \wedge \neg p$ | $\neg q$ | $[(p \rightarrow q) \wedge \neg p] \rightarrow \neg q$ |
|---|---|---|---|---|---|---|
| T | T | T | F | F | F | T |
| T | F | F | F | F | T | T |
| F | T | T | T | T | F | F |
| F | F | T | T | T | T | T |

**Fig. 10-15**

**10.12.** Determine the validity of the following argument: $p \rightarrow q, \neg q \vdash \neg p$.

Construct the truth table for $[(p \rightarrow q) \wedge \neg q] \rightarrow \neg p$ as in Fig. 10.16. Since the proposition $[(p \rightarrow q) \vee \neg q] \rightarrow \neg p$ is a tautology, the argument is valid.

| $p$ | $q$ | $[(p$ | $\rightarrow$ | $q)$ | $\wedge$ | $\neg$ | $q]$ | $\rightarrow$ | $\neg$ | $p$ |
|---|---|---|---|---|---|---|---|---|---|---|
| T | T | T | T | T | F | F | T | T | F | T |
| T | F | T | F | F | F | T | F | T | F | T |
| F | T | F | T | T | F | F | T | T | T | F |
| F | F | F | T | F | T | T | F | T | T | F |
| Step | | 1 | 2 | 1 | 3 | 2 | 1 | 4 | 2 | 1 |

**Fig. 10-16**

**10.13.** Prove that the following argument is valid: $p \to \neg q, r \to q, r \vdash \neg p$.

Construct the truth tables of the premises and conclusion as in Fig. 10.17. Now, $p \to \neg q$, $r \to q$, and $r$ are true simultaneously only in the fifth line of the table, where $\neg p$ is also true. Hence, the argument is valid.

| | $p$ | $q$ | $r$ | $p \to \neg q$ | $r \to q$ | $\neg q$ |
|---|---|---|---|---|---|---|
| 1 | T | T | T | F | T | F |
| 2 | T | T | F | F | T | F |
| 3 | T | F | T | T | F | F |
| 4 | T | F | F | T | T | F |
| 5 | F | T | T | T | T | T |
| 6 | F | T | F | T | T | T |
| 7 | F | F | T | T | F | T |
| 8 | F | F | F | T | T | T |

**Fig. 10-17**

**10.14.** Test the validity of the following argument:

If two sides of a triangle are equal, then the opposite angles are equal.
Two sides of a triangle are not equal.

---

The opposite angles are not equal.

First translate the argument into the symbolic form $p \to q, \neg p \vdash \neg q$, where $p$ is "Two sides of a triangle are equal" and $q$ is "The opposite angles are equal". By problem 10.11, this argument is a fallacy.

**Remark**: Although the conclusion *does* follow from the second premise and axioms of Euclidean geometry, the above argument does not constitute such a proof since the argument is a fallacy.

**10.15.** Determine the validity of the following argument:

If 7 is less than 4, then 7 is not a prime number.
7 is not less than 4.

---

7 is a prime number.

First translate the argument into symbolic form. Let $p$ be "7 is less than 4" and $q$ be "7 is a prime number". Then the argument is of the form

$$p \to \neg q, \neg p \vdash q$$

Now, we construct a truth table as shown in Fig. 10.18. The above argument is shown to be a fallacy since, in the fourth line of the truth table, the premises $p \to \neg q$ and $\neg p$ are true, but the conclusion $q$ is false.

**Remark**: The fact that the conclusion of the argument happens to be a true statement is irrelevant to the fact that the argument presented is a fallacy

| $p$ | $q$ | $\neg q$ | $p \to \neg q$ | $\neg p$ |
|---|---|---|---|---|
| T | T | F | F | F |
| T | F | T | T | F |
| F | T | F | T | T |
| F | F | T | T | T |

**Fig. 10-18**

**10.16.** Show that $p \wedge q$ logically implies $p \leftrightarrow q$.

Consider the truth tables of $p \wedge q$ and $p \leftrightarrow q$ shown in Fig. 10.19. Now $p \wedge q$ is true only in the first line of the table and, in this case, the proposition $p \leftrightarrow q$ is also true. Thus $p \wedge q$ logically implies $p \leftrightarrow q$.

| $p$ | $q$ | $p \wedge q$ | $p \leftrightarrow q$ |
|---|---|---|---|
| T | T | T | T |
| T | F | F | F |
| F | T | F | F |
| F | F | F | T |

**Fig. 10-19**

## QUANTIFIERS AND PROPOSITIONAL FUNCTIONS

**10.17.** Let $A = \{1, 2, 3, 4, 5\}$. Determine the truth value of each of the following statements:

(*a*) $(\exists x \in A)(x + 3 = 10)$ (*c*) $(\exists x \in A)(x + 3 < 5)$
(*b*) $(\forall x \in A)(x + 3 < 10)$ (*d*) $(\forall x \in A)(x + 3 \leq 7)$

(*a*) False. For no number in $A$ is a solution to $x + 3 = 10$.

(*b*) True. For every number in $A$ satisfies $x + 3 < 10$.

(*c*) True. For if $x_0 = 1$, then $x_0 + 3 < 5$, i.e., 1 is a solution.

(*d*) False. For if $x_0 = 5$, then $x_0 + 3$ is not less than or equal 7. In other words, 5 is not a solution to the given condition.

**10.18.** Determine the truth value of each of the following statements where $\mathbf{U} = \{1, 2, 3\}$ is the universal set:

(*a*) $\exists x \forall y,\ x^2 < y + 1$; (*b*) $\exists x \forall y,\ x^2 + y^2 < 12$; (*c*) $\forall x \forall y,\ x^2 + y^2 < 12$.

(*a*) True. For if $x = 1$, then 1, 2, and 3 are all solutions to $1 < y + 1$.

(*b*) True. For each $x_0$, let $y = 1$; then $x_0^2 + 1 < 12$ is a true statement.

(*c*) False. For if $x_0 = 2$ and $y_0 = 3$, then $x_0^2 + y_0^2 < 12$ is not a true statement.

**10.19.** Negate each of the following statements:

(*a*) $\exists x \forall y,\ p(x, y)$; (*b*) $\forall x \forall y,\ p(x, y)$; (*c*) $\exists y \exists x \forall z,\ p(x, y, z)$.

Use $\neg \forall x p(x) \equiv \exists x \neg p(x)$ and $\neg \exists x p(x) \equiv \forall x \neg p(x)$:

(*a*) $\neg(\exists x \forall y,\ p(x, y)) \equiv \forall x \exists y \neg p(x, y)$.

(*b*) $\neg(\forall x \forall y,\ p(x, y)) \equiv \exists x \exists y \neg p(x, y)$.

(*c*) $\neg(\exists y \exists x \forall z,\ p(x, y, z)) \equiv \forall y \forall x \exists z \neg p(x, y, z)$.

**10.20.** Let $p(x)$ denote the sentence "$x + 2 > 5$". State whether or not $p(x)$ is a propositional function on each of the following sets: (*a*) $\mathbf{P}$, the set of positive integers; (*b*) $M = \{-1, -2, -3, \ldots\}$; (*c*) $\mathbf{C}$, the set of complex numbers.

(*a*) Yes.

(*b*) Although $p(x)$ is false for every element in $M$, $p(x)$ is still a propositional function on $M$.

(*c*) No. Note that $2i + 2 > 5$ does not have any meaning. In other words, inequalities are not defined for complex numbers.

**10.21.** Negate each of the following statements: (*a*) All students live in the dormitories. (*b*) All mathematics majors are males. (*c*) Some students are 25 (years) or older.

Use Theorem 4.5 to negate the quantifiers.

(*a*) At least one student does not live in the dormitories. (Some students do not live in the dormitories.)

(*b*) At least one mathematics major is female. (Some mathematics majors are female.)

(*c*) None of the students is 25 or older. (All the students are under 25.)

## Supplementary Problems

### PROPOSITION AND LOGICAL OPERATIONS

**10.22.** Let $p$ be "Audrey speaks French" and let $q$ be "Audrey speaks Danish". Give a simple verbal sentence which describes each of the following:

(*a*) $p \vee q$; (*b*) $p \wedge q$; (*c*) $p \wedge \neg q$; (*d*) $\neg p \vee \neg q$; (*e*) $\neg\neg p$; (*f*) $\neg(\neg p \wedge \neg q)$.

**10.23.** Let $p$ denote "He is rich" and let $q$ denote "He is happy". Write each statement in symbolic form using $p$ and $q$. Note that "He is poor" and "He is unhappy" are equivalent to $\neg p$ and $\neg q$, respectively.

(*a*) If he is rich, then he is unhappy.

(*b*) He is neither rich nor happy.

(*c*) It is necessary to be poor in order to be happy.

(*d*) To be poor is to be unhappy.

**10.24.** Find the truth table for: (*a*) $p \vee \neg q$; (*b*) $\neg p \wedge \neg q$.

**10.25.** Verify that the proposition $(p \wedge q) \wedge \neg(p \vee q)$ is a contradiction.

### ARGUMENTS

**10.26.** Test the validity of each argument:

(*a*) If it rains, Erik will be sick.
It did not rain.
______________________
Erik was not sick.

(*b*) If it rains, Erik will be sick.
Erik was not sick.
______________________
It did not rain.

**10.27.** Test the validity of the following argument:

If I study, then I will not fail mathematics.
If I do not play basketball, then I will study.
But I failed mathematics.
______________________
Therefore I must have played basketball.

**10.28.** Show that $p \leftrightarrow \neg q$ does not logically imply $p \rightarrow q$.

**QUANTIFIERS**

**10.29.** Let $A = \{1, 2, \ldots, 9, 10\}$. Consider each of the following sentences. If it is a statement, then determine its truth value. If it is a propositional function, determine its truth set.

(*a*) $(\forall x \in A)(\exists y \in A)(x + y < 14)$ (*c*) $(\forall x \in A)(\forall y \in A)(x + y < 14)$

(*b*) $(\forall y \in A)(x + y < 14)$ (*d*) $(\exists y \in A)(x + y < 14)$

**10.30.** Negate each of the following statements:

(*a*) If the teacher is absent, then some students do not complete their homework.

(*b*) All the students completed their homework and the teacher is present.

(*c*) Some of the students did not complete their homework or the teacher is absent.

**10.31.** Negate each of the statements in Problem 10.17.

**10.32.** Find a counterexample for each statement where $\mathbf{U} = \{3, 5, 7, 9\}$ is the universal set:

(*a*) $\forall x,\ x + 3 \geq 7$; (*b*) $\forall x$, $x$ is odd; (*c*) $\forall x$, $x$ is prime; (*d*) $\forall x, |x| = x$.

## Answers to Supplementary Problems

**10.22.** In each case, translate $\wedge$, $\vee$, and $\neg$ to read "and", "or", and "It is false that" or "not", respectively; and then simplify the English sentence.

**10.23.** (*a*) $p \to \neg q$; (*b*) $\neg p \wedge \neg q$; (*c*) $q \to \neg p$; (*d*) $\neg p \leftrightarrow \neg q$

**10.24.** The truth tables appear in Fig. 10-20.

| $p$ | $q$ | $\neg q$ | $p \vee \neg q$ |
|---|---|---|---|
| T | T | F | T |
| T | F | T | T |
| F | T | F | F |
| F | F | T | T |

(*a*)

| $p$ | $q$ | $\neg p$ | $\neg q$ | $\neg p \wedge \neg q$ |
|---|---|---|---|---|
| T | T | F | F | F |
| T | F | F | T | F |
| F | T | T | F | F |
| F | F | T | T | T |

(*b*)

**Fig. 10-20**

**10.25.** It is a contradiction since its truth table in Fig. 10-21 is false for all values of $p$ and $q$.

| $p$ | $q$ | $p \wedge q$ | $p \vee q$ | $\neg(p \vee q)$ | $(p \wedge q) \wedge \neg(p \vee q)$ |
|---|---|---|---|---|---|
| T | T | T | T | F | F |
| T | F | F | T | F | F |
| F | T | F | T | F | F |
| F | F | F | F | T | F |

**Fig. 10-21**

**10.26.** First translate the arguments into symbolic form: (*a*) $p \to q,\ \neg p \vdash \neg q$, (*b*) $p \to q,\ \neg q \vdash \neg p$. By Problem 10.11, argument (*a*) is a fallacy. By Problem 10.12, argument (*b*) is valid.

**10.27.** Translate the argument into the following symbolic form where $p$ is "I study", $q$ is "I fail mathematics", and $r$ is "I play basketball":

$$p \to \neg q, \ \neg r \to p, \ q \vdash r$$

Construct the truth tables as in Fig. 10.22 where the premises $p \to \neg q$, $\neg r \to p$, and $q$ are true simultaneously only in the fifth row of the table, and in that case the conclusion $r$ is also true. Hence the argument is valid.

| $p$ | $q$ | $r$ | $\neg q$ | $p \to \neg q$ | $\neg r$ | $\neg r \to p$ |
|---|---|---|---|---|---|---|
| T | T | T | F | F | F | T |
| T | T | F | F | F | T | T |
| T | F | T | T | T | F | T |
| T | F | F | T | T | T | T |
| F | T | T | F | T | F | T |
| F | T | F | F | T | T | F |
| F | F | T | T | T | F | T |
| F | F | F | T | T | T | F |

**Fig. 10-22**

| $p$ | $q$ | $\neg q$ | $p \leftrightarrow \neg q$ | $p \to q$ |
|---|---|---|---|---|
| T | T | F | F | T |
| T | F | T | T | F |
| F | T | F | T | T |
| F | F | T | F | T |

**Fig. 10-23**

**10.28. Method 1.** Construct the truth tables of $p \leftrightarrow \neg q$ and $p \to q$ as in Fig. 10.23. Note that $p \leftrightarrow \neg q$ is true in line 2 of the truth table whereas $p \to q$ is false.

**Method 2.** Construct the truth table of the proposition $(p \leftrightarrow \neg q) \to (p \to q)$. It will not be a tautology; hence, by Theorem 10.4, $p \leftrightarrow \neg q$ does not logically imply $p \to q$.

**10.29.** (*a*) The open sentence in two variables is preceded by two quantifiers; hence it is a statement. Moreover, the statement is true.

(*b*) The open sentence is preceded by one quantifier; hence it is a propositional function of the other variable. Note that for every $y \in A$, $x_0 + y < 14$ if and only if $x_0 = 1, 2$, or $3$. Hence the truth set is $\{1, 2, 3\}$.

(*c*) It is a statement and it is false: if $x_0 = 8$ and $y_0 = 9$, then $x_0 + y_0 < 14$ is not true.

(*d*) It is an open sentence in $x$. The truth set is $A$ itself.

**10.30.** (*a*) The teacher is absent and all the students completed their homework.

(*b*) Some of the students did not complete their homework or the teacher is absent.

(*c*) All the students completed their homework and the teacher is present.

**10.31.** (*a*) $(\forall x \in A)(x + 3 \neq 10)$ (*c*) $(\forall x \in A)(x + 3 \geq 5)$

(*b*) $(\exists x \in A)(x + 3 \geq 10)$ (*d*) $(\exists x \in A)(x + 3 > 7)$

**10.32.** (*a*) Here 5, 7, and 9 are counterexamples.

(*b*) The statement is true; hence no counterexample exists.

(*c*) Here 9 is the only counterexample.

(*d*) The statement is true; hence there is no counterexample.

# Chapter 11

# Boolean Algebra

## 11.1 INTRODUCTION

Both sets and propositions satisfy similar laws which are listed in Tables 1-1 and 10-1 (appearing in Chapters 1 and 10, respectively). These laws are used to define an abstract mathematical structure called a *Boolean algebra*, which is named after the mathematician George Boole (1813–1864).

## 11.2 BASIC DEFINITIONS

Let $B$ be a nonempty set with two binary operations $+$ and $*$, a unary operation $'$, and two distinct elements 0 and 1. Then $B$ is called a *Boolean algebra* if the following axioms hold where $a, b, c$ are any elements in $B$:

[$\mathbf{B}_1$] ***Commutative laws***:
(1*a*) $a + b = b + a$ (1*b*) $a * b = b * a$

[$\mathbf{B}_2$] ***Distributive laws***:
(2*a*) $a + (b * c) = (a + b) * (a + c)$ (2*b*) $a * (b + c) = (a * b) + (a * c)$

[$\mathbf{B}_3$] ***Identity laws***:
(3*a*) $a + 0 = a$ (3*b*) $a * 1 = a$

[$\mathbf{B}_4$] ***Complement laws***:
(4*a*) $a + a' = 1$ (4*b*) $a * a' = 0$

We will sometimes designate a Boolean algebra by $\langle B, +, *, ', 0, 1\rangle$ when we want to emphasize its six parts. We say 0 is the *zero* element, 1 is the *unit* element and $a'$ is the *complement* of $a$. We will usually drop the symbol $*$ and use juxtaposition instead. Then (2*b*) is written $a(b + c) = ab + ac$ which is the familiar algebraic identity of rings and fields. However, (2*a*) becomes $a + bc = (a + b)(a + c)$, which is certainly not a usual identity in algebra.

The operations $+$, $*$ and $'$ are called sum, product, and complement respectively. We adopt the usual convention that, unless we are guided by parentheses, $'$ has precedence over $*$, and $*$ has precedence over $+$. For example,

$$a + b * c \text{ means } a + (b * c) \text{ and not } (a + b) * c \qquad a * b' \text{ means } a * (b') \text{ and not } (a * b)'$$

Of course when $a + b * c$ is written $a + bc$ then the meaning is clear.

**EXAMPLE 11.1**

(*a*) Let $\mathbf{B} = \{0, 1\}$, the set of *bits* (binary digits), with the binary operations of $+$ and $*$ and the unary operation $'$ defined by Fig. 11-1. Then $\mathbf{B}$ is a Boolean algebra. (Note $'$ simply changes the bit, i.e., $1' = 0$ and $0' = 1$.)

| + | 1 | 0 |
|---|---|---|
| 1 | 1 | 1 |
| 0 | 1 | 0 |

| * | 1 | 0 |
|---|---|---|
| 1 | 1 | 0 |
| 0 | 0 | 0 |

| ' | 1 | 0 |
|---|---|---|
| | 0 | 1 |

**Fig. 11-1**

(*b*) Let $\mathbf{B}^n = \mathbf{B} \times \mathbf{B} \times \cdots \times \mathbf{B}$ ($n$ factors) where the operations of $+$, $*$ and $'$ are defined componentwise using Fig. 11-1. For notational convenience, we write the elements of $\mathbf{B}^n$ as $n$-bit sequences without commas, e.g., $x = 110011$ and $y = 111000$ belong to $\mathbf{B}^6$. Hence

$$x + y = 111011, \qquad x * y = 110000, \qquad x' = 001100$$

Then $\mathbf{B}^n$ is a Boolean algebra. Here $0 = 000\cdots0$ is the zero element, and $1 = 111\cdots1$ is the unit element. We note that $\mathbf{B}^n$ has $2^n$ elements.

(*c*) Let $\mathbf{D}_{70} = \{1, 2, 5, 7, 10, 14, 35, 79\}$, the divisors of 70. Define $+$, $*$ and $'$ by

$$a + b = \text{lcm}(a, b), \qquad a * b = \text{gcd}(a, b), \qquad a' = \frac{70}{a}$$

Then $\mathbf{D}_{70}$ is a Boolean algebra with 1 the zero element and 70 the unit element.

(*d*) Let $\mathscr{C}$ be a collection of sets closed under the set operations of union, intersection, and complement. Then $\mathscr{C}$ is a Boolean algebra with the empty set $\varnothing$ as the zero element and the universal set $\mathbf{U}$ as the unit element.

### Subalgebras, Isomorphic Boolean Algebras

Suppose $C$ is a nonempty subset of a Boolean algebra $B$. We say $C$ is a *subalgebra* of $B$ if $C$ itself is a Boolean algebra (with respect to the operations of $B$). We note that $C$ is a subalgebra of $B$ if and only if $C$ is closed under the three operations of $B$, i.e., $+$, $*$, and $'$. For example, $\{1, 2, 35, 70\}$ is a subalgebra of $D_{70}$ in Example 11.1(*c*).

Two Boolean algebras $B$ and $B'$ are said to be *isomorphic* if there is a one-to-one correspondence $f\colon B \to B'$ which preserves the three operations, i.e., such that

$$f(a + b) = f(a) + f(b), \qquad f(a * b) = f(a) * f(b) \qquad \text{and} \qquad f(a') = f(a)'$$

for any elements $a, b$ in $B$.

## 11.3 DUALITY

The *dual* of any statement in a Boolean algebra $B$ is the statement obtained by interchanging the operations $+$ and $*$, and interchanging their identity elements 0 and 1 in the original statement. For example, the dual of

$$(1 + a) * (b + 0) = b \qquad \text{is} \qquad (0 * a) + (b * 1) = b$$

Observe the symmetry in the axioms of a Boolean algebra $B$. That is, the dual of the set of axioms of $B$ is the same as the original set of axioms. Accordingly, the important principle of duality holds in $B$. Namely,

**Theorem 11.1 (Principle of Duality):** The dual of any theorem in a Boolean algebra is also a theorem.

In other words, if any statement is a consequence of the axioms of a Boolean algebra, then the dual is also a consequence of those axioms since the dual statement can be proven by using the dual of each step of the proof of the original statement.

## 11.4 BASIC THEOREMS

Using the axioms [$\mathbf{B}_1$] through [$\mathbf{B}_4$], we prove (Problem 11.5) the following theorem.

**Theorem 11.2**: Let $a, b, c$ be any elements in a Boolean algebra $B$.

(i) ***Idempotent laws***:

(5$a$) $a + a = a$ (5$b$) $a * a = a$

(ii) ***Boundedness laws***:

(6$a$) $a + 1 = 1$ (6$b$) $a * 0 = 0$

(iii) ***Absorption laws***:

(7$a$) $a + (a * b) = a$ (7$b$) $a * (a + b) = a$

(iv) ***Associative laws***:

(8$a$) $(a + b) + c = a + (b + c)$ (8$b$) $(a * b) * c = a * (b * c)$

Theorem 11.2 and our axioms still do not contain all the properties of sets listed in Table 1-1. The next two theorems (proved in Problems 11.6 and 11.7) give us the remaining properties.

**Theorem 11.3**: Let $a$ be any element of a Boolean algebra $B$.

(i) (Uniqueness of Complement)

If $a + x = 1$ and $a * x = 0$, then $x = a'$.

(ii) (Involution law) $(a')' = a$

(iii) (9$a$) $0' = 1$, (9$b$) $1' = 0$

**Theorem 11.4 (DeMorgan's laws):** (10$a$) $(a + b)' = a' * b'$. (10$b$) $(a * b)' = a' + b'$.

## 11.5 BOOLEAN ALGEBRAS AS LATTICES

By Theorem 11.2 and axiom [$\mathbf{B}_1$], every Boolean algebra $B$ satisfies the associative, commutative, and absorption laws and hence is a lattice where $+$ and $*$ are the join and meet operations, respectively. With respect to this lattice, $a + 1 = 1$ implies $a \leq 1$ and $a * 0 = 0$ implies $0 \leq a$, for any element $a \in B$. Thus $B$ is a bounded lattice. Furthermore, axioms [$\mathbf{B}_2$] and [$\mathbf{B}_4$] show that $B$ is also distributive and complemented. Conversely, every bounded, distributive, and complemented lattice $L$ satisfies the axioms [$\mathbf{B}_1$] through [$\mathbf{B}_4$]. Accordingly, we have the following

**Alternate Definition**: A Boolean algebra $B$ is a bounded, distributive, and complemented lattice.

Since a Boolean algebra $B$ is a lattice, it has a natural partial ordering (and so its diagram can be drawn). Recall (Chapter 7) that we define $a \leq b$ when the equivalent conditions $a + b = b$ and $a * b = a$ hold. Since we are in a Boolean algebra, we can actually say much more. Specifically, the following theorem (proved in Problem 11.8) applies.

**Theorem 11.5**: The following are equivalent in a Boolean algebra:

(1) $a + b = b$, (2) $a * b = a$, (3) $a' + b = 1$, (4) $a * b' = 0$.

Thus in a Boolean algebra we can write $a \leq b$ whenever any of the above four conditions is known to be true.

**EXAMPLE 11.2**

($a$) Consider a Boolean algebra of sets. Then set $A$ precedes set $B$ if $A$ is a subset of $B$. Theorem 11.4 states that if $A \subseteq B$, as illustrated in the Venn diagram in Fig. 11-2, then the following conditions hold:

(1) $A \cup B = B$, (2) $A \cap B = A$, (3) $A^c \cup B = U$, (4) $A \cap B^c = \varnothing$.

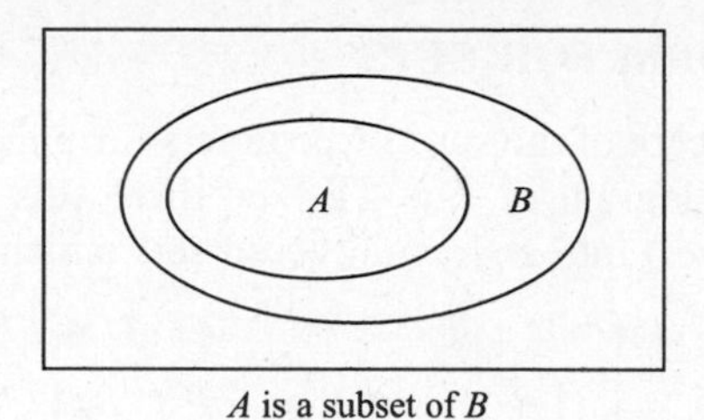


*A* is a subset of *B*

**Fig. 11-2**

(*b*) Consider the Boolean algebra of the proposition calculus. Then the proposition $P$ precedes the proposition $Q$ if $P$ logically implies $Q$, i.e., if $P \Rightarrow Q$.

## 11.6 REPRESENTATION THEOREM

Let $B$ be a finite Boolean algebra. Recall (Section 7.9) that an element $a$ in $B$ is an atom if $a$ immediately succeeds 0, that is if $0 << a$. Let $A$ be the set of atoms of $B$ and let $P(A)$ be the Boolean algebra of all subsets of the set $A$ of atoms. By Theorem 7.15, each $x \neq 0$ in $B$ can be expressed uniquely (except for order) as the sum (join) of atoms, i.e. elements of $A$ Say,

$$x = a_1 + a_2 + \cdots + a_r$$

is such a representation. Consider the function $f: B \to P(A)$ defined by

$$f(x) = \{a_1, a_2, \ldots, a_r\}$$

The mapping is well-defined since the representation is unique.

**Theorem 11.6**: The above mapping $f: B \to \mathcal{P}(A)$ is an isomorphism.

Thus we see the intimate relationship between set theory and abstract Boolean algebras in the sense that every finite Boolean algebra is structurally the same as a Boolean algebra of sets.

If a set $A$ has $n$ elements, then its power set $\mathcal{P}(A)$ has $2^n$ elements. Thus the above theorem gives us our next result.

**Corollary 11.7**: A finite Boolean algebra has $2^n$ elements for some positive integer $n$.

**EXAMPLE 11.3** Consider the Boolean algebra $\mathbf{D}_{70} = \{1, 2, 5, \ldots, 70\}$ whose diagram is given in Fig. 11-3(*a*). Note that $A = \{2, 5, 7\}$ is the set of atoms of $\mathbf{D}_{70}$. The following is the unique representation of each non-atom by atoms:

$$10 = 2 \vee 5, \qquad 14 = 2 \vee 7, \qquad 35 = 5 \vee 7, \qquad 70 = 2 \vee 5 \vee 7$$

Figure 11-3(*b*) gives the diagram of the Boolean algebra of the power set $\mathcal{P}(A)$ of the set $A$ of atoms. Observe that the two diagrams are structurally the same.

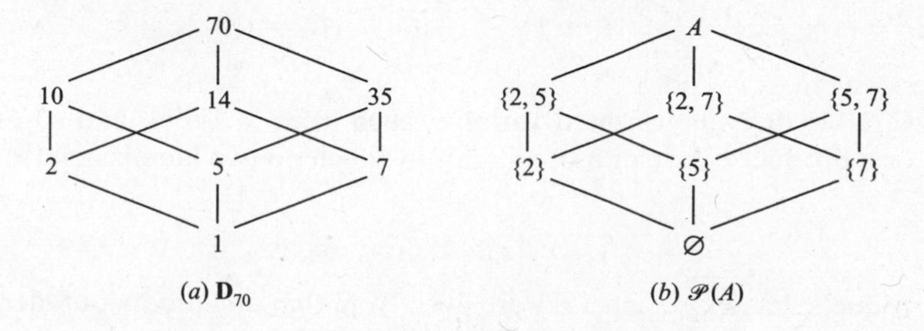


(*a*) $\mathbf{D}_{70}$ (*b*) $\mathcal{P}(A)$

**Fig. 11-3**

## 11.7 SUM-OF-PRODUCTS FORM FOR SETS

This section motivates the concept of the sum-of-products form in Boolean algebra by an example of set theory. Consider the Venn diagram in Fig. 11-4 of three sets $A, B, C$. Observe that these sets partition the rectangle (universal set) into eight numbered sets which can be represented as follows:

$$\begin{array}{llll} (1)\ A \cap B \cap C & (3)\ A \cap B^c \cap C & (5)\ A \cap B^c \cap C^c & (7)\ A^c \cap B^c \cap C \\ (2)\ A \cap B \cap C^c & (4)\ A^c \cap B \cap C & (6)\ A^c \cap B \cap C^c & (8)\ A^c \cap B^c \cap C^c \end{array}$$

Each of these eight sets is of the form $A^* \cap B^* \cap C^*$, where

$$A^* = A \text{ or } A^c, \qquad B^* = B \text{ or } B^c, \qquad C^* = C \text{ or } C^c$$

Consider any nonempty set expression $E$ involving the sets $A, B$, and $C$, say,

$$E = [(A \cap B^c)^c \cup (A^c \cap C^c)] \cap [B^c \cup C)^c \cap (A \cup C^c)]$$

Then $E$ will represent some area in Fig. 11-4 and hence will uniquely equal the union of one or more of the eight sets.

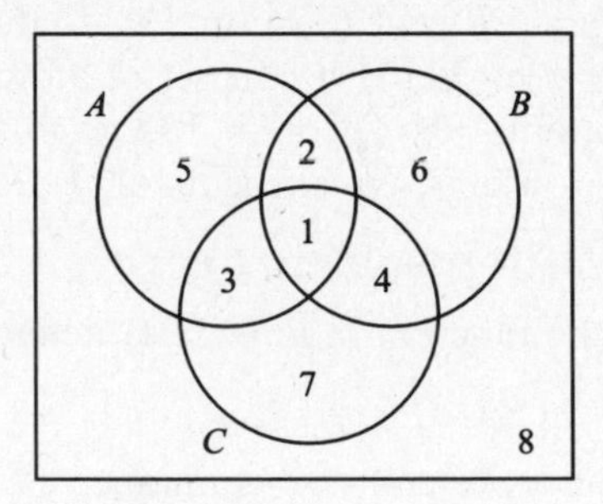


**Fig. 11-4**

Suppose we now interpret a union as a sum and an intersection as a product. Then the above eight sets are products, and the unique representation of $E$ will be a sum (union) of products. This unique representation of $E$ is the same as the complete sum-of-products expansion in Boolean algebras which we discuss below.

## 11.8 SUM-OF-PRODUCTS FORM FOR BOOLEAN ALGEBRAS

Consider a set of variables (or letters or symbols), say, $x_1, x_2, \ldots, x_n$. A *Boolean expression* $E$ in these variables, sometimes written $E(x_1, \ldots, x_n)$, is any variable or any expression built up from the variables using the Boolean operations $+$, $*$ and $'$. (Naturally, the expression $E$ must be *well-formed*, that is, where $+$ and $*$ are used as binary operations, and $'$ is used as a unary operation.) For example,

$$E_1 = (x + y'z)' + (xyz' + x'y)' \qquad \text{and} \qquad E_2 = ((xy'z' + y)' + x'z)'$$

are Boolean expressions in $x, y$, and $z$.

A *literal* is a variable or complemented variable, such as $x$, $x'$, $y$, $y'$, and so on. A *fundamental product* is a literal or a product of two or more literals in which no two literals involve the same variable. Thus

$$xz', \quad xy'z, \quad x, \quad y', \quad x'yz$$

are fundamental products, but $xyx'z$ and $xyzy$ are not. Note that any product of literals can be reduced to either 0 or a fundamental product, e.g., $xyx'z = 0$ since $xx' = 0$ (complement law), and $xyzy = xyz$ since $yy = y$ (idempotent law).

A fundamental product $P_1$ is said to be *contained in* (or *included in*) another fundamental product $P_2$ if the literals of $P_1$ are also literals of $P_2$. For example, $x'z$ is contained in $x'yz$, but $x'z$ is not contained in $xy'z$ since $x'$ is not a literal of $xy'z$. Observe that if $P_1$ is contained in $P_2$, say $P_2 = P_1 * Q$, then, by the absorption law,

$$P_1 + P_2 = P_1 + P_1 * Q = P_1$$

Thus, for instance, $x'z + x'yz = x'z$.

**Definition:** A Boolean expression $E$ is called a *sum-of-products* expression if $E$ is a fundamental product or the sum of two or more fundamental products none of which is contained in another.

**Definition:** Let $E$ be any Boolean expression. A *sum-of-products form* of $E$ is an equivalent Boolean sum-of-products expression.

**EXAMPLE 11.4** Consider the expressions

$$E_1 = xz' + y'z + xyz' \quad \text{and} \quad E_2 = xz' + x'yz' + xy'z$$

Although the first expression $E_1$ is a sum of products, it is not a sum-of-products expression. Specifically, the product $xz'$ is contained in the product $xyz'$. However, by the absorption law, $E_1$ can be expressed as

$$E_1 = xz' + y'z + xyz' = xz' + xyz' + y'z = xz' + y'z$$

This yields a sum-of-products form for $E_1$. The second expression $E_2$ is already a sum-of-products expression.

### Algorithm for Finding Sum-of-Products Forms

The following four-step algorithm uses the Boolean algebra laws to transform any Boolean expression $E$ into an equivalent sum-of-products expression:

**Algorithm 11.8A:** The input is a Boolean expression $E$. The output is a sum-of-products expression equivalent to $E$.

***Step 1.*** Use DeMorgan's laws and involution to move the complement operation into any parenthesis until finally the complement operation only applies to variables. Then $E$ will consist only of sums and products of literals.

***Step 2.*** Use the distributive operation to next transform $E$ into a sum of products.

***Step 3.*** Use the commutative, idempotent, and complement laws to transform each product in $E$ into 0 or a fundamental product.

***Step 4.*** Use the absorption and identity laws to finally transform $E$ into a sum-of-products expression.

**EXAMPLE 11.5** Suppose Algorithm 11.8A is applied to the following Boolean expression:

$$E = ((xy)'z)'((x' + z)(y' + z'))'$$

***Step 1.*** Using DeMorgan's laws and involution, we obtain

$$E = ((xy)'' + z')((x' + z)' + (y' + z')') = (xy + z')(xz' + yz)$$

$E$ now consists only of sums and products of literals.

***Step 2.*** Using the distributive laws, we obtain

$$E = xyxz' + xyyz + xz'z' + yzz'$$

$E$ now is a sum of products.

***Step 3.*** Using the commutative, idempotent, and complement laws, we obtain

$$E = xyz' + xyz + xz' + 0$$

Each term in $E$ is a fundamental product or 0.

***Step 4.*** The product $ac'$ is contained in $abc'$; hence, by the absorption law,

$$xz' + (xz' * y) = xz'$$

Thus we may delete $abc'$ from the sum. Also, by the identity law for 0, we may delete 0 from the sum. Accordingly,

$$E = xyz + xz'$$

$E$ is now represented by a sum-of-products expression.

### Complete Sum-of-Products Forms

A Boolean expression $E = E(x_1, x_2, \ldots, x_n)$ is said to be a *complete sum-of-products* expression if $E$ is a sum-of-products expression where each product $P$ involves all the $n$ variables. Such a fundamental product $P$ which involves all the variables is called a *minterm*, and there is a maximum of $2^n$ such products for $n$ variables. The following theorem applies.

**Theorem 11.8**: Every nonzero Boolean expression $E = E(x_1, x_2, \ldots, x_n)$ is equivalent to a complete sum-of-products expression and such a representation is unique.

The above unique representation of $E$ is called the *complete sum-of-products form* of $E$. Recall that Algorithm 11.8A tells us how to transform $E$ into a sum-of-products form. The following algorithm shows how to transform a sum-of-products form into a complete sum-of-products form.

**Algorithm 11.8B:** The input is a Boolean sum-of-products expression $E = E(x_1, x_2, \ldots, x_n)$. The output is a complete sum-of-products expression equivalent to $E$.

***Step 1.*** Find a product $P$ in $E$ which does not involve the variable $x_i$, and then multiply $P$ by $x_i + x_i'$, deleting any repeated products. (This is possible since $x_i + x_i' = 1$, and $P + P = P$.)

***Step 2.*** Repeat Step 1 until every product $P$ in $E$ is a minterm, i.e., every product $P$ involves all the variables.

**EXAMPLE 11.6** Express $E(x, y, z) = x(y'z)'$ in its complete sum-of-products form.

(*a*) Apply Algorithm 11.8A to $E$ to obtain

$$E = x(y'z)' = x(y + z') = xy + xz'$$

Now $E$ is represented by a sum-of-products expression.

(*b*) Apply Algorithm 11.8B to obtain

$$E = xy(z + z') + xz'(y + y') = xyz + xyz' + xyz' + xy'z'$$
$$= xyz + xyz' + xy'z'$$

Now $E$ is represented by its complete sum-of-products form.

**Warning**: The terminology in this section has not been standardized. The sum-of-products form for a Boolean expression $E$ is also called the *disjunctive normal form* or DNF of $E$. The complete sum-of-products form for $E$ is also called the *full disjunctive normal form*, or the *disjunctive canonical form*, or the *minterm canonical form* of $E$.

## 11.9 MINIMAL BOOLEAN EXPRESSIONS, PRIME IMPLICANTS

There are many ways of representing the same Boolean expression $E$. Here we define and investigate a minimal sum-of-products form for $E$. We must also define and investigate prime implicants of $E$ since the minimal sum-of-products involves such prime implicants. Other minimal forms exist, but their investigation lies beyond the scope of this text.

### Minimal Sum-of-Products

Consider a Boolean sum-of-products expression $E$. Let $E_L$ denote the number of literals in $E$ (counted according to multiplicity), and let $E_S$ denote the number of summands in $E$. For instance, suppose

$$E = xyz' + x'y't + xy'z't + x'yzt$$

Then $E_L = 3 + 3 + 4 + 4 = 14$ and $E_S = 4$.

Suppose $E$ and $F$ are equivalent Boolean sum-of-products expressions. We say $E$ is *simpler* than $F$ if

$$\text{(i) } E_L < F_L \text{ and } E_S \leq F_L, \quad \text{or} \quad \text{(ii) } E_L \leq F_L \text{ and } E_S < F_L$$

We say $E$ is *minimal* if there is no equivalent sum-of-products expression which is simpler than $E$. We note that there can be more than one equivalent minimal sum-of-products expression.

### Prime Implicants

A fundamental product $P$ is called a *prime implicant* of a Boolean expression $E$ if

$$P + E = E$$

but no other fundamental product contained in $P$ has this property. For instance, suppose

$$E = xy' + xyz' + x'yz'$$

One can show (Problem 11.15) that

$$xz' + E = E \quad \text{but} \quad x + E \neq E \quad \text{and} \quad z' + E \neq E$$

Thus $xz'$ is a prime implicant of $E$.

The following theorem applies.

**Theorem 11.9**: A minimal sum-of-products form for a Boolean expression $E$ is a sum of prime implicants of $E$.

The following subsections give a method for finding the prime implicants of $E$ based on the notion of the consensus of fundamental products. This method can then be used to find a minimal sum-of-products form for $E$. Section 11.10 gives a geometric method for finding such prime implicants.

### Consensus of Fundamental Products

Let $P_1$ and $P_2$ be fundamental products such that exactly one variable, say $x_k$, appears uncomplemented in one of $P_1$ and $P_2$ and complemented in the other. Then the *consensus* of $P_1$ and $P_2$ is the product (without repetitions) of the literals of $P_1$ and the literals of $P_2$ after $x_k$ and $x_k'$ deleted. (We do not define the consensus of $P_1 = x$ and $P_2 = x'$.)

The following lemma (proved in Problem 11.19) applies.

**Lemma 11.10**: Suppose $Q$ is the consensus of $P_1$ and $P_2$. Then $P_1 + P_2 + Q = P_1 + P_2$.

**EXAMPLE 11.7** Find the consensus $Q$ of $P_1$ and $P_2$ where:

(*a*) $P_1 = xyz's$ and $P_2 = xy't$.

Delete $y$ and $y'$ and then multiply the literals of $P_1$ and $P_2$ (without repetition) to obtain $Q = xz'st$.

(*b*) $P_1 = xy'$ and $P_2 = y$.

Deleting $y$ and $y'$ yields $Q = x$.

(*c*) $P_1 = x'yz$ and $P_2 = x'yt$.

No variable appears uncomplemented in one of the products and complemented in the other. Hence $P_1$ and $P_2$ have no consensus.

(*d*) $P_1 = x'yz$ and $P_2 = xyz'$.

Each of $x$ and $z$ appear complemented in one of the products and uncomplemented in the other. Hence $P_1$ and $P_2$ have no consensus.

### Consensus Method for Finding Prime Implicants

The following algorithm, called the *consensus method*, is used to find the prime implicants of a Boolean expression.

**Algorithm 11.9A (Consensus Method):** The input is a Boolean expression

$$E = P_1 + P_2 + \cdots + P_m$$

where the $P$'s are fundamental products. The output expresses $E$ as a sum of its prime implicants (Theorem 11.11).

***Step 1.*** Delete any fundamental product $P_i$ which includes any other fundamental product $P_j$. (Permissible by the absorption law.)

***Step 2.*** Add the consensus of any $P_i$ and $P_j$ providing $Q$ does not include any of the $P$'s. (Permissible by Lemma 11.10.)

***Step 3.*** Repeat Step 1 and/or Step 2 until neither can be applied.

The following theorem gives the basic property of the above algorithm.

**Theorem 11.11:** The consensus method will eventually stop, and then $E$ will be the sum of its prime implicants.

**EXAMPLE 11.8** Let $E = xyz + x'z' + xyz' + x'y'z + x'yz'$. Then:

$$\begin{aligned}
E &= xyz + x'z' + xyz' + x'y'z && (x'yz' \text{ includes } x'z')\\
&= xyz + x'y' + xyz' + x'y'z + xy && (\text{Consensus of } xyz \text{ and } xyz')\\
&= x'z' + x'y'z + xy && (xyz \text{ and } xyz' \text{ include } xy)\\
&= x'z' + x'y'z + xy + x'y' && (\text{Consensus of } x'z' \text{ and } x'y'z)\\
&= x'z' + xy + x'y' && (x'y'z \text{ includes } x'y')\\
&= x'z' + xy + x'y' + yz' && (\text{Consensus of } x'z' \text{ and } xy)
\end{aligned}$$

Now neither step in the consensus method will change $E$. Thus $E$ is the sum of its prime implicants, which are $x'z'$, $xy$, $x'y'$, and $yz'$.

### Finding a Minimal Sum-of-Products Form

The consensus method (Algorithm 11.9A) can be used to express a Boolean expression $E$ as a sum of all its prime implicants. Using such a sum, one may find a minimal sum-of-products form for $E$ as follows.

**Algorithm 11.9B:** The input is a Boolean expression $E = P_1 + P_2 + \cdots + P_m$ where the $P$'s are all the prime implicants of $E$. The output expresses $E$ as a minimal sum-of-products.

***Step 1.*** Express each prime implicant $P$ as a complete sum-of-products.

***Step 2.*** Delete one by one those prime implicants whose summands appear among the summands of the remaining prime implicants.

**EXAMPLE 11.9** We apply Algorithm 11.9B to

$$E = x'z' + xy + x'y' + yz'$$

(By Example 11.8, $E$ is now expressed as the sum of all its prime implicants.)

***Step 1.*** Express each prime implicant of $E$ as a complete sum-of-products to obtain:

$$x'z' = x'z'(y + y') = x'yz' + x'y'z'$$
$$xy = xy(z + z') = xyz + xyz'$$
$$x'y' = x'y'(z + z') = x'y'z + x'y'z'$$
$$yz' = yz'(x + x') = xyz' + x'yz'$$

***Step 2.*** The summands of $x'z'$ are $x'yz$ and $x'y'z'$ which appear among the other summands. Thus delete $x'z'$ to obtain

$$E = xy + x'y' + yz'$$

The summands of no other prime implicant appear among the summands of the remaining prime implicants, and hence this is a minimal sum-of-products form for $E$. In other words, none of the remaining prime implicants is *superfluous*, that is, none can be deleted without changing $E$.

## 11.10 KARNAUGH MAPS

Karnaugh maps, where minterms involving the same variables are represented by squares, are pictorial devices for finding prime implicants and minimal forms for Boolean expressions involving at most six variables. We will only treat the cases of two, three, and four variables. In the context of Karnaugh maps, we will sometimes use the terms "squares" and "minterm" interchangeably. Recall that a minterm is a fundamental product which involves all the variables, and that a complete sum-of-products expression is a sum of minterms.

First we need to define the notion of adjacent products. Two fundamental products $P_1$ and $P_2$ are said to be *adjacent* if $P_1$ and $P_2$ have the same variables and if they differ in exactly one literal. Thus there must be an uncomplemented variable in one product and complemented in the other. In particular, the sum of two such adjacent products will be a fundamental product with one less literal (Problem 11.51).

**EXAMPLE 11.10** Find the sum of the following adjacent products $P_1$ and $P_2$:

(*a*) $P_1 = xyz'$ and $P_2 = xy'z'$.

$$P_1 + P_2 = xyz' + xy'z' = xz'(y + y') = xz'(1) = xz'$$

(*b*) $P_1 = x'yzt$ and $P_2 = x'yz't$.

$$P_1 + P_2 = x'yzt + x'yz't = x'yt(z + z') = x'yt(1) = x'yt$$

(c) $P_1 = x'yzt$ and $P_2 = xyz't$.

Here $P_1$ and $P_2$ are not adjacent since they differ in two literals. In particular,

$$P_1 + P_2 = x'yzt + xyz't = (x' + x)y(z + z')t = (1)y(1)t = yt$$

(d) $P_1 = xyz'$ and $P_2 = xyzt$.

Here $P_1$ and $P_2$ are not adjacent since they have different variables. Thus, in particular, they will not appear as squares in the same Karnaugh map.

### Case of Two Variables

The Karnaugh map corresponding to Boolean expressions $E = E(x, y)$ with two variables $x$ and $y$ is shown in Fig. 11-5(*a*). The Karnaugh map may be viewed as a Venn diagram where $x$ is represented by the points in the upper half of the map, shaded in Fig. 11-5(*b*), and $y$ is represented by the points in the left half of the map, shaded in Fig. 11-5(*c*). Thus $x'$ is represented by the points in the lower half of the map, and $y'$ is represented by the points in the right half of the map. Accordingly, the four possible minterms with two literals,

$$xy, \quad xy', \quad x'y, \quad x'y'$$

are represented by the four squares in the map, as labeled in Fig. 11-5(*d*). Note that two such squares are adjacent, as defined above, if and only if the squares are geometrically adjacent (have a side in common).

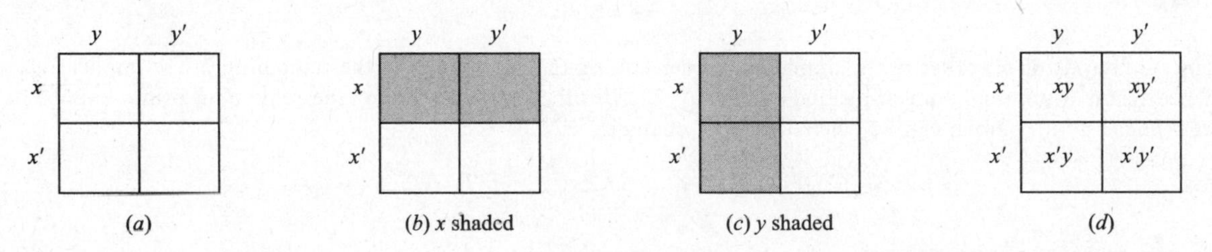

(*a*) (*b*) *x* shaded (*c*) *y* shaded (*d*)

**Fig. 11-5**

Any complete sum-of-products Boolean expression $E(x, y)$ is a sum of minterms and hence can be represented in the Karnaugh map by placing checks in the appropriate squares. A prime implicant of $E(x, y)$ will be either a pair of adjacent squares in $E$ or an *isolated* square, i.e., a square which is not adjacent to any other square of $E(x, y)$. A minimal sum-of-products form for $E(x, y)$ will consist of a minimal number of prime implicants which cover all the squares of $E(x, y)$ as illustrated in the next example.

**EXAMPLE 11.11** Find the prime implicants and a minimal sum-of-products form for each of the following complete sum-of-products Boolean expressions:

(*a*) $E_1 = xy + xy'$; (*b*) $E_1 = xy + x'y + x'y'$; (*c*) $E_1 = xy + x'y'$.

This can be solved by using Karnaugh maps as follows:

(*a*) Check the squares corresponding to $xy$ and $xy'$ as in Fig. 11-6(*a*). Note that $E_1$ consists of one prime implicant, the two adjacent squares designated by the loop in Fig. 11-6(*a*). This pair of adjacent squares represents the variable $x$, so $x$ is a (the only) prime implicant of $E_1$. Consequently, $E_1 = x$ is its minimal sum.

(*b*) Check the squares corresponding to $xy$, $x'y$, and $x'y'$ as in Fig. 11-6(*b*). Note that $E_2$ contains two pairs of adjacent squares (designated by the two loops) which include all the squares of $E_2$. The vertical pair represents $y$ and the horizontal pair represents $x'$; hence $y$ and $x'$ are the prime implicants of $E_2$. Thus $E_2 = x' + y$ is its minimal sum.

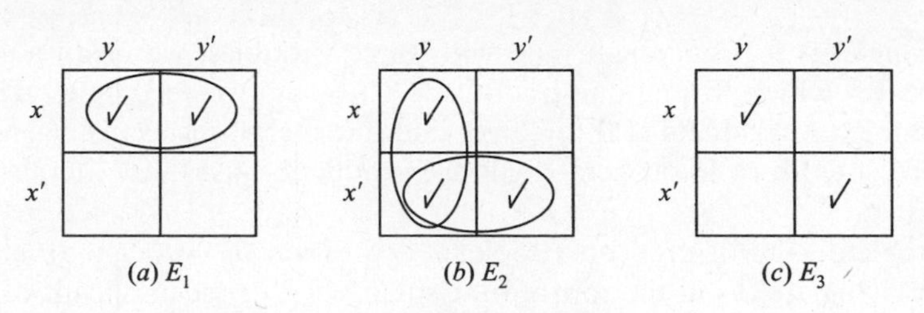


(a) $E_1$ (b) $E_2$ (c) $E_3$

**Fig. 11-6**

(c) Check the squares corresponding to $xy$ and $x'y'$ as in Fig. 11-6(c). Note that $E_3$ consists of two isolated squares which represent $xy$ and $x'y'$; hence $xy$ and $x'y'$ are the prime implicants of $E_3$ and $E_3 = xy + x'y'$ is its minimal sum.

### Case of Three Variables

The Karnaugh map corresponding to Boolean expressions $E = E(x, y, z)$ with three variables $x, y, z$ is shown in Fig. 11-7(a). Recall that there are exactly eight minterms with three variables:

$$xyz, \quad xyz', \quad xy'z', \quad xy'z, \quad x'yz, \quad x'yz', \quad x'y'z', \quad x'y'z$$

These minterms are listed so that they correspond to the eight squares in the Karnaugh map in the obvious way.

Furthermore, in order that every pair of adjacent products in Fig. 11-7(a) are geometrically adjacent, the right and left edges of the map must be identified. This is equivalent to cutting out, bending, and gluing the map along the identified edges to obtain the cylinder pictured in Fig. 11-7(b), where adjacent products are now represented by squares with one edge in common.

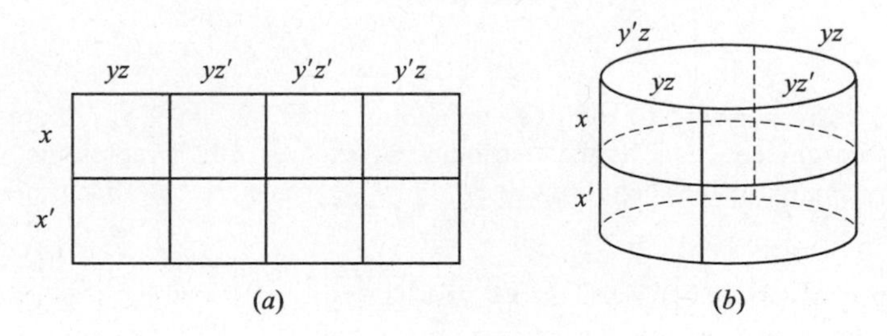


(a) (b)

**Fig. 11-7**

Viewing the Karnaugh map in Fig. 11-7(a) as a Venn diagram, the areas represented by the variables $x, y$, and $z$ are shown in Fig. 11-8. Specifically, the variable $x$ is still represented by the points in the upper half of the map, as shaded in Fig. 11-8(a), and the variable $y$ is still represented by the points in the left half of the map, as shaded in Fig. 11-8(b). The new variable $z$ is represented by the points in the left and right quarters of the map, as shaded in Fig. 11-8(c). Thus $x'$, $y'$, and $z'$ are represented, respectively, by points in the lower half, right half, and middle two quarters of the map.

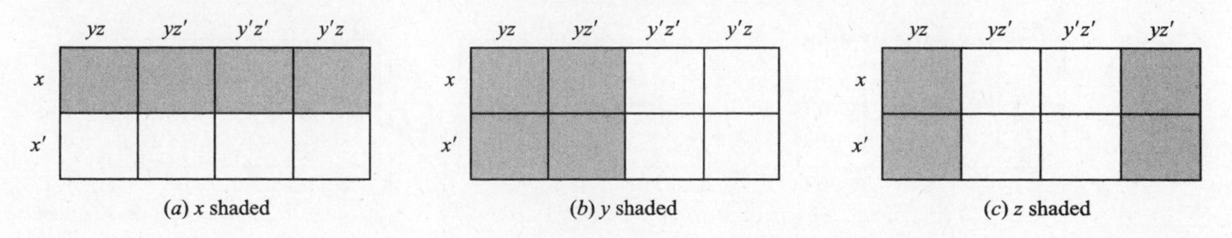


(a) $x$ shaded (b) $y$ shaded (c) $z$ shaded

**Fig. 11-8**

By a *basic rectangle* in the Karnaugh map with three variables, we mean a square, two adjacent squares, or four squares which form a one-by-four or a two-by-two rectangle. These basic rectangles correspond to fundamental products of three, two, and one literal, respectively. Moreover, the fundamental product represented by a basic rectangle is the product of just those literals that appear in every square of the rectangle.

Suppose a complete sum-of-products Boolean expression $E = E(x, y, z)$ is represented in the Karnaugh map by placing checks in the appropriate squares. A prime implicant of $E$ will be a *maximal basic rectangle* of $E$, i.e., a basic rectangle contained in $E$ which is not contained in any larger basic rectangle in $E$. A minimal sum-of-products form for $E$ will consist of a *minimal cover* of $E$, that is, a minimal number of maximal basic rectangles of $E$ which together include all the squares of $E$.

**EXAMPLE 11.12** Find the prime implicants and a minimal sum-of-products form for each of the following complete sum-of-products Boolean expressions:

$$(a)\quad E_1 = xyz + xyz' + x'yz' + x'y'z$$

$$(b)\quad E_2 = xyz + xyz' + xy'z + x'yz + x'y'z$$

$$(c)\quad E_3 = xyz + xyz' + x'yz' + x'y'z + x'y'z$$

This can be solved by using Karnaugh maps as follows:

(*a*) Check the squares corresponding to the four summands as in Fig. 11-9(*a*). Observe that $E_1$ has three prime implicants (maximal basic rectangles), which are circled; these are $xy$, $yz'$, and $x'y'z$. All three are needed to cover $E_1$; hence the minimal sum for $E_1$ is

$$E_1 = xy + yz' + x'y'z$$

(*b*) Check the squares corresponding to the five summands as in Fig. 11-9(*b*). Note that $E_2$ has two prime implicants, which are circled. One is the two adjacent squares which represents $xy$, and the other is the two-by-two square (spanning the identified edges) which represents $z$. Both are needed to cover $E_2$, so the minimal sum for $E_2$ is

$$E_2 = xy + z$$

(*c*) Check the squares corresponding to the five summands as in Fig. 11-9(*c*). As indicated by the loops, $E_3$ has four prime implicants, $xy$, $yz'$, $x'z'$, and $x'y'$. However, only one of the two dashed ones, i.e., one of $yz'$ or $x'z'$, is needed in a minimal cover of $E_3$. Thus $E_3$ has two minimal sums:

$$E_3 = xy + yz' + x'y' = xy + x'z' + x'y'$$

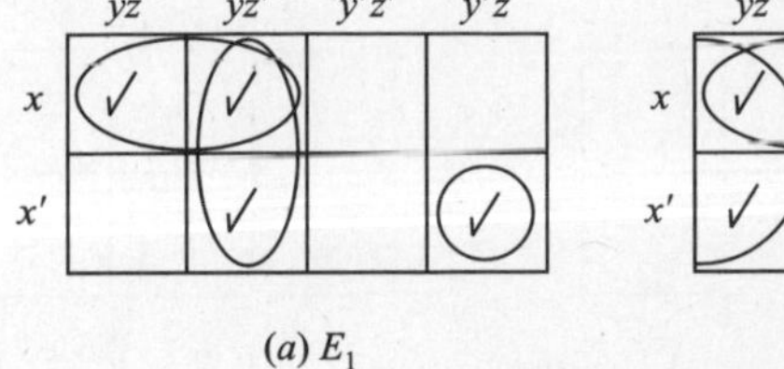

(*a*) $E_1$

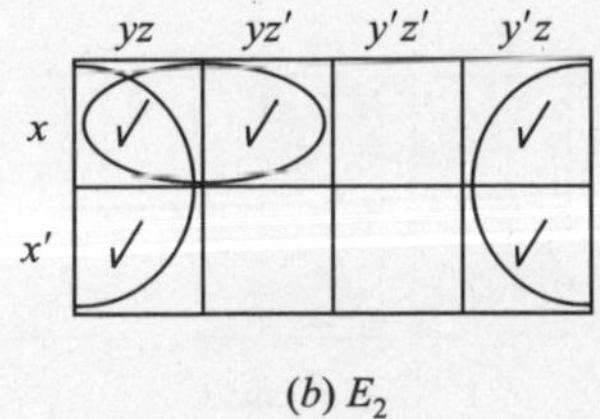

(*b*) $E_2$

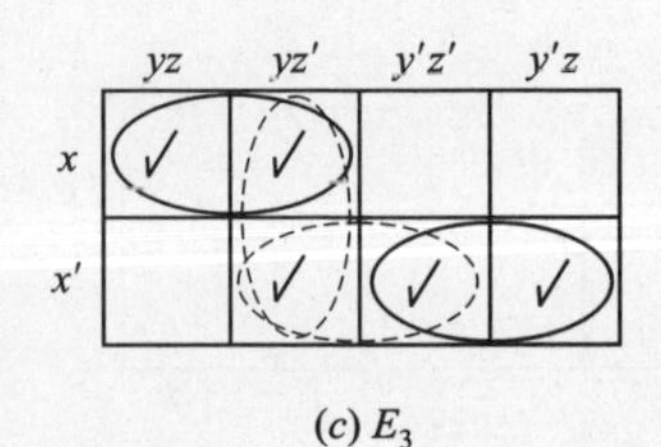

(*c*) $E_3$

**Fig. 11-9**

### Case of Four Variables

The Karnaugh map corresponding to Boolean expressions $E = E(x, y, z, t)$ with four variables $x, y, z, t$ is shown in Fig. 11-10. Each of the 16 squares corresponds to one of the 16 minterms with four variables,

$$xyzt, \quad xyzt', \quad xyz't', \quad xyz't, \quad \ldots, \quad x'yz't$$

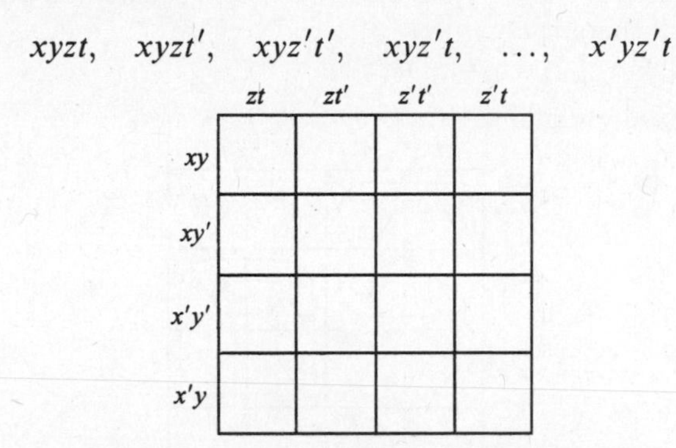


**Fig. 11-10**

This is indicated by the labels of the row and column of the square. Observe that the top line and the left side are labeled so that adjacent products differ in precisely one literal. Again, we must identify the left edge with the right edge (as we did with three variables) but we must also identify the top edge with the bottom edge. (These identifications give rise to a donut-shaped surface called a *torus*, and we may view our map as really being a torus.)

A *basic rectangle* in a four-variable Karnaugh map is a square, two adjacent squares, four squares which form a one-by-four or two-by-two rectangle, or eight squares which form a two-by-four rectangle. These basic rectangles correspond to fundamental products of four, three, two, and one literal, respectively. Again, maximal basic rectangles are the prime implicants. The minimizing technique for a Boolean expression $E(x, y, z, t)$ is the same as before.

**EXAMPLE 11.13** Find the fundamental product $P$ represented by the basic rectangle in the Karnaugh maps shown in Fig. 11-11.

In each case, find the literals which appear in all the squares of the basic rectangle; $P$ is the product of such literals.

(*a*) $x, y$, and $z'$ appear in both squares; hence $P = xy'z'$.

(*b*) Only $y$ and $z$ appear in all four squares; hence $P = yz$.

(*c*) Only $t$ appears in all eight squares; hence $P = t$.

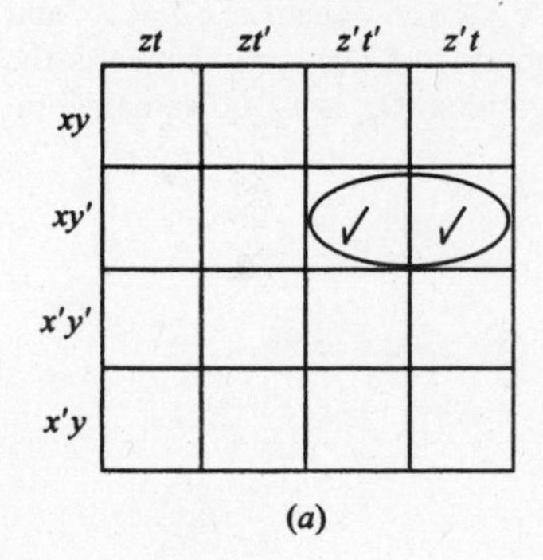


(*a*)

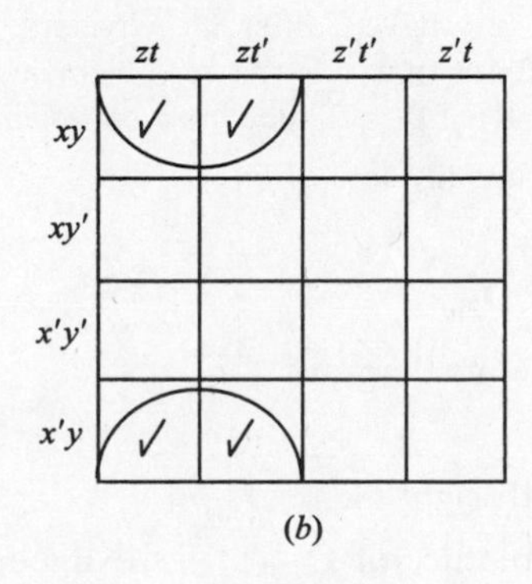


(*b*)

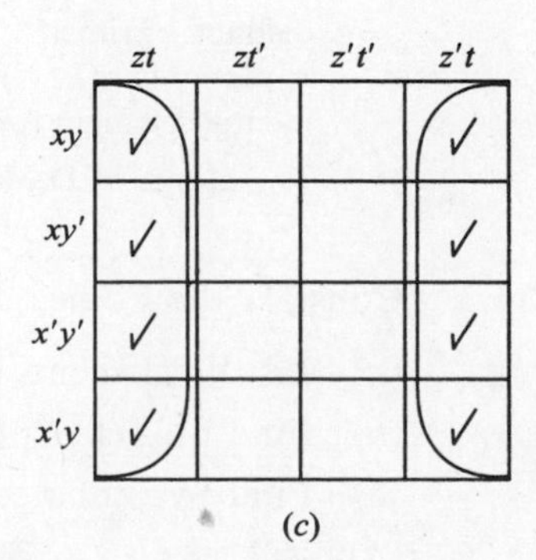


(*c*)

**Fig. 11-11**

**EXAMPLE 11.14** Find a minimal sum-of-products form for $E = xy' + xyz + x'y'z' + x'yzt'$.

Check all the squares representing each fundamental product. That is, check all four squares representing $xy'$, the two squares representing $xyz$, the two squares representing $x'y'z'$ and the one square representing $x'yzt'$, as in Fig. 11-12. A minimal cover of the map consists of the three designated basic rectangles. The two-by-two squares represent the fundamental products $xz$ and $y'z'$, and the two adjacent squares (on top and bottom) represents $yzt'$. Hence the following is a minimal sum for $E$:

$$E = xz + y'z' + yzt'$$

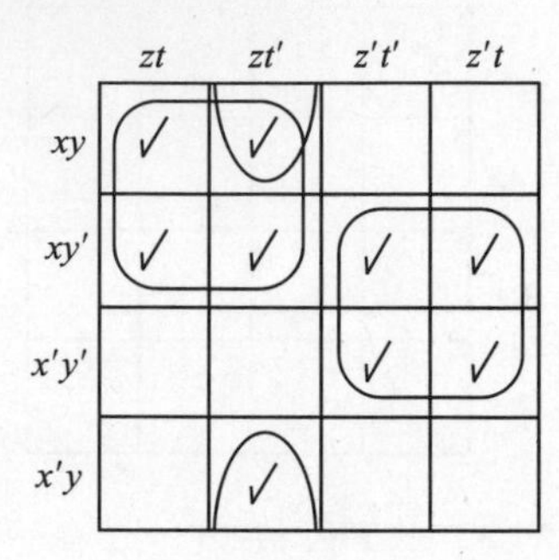


Fig. 11-12

## Solved Problems

### BOOLEAN ALGEBRAS

**11.1.** Write the dual of each Boolean equation: (a) $(a * 1) * (0 + a') = 0$; (b) $a + a'b = a + b$.

(a) To obtain the dual equation, interchange + and $*$, and interchange 0 and 1. This yields

$$(a + 0) + (1 * a') = 1$$

(b) First write the equation using $*$: $a + (a' * b) = a + b$. Then the dual is $a * (a' + b) = a * b$, which can we written as

$$a(a' + b) = ab$$

**11.2.** Recall (Chapter 7) that the set $\mathbf{D}_m$ of divisors of $m$ is a bounded, distributive lattice with $a + b = a \vee b = \operatorname{lcm}(a, b)$ and $a * b = a \wedge b = \gcd(a, b)$. (a) Show that $\mathbf{D}_m$ is a Boolean algebra if $m$ is square free, i.e., if $m$ is a product of distinct primes. (b) Find the atoms of $\mathbf{D}_m$.

(a) We need only show that $\mathbf{D}_m$ is complemented. Let $x$ be in $\mathbf{D}_m$ and let $x' = m/x$. Since $m$ is a product of distinct primes, $x$ and $x'$ have different divisors. Hence $x * x' = \gcd(x, x') = 1$ and $x + x' = \operatorname{lcm}(x, x') = m$. Recall that 1 is the zero element (lower bound) of $\mathbf{D}_m$, and that $m$ is the identity element (upper bound) of $\mathbf{D}_m$. Thus $x'$ is a complement of $x$, and so $\mathbf{D}_m$ is a Boolean algebra.

(b) The atoms of $\mathbf{D}_m$ are the prime divisors of $m$.

**11.3.** Consider the Boolean algebra $\mathbf{D}_{210}$.

(a) List its elements and draw its diagram.

(b) Find the set $A$ of atoms.

(c) Find two subalgebras with eight elements.

(d) Is $X = \{1, 2, 6, 210\}$ a sublattice of $\mathbf{D}_{210}$? A subalgebra?

(e) Is $Y = \{1, 2, 3, 6\}$ a sublattice of $\mathbf{D}_{210}$? A subalgebra?

(*a*) The divisors of 210 are 1, 2, 3, 5, 6, 7, 10, 14, 15, 21, 30, 35, 42, 70, 105 and 210. The diagram of $\mathbf{D}_{210}$ appears in Fig. 11-13.

(*b*) $A = \{2, 3, 5, 7\}$, the set of prime divisors is 210.

(*c*) $B = \{1, 2, 3, 35, 6, 70, 105, 210\}$ and $C = \{1, 5, 6, 7, 30, 35, 42, 210\}$ are subalgebras of $\mathbf{D}_{210}$.

(*d*) $X$ is a sublattice since it is linearly ordered. However, $X$ is not a subalgebra since 35 is the complement of 2 in $\mathbf{D}_{210}$ but 35 does not belong to $X$. (In fact, no Boolean algebra with more than two elements is linearly ordered.)

(*e*) $Y$ is a sublattice of $\mathbf{D}_{210}$ since it is closed under $+$ and $*$. However, $Y$ is not a subalgebra of $\mathbf{D}_{210}$ since it is not closed under complements in $\mathbf{D}_{210}$, e.g., $35 = 2'$ does not belong to $Y$. (We note that $Y$ itself is a Boolean algebra, in fact, $Y = \mathbf{D}_6$.)

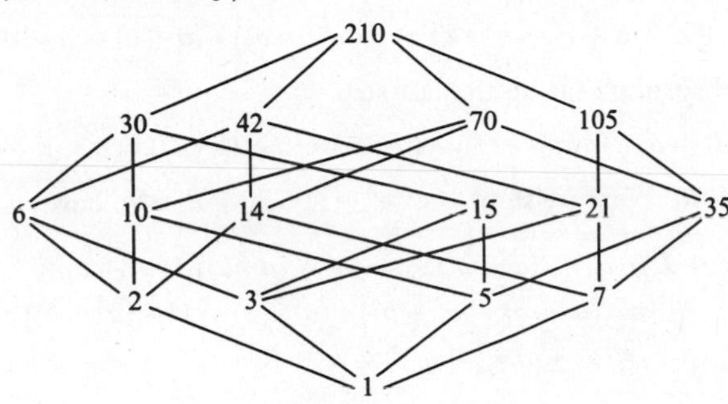


**Fig. 11-13**

**11.4.** Find the number of subalgebras of $\mathbf{D}_{210}$.

A subalgebra of $\mathbf{D}_{210}$ must contain two, four, eight or sixteen elements.

(i) There can be only one two-element subalgebra which consists of the upper and lower bounds, i.e., $\{1,210\}$.

(ii) Since $\mathbf{D}_{210}$ contains sixteen elements, the only sixteen-element subalgebra is $\mathbf{D}_{210}$ itself.

(iii) Any four-element subalgebra is of the form $\{1, x, x', 210\}$, i.e., consists of the upper and lower bounds and a nonbound element and its complement. There are fourteen nonbound elements in $\mathbf{D}_{210}$ and so there are $14/2 = 7$ pairs $\{x, x'\}$. Thus $\mathbf{D}_{210}$ has seven 4-element subalgebras.

(iv) Any eight-element subalgebra $S$ will itself contain three atoms $s_1, s_2, s_3$. We can choose $s_1$ and $s_2$ to be any two of the four atoms of $\mathbf{D}_{210}$ and then $s_3$ must be the product of the other two atoms, e.g., we can let $s_1 = 2$, $s_2 = 3$, $s_3 = 5 \cdot 7 = 35$ (which determines the subalgebra $B$ above), or we can let $s_1 = 5$, $s_2 = 7$, $s_3 = 2 \cdot 3 = 6$ (which determines the subalgebra $C$ above). There are $\binom{4}{2} = 6$ ways to choose $s_1$ and $s_2$ from the four atoms of $\mathbf{D}_{210}$ and so $\mathbf{D}_{210}$ has six 8-element subalgebras.

Accordingly, $\mathbf{D}_{210}$ has $1 + 1 + 7 + 6 = 15$ subalgebras.

**11.5.** Prove Theorem 11.2: Let $a, b, c$ be any elements in a Boolean algebra $B$.

(i) Idempotent laws:

(5*a*) $a + a = a$ (5*b*) $a * a = a$

(ii) Boundedness laws:

(6*a*) $a + 1 = 1$ (6*b*) $a * 0 = 0$

(iii) Absorption laws:

(7*a*) $a + (a * b) = a$ (7*b*) $a * (a + b) = a$

(iv) Associative laws:

(8*a*) $(a + b) + c = a + (b + c)$ (8*b*) $(a * b) * c = a * (b * c)$

The proofs follow:

(5*b*) $a = a * 1 = a * (a + a') = (a * a) + (a * a') = (a * a) + 0 = a * a$

(5*a*) Follows from (5*b*) and duality.

(6*b*) $a * 0 = (a * 0) + 0 = (a * 0) + (a * a') = a * (0 + a') = a * (a' + 0) = a * a' = 0$

(6*a*) Follows from (6*b*) and duality.

(7*b*) $a * (a + b) = (a + 0) * (a + b) = a + (0 * b) = a + (b * 0) = a + 0 = a$

(7*a*) Follows from (7*b*) and duality.

(8*b*) Let $L = (a * b) * c$ and $R = a * (b * c)$. We need to prove that $L = R$. We first prove that $a + L = a + R$. Using the absorption laws in the last two steps,

$$a + L = a + ((a * b) * c) = (a + (a * b)) * (a + c) = a * (a + c) = a$$

Also, using the absorption law in the last step,

$$a + R = a + (a * (b * c)) = (a + a) * (a + (b * c)) = a * (a + (b * c)) = a$$

Thus $a + L = a + R$. Next we show that $a' + L = a' + R$. We have

$$\begin{aligned} a' + L &= a' + ((a * b) * c) = (a' + (a * b)) * (a' + c) \\ &= ((a' + a) * (a' + b)) * (a' + c) = (1 * (a' + b)) * (a' + c) \\ &= (a' + b) * (a' + c) = a' + (b * c) \end{aligned}$$

Also,

$$\begin{aligned} a' + R &= a' + (a * (b * c)) = (a' + a) * (a' + (b * c)) \\ &= 1 * (a' + (b * c)) = a' + (b * c) \end{aligned}$$

Thus $a' + L = a' + R$. Consequently

$$\begin{aligned} L &= 0 + L = (a * a') + L = (a + L) * (a' + L) = (a + R) * (a' + R) \\ &= (a * a') + R = 0 + R = R \end{aligned}$$

(8*a*) Follows from (8*b*) and duality.

**11.6.** Prove Theorem 11.3: Let $a$ be any element of a Boolean Algebra $B$.

(i) (Uniqueness of complement) If $a + x = 1$ and $a * x = 0$, then $x = a'$.

(ii) (Involution law) $(a')' = a$

(iii) (9*a*) $0' = 1$. (9*b*) $1' = 0$.

(i) We have

$$a' = a' + 0 = a' + (a * x) = (a' + a) * (a' + x) = 1 * (a' + x) = a' + x$$

Also,

$$x = x + 0 = x + (a * a') = (x + a) * (x + a') = 1 * (x + a') = x + a'$$

Hence $x = x + a' = a' + x = a'$.

(ii) By definition of complement, $a + a' = 1$ and $a * a' = 0$. By commutativity, $a' + a = 1$ and $a' * a = 0$. By uniqueness of complement, $a$ is the complement of $a'$, that is, $a = (a')'$.

(iii) By boundedness law (6*a*), $0 + 1 = 1$, and by identity axiom (3*b*), $0 * 1 = 0$. By uniqueness of complement, 1 is the complement of 0, that is, $1 = 0'$. By duality, $0 = 1'$.

**11.7.** Prove Theorem 11.4 (DeMorgan's laws):

(10$a$) $(a+b)' = a' * b'$. (10$b$) $(a*b)' = a' + b'$

(10$a$) We need to show that $(a+b)+(a'*b') = 1$ and $(a+b)*(a'*b') = 0$; then by uniqueness of complement, $a'*b' = (a+b)'$. We have

$$\begin{aligned}(a+b)+(a'*b') &= b+a+(a'*b') = b+(a+a')*(a+b') \\ &= b+1*(a+b') = b+a+b' = b+b'+a = 1+a = 1\end{aligned}$$

Also,

$$\begin{aligned}(a+b)*(a'*b') &= ((a+b)*a')*b' \\ &= ((a*a')+(b*a'))*b' = (0+(b*a'))*b' \\ &= (b*a')*b' = (b*b')*a' = 0*a' = 0\end{aligned}$$

Thus $a'*b' = (a+b)'$.

(10$b$) Principle of duality (Theorem 11.1).

**11.8.** Prove Theorem 11.5: The following are equivalent in a Boolean algebra:

(1) $a+b=b$, (2) $a*b=a$, (3) $a'+b=1$, (4) $a*b'=0$.

By Theorem 7.8, (1) and (2) are equivalent. We show that (1) and (3) are equivalent. Suppose (1) holds. Then

$$a'+b = a'+(a+b) = (a'+a)+b = 1+b = 1$$

Now suppose (3) holds. Then

$$a+b = 1*(a+b) = (a'+b)*(a+b) = (a'*a)+b = 0+b = b$$

Thus (1) and (3) are equivalent.

We next show that (3) and (4) are equivalent. Suppose (3) holds. By DeMorgan's law and involution,

$$0 = 1' - (a'+b)' = a''*b' = a*b'$$

Conversely, if (4) holds then

$$1 = 0' = (a*b')' = a'+b'' = a'+b$$

Thus (3) and (4) are equivalent. Accordingly, all four are equivalent.

**11.9.** Prove Theorem 11.6: The mapping $f\colon B \to \mathcal{P}(A)$ is an isomorphism where $B$ is a Boolean algebra, $\mathcal{P}(A)$ is the power set of the set $A$ of atoms, and

$$f(x) = \{a_1, a_2, \ldots, a_n\}$$

where $x = a_1 + \cdots + a_n$ is the unique representation of $a$ as a sum of atoms.

Recall (Chapter 7) that if the $a$'s are atoms then $a_i^2 = a_i$ but $a_ia_j = 0$ for $a_i \neq a_j$. Suppose $x, y$ are in $B$ and suppose

$$\begin{aligned}x &= a_1 + \cdots + a_r + b_1 + \cdots + b_s \\ y &= b_1 + \cdots + b_s + c_1 + \cdots + c_t\end{aligned}$$

where

$$A = \{a_1, \ldots, a_r, b_1, \ldots, b_s, c_1, \ldots, c_t, d_1, \ldots, d_k\}$$

is the set of atoms of $B$. Then

$$\begin{aligned}x+y &= a_1 + \cdots + a_r + b_1 + \cdots + b_s + c_1 + \cdots + c_t \\ xy &= b_1 + \cdots + b_s\end{aligned}$$

Hence

$$\begin{aligned} f(x+y) &= \{a_1, \ldots, a_r, b_1, \ldots, b_s, c_1, \ldots, c_t\} \\ &= \{a_1, \ldots, a_r, b_1, \ldots, b_s\} \cup \{b_1, \ldots, b_s, c_1, \ldots, c_t\} \\ &= f(x) \cup f(y) \\ f(xy) &= \{b_1, \ldots, b_s\} \\ &= \{a_1, \ldots, a_r, b_1, \ldots, b_s\} \cap \{b_1, \ldots, b_s, c_1, \ldots, c_t\} \\ &= f(x) \cap f(y) \end{aligned}$$

Let $y = c_1 + \cdots + c_t + d_1 + \cdots + d_k$. Then $x + y = 1$ and $xy = 0$, and so $y = x'$. Thus

$$f(x') = \{c_1, \ldots, c_t, d_1, \ldots, d_k\} = \{a_1, \ldots, a_r, b_1, \ldots, b_2\}^c = (f(x))^c$$

Since the representation is unique, $f$ is one-to-one and onto. Hence $f$ is a Boolean algebra isomorphism.

## BOOLEAN EXPRESSIONS

**11.10.** Reduce the following Boolean products to either 0 or a fundamental product:

(*a*) $xyx'z$; (*b*) $xyzy$; (*c*) $xyz'yx$; (*d*) $xyz'yx'z'$.

Use the commutative law $x * y = y * x$, the complement law $x * x' = 0$, and the idempotent law $x * x = x$:

(*a*) $zyx'z = xx'yz = 0yz = 0$

(*b*) $xyzy = xyyz = xyz$

(*c*) $xyz'yx = xxyyz' = xyz'$

(*d*) $xyz'yx'z' = xx'yyz'z' = 0yz' = 0$

**11.11.** Express each Boolean expression $E(x, y, z)$ as a sum-of-products and then in its complete sum-of-products form:

(*a*) $E = x(xy' + x'y + y'z)$; (*b*) $E = z(x' + y) + y'$.

First use Algorithm 11.8A to express $E$ as a sum-of-products, and then use Algorithm 11.8B to express $E$ as a complete sum-of-products.

(*a*) First we have $E = xxy' + xx'y + xy'z = xy' + xy'z$. Then

$$E = xy'(z + z') + xy'z = xy'z + xy'z' + xy'z = xy'z + xy'z'$$

(*b*) First we have

$$E = z(x' + y) + y' = x'z + yz + y'$$

Then

$$\begin{aligned} E = x'z + yz + y' &= x'z(y + y') + yz(x + x') + y'(x + x')(z + z') \\ &= x'yz + x'y'z + xyz + x'yz + xy'z + xy'z' + x'y'z + x'y'z' \\ &= xyz + xy'z + xy'z' + x'yz + x'y'z + x'y'z' \end{aligned}$$

**11.12.** Express $E(x, y, z) = (x' + y)' + x'y$ in its complete sum-of-products form.

We have $E = (x' + y)' + x'y = +xy' + x'y$, which would be the complete sum-of-products form of $E$ if $E$ were a Boolean expression in $x$ and $y$. However, it is specified that $E$ is a Boolean expression in the three variables $x$, $y$, and $z$. Hence,

$$E = xy' + x'y = xy'(z + z') + x'y(z + z') = xy'z + xy'z' + x'yz + x'yz'$$

is the complete sum-of-products form of $E$.

**11.13.** Express each Boolean expression $E(x, y, z)$ as a sum-of-products and then in its complete sum-of-products form:

(*a*) $E = y(x + yz)'$; (*b*) $E = x(xy + y' + x'y)$.

(*a*) $E = y(x'(yz)') = yx'(y' + z') = yx'y' + x'yz' = x'yz'$

which already is in its complete sum-of-products form.

(*b*) First we have $E = xxy + xy' + xx'y = xy + xy'$. Then

$$E = xy(z + z') + xy'(z + z') = xyz + xyz' + xy'z + xy'z'$$

**11.14.** Express each set expression $E(A, B, C)$ involving sets $A, B, C$ as a union of intersections:

(*a*) $E = (A \cup B)^c \cap (C^c \cup B)$; (*b*) $E = (B \cap C)^c \cap (A^c \cup C)^c$

Use Boolean notation, $'$ for complement, $+$ for union and $*$ (or juxtaposition) for intersection, and then express $E$ as a sum of products (union of intersections).

(*a*) $E = (A + B)'(C' + B) = A'B'(C' + B) = A'B'C' + A'B'B = A'B'C'$ or $E = A^c \cap B^c \cap C^c$

(*b*) $E = (BC)'(A' + C)' = (B' + C')(AC') = AB'C' + AC'$ or $E = (A \cap B^c \cap C^c) \cup (A \cap C^c)$

**11.15.** Let $E = xy' + xyz' + x'yz'$. Prove that (*a*) $xz' + E = E$; (*b*) $x + E \neq E$, (*c*) $z' + E \neq E$.

Since the complete sum-of-products form is unique, $A + E = E$, where $A \neq 0$, if and only if the summands in the complete sum-of-products form for $A$ are among the summands in the complete sum-of-products form for $E$. Hence, first find the complete sum-of-products form for $E$:

$$E = xy'(z + z') + xyz' + x'yz' = xy'z + xy'z' + xyz' + x'yz'$$

(*a*) Express $xz'$ in complete sum-of-products form:

$$xz' = xz'(y + y') = xyz' + xy'z'$$

Since the summands of $xz'$ are among those of $E$, we have $xz' + E = E$.

(*b*) Express $x$ in complete sum-of-products form:

$$x = x(y + y')(z + z') = xyz + xyz' + xy'z + xy'z'$$

The summand $xyz$ of $x$ is not a summand of $E$; hence $x + E \neq E$.

(*c*) Express $z'$ in complete sum-of-products form:

$$z' = z'(x + x')(y + y') = xyz' + xy'z' + x'yz' + x'y'z'$$

The summand $x'y'z'$ of $z'$ is not a summand of $E$; hence $z' + E \neq E$.

## MINIMAL BOOLEAN EXPRESSIONS, PRIME IMPLICANTS

**11.16.** For any Boolean sum-of-products expression $E$, we let $E_L$ denote the number of literals in $E$ (counting multiplicity) and $E_S$ denote the number of summands in $E$. Find $E_L$ and $E_S$ for each of the following:

(*a*) $E = xy'z + x'z' + yz' + x$

(*b*) $E = x'y'z + xyz + y + yz' + x'z$

(*c*) $E = xyt' + x'y'zt + xz't$

(*d*) $E = (xy' + z)' + xy'$

Simply add up the number of literals and the number of summands in each expression:

(*a*) $E_L = 3 + 2 + 2 + 1 = 8$, $E_S = 4$.

(*b*) $E_L = 3 + 3 + 1 + 2 + 2 = 11$, $E_S = 5$.

(*c*) $E_L = 3 + 4 + 3 = 10$, $E_S = 3$.

(*d*) Because $E$ is not written as a sum of products, $E_L$ and $E_S$ are not defined.

**11.17.** Given $E$ and $F$ are equivalent Boolean sum-of-products, define:

(*a*) $E$ is simpler than $F$; (*b*) $E$ is minimal.

(*a*) $E$ is simpler than $F$ if $E_L < F_L$ and $E_S \leq F_S$, or if $E_L \leq F_L$ and $E_S < F_S$.

(*b*) $E$ is minimal if there is no equivalent sum-of-products expression which is simpler than $E$.

**11.18.** Find the consensus $Q$ of the fundamental products $P_1$ and $P_2$ where:

(*a*) $P_1 = xy'z'$, $P_2 = xyt$ (*c*) $P_1 = xy'z'$, $P_2 = x'y'zt$

(*b*) $P_1 = xyz't$, $P_2 = xzt$ (*d*) $P_1 = xyz'$, $P_2 = xz't$

The consensus $Q$ of $P_1$ and $P_2$ exists if there is exactly one variable, say $x_k$, which is complemented in one of $P_1$ and $P_2$ and uncomplemented in the other. Then $Q$ is the product (without repetition) of the literals in $P_1$ and $P_2$ after $x_k$ and $x_k'$ have been deleted.

(*a*) Delete $y'$ and $y$ and then multiply the literals of $P_1$ and $P_2$ (without repetition) to obtain $Q = xz't$.

(*b*) Deleting $z'$ and $z$ yields $Q = xyt$.

(*c*) They have no consensus since both $x$ and $z$ appear complemented in one of the products and uncomplemented in the other.

(*d*) They have no consensus since no variable appears complemented in one of the products and uncomplemented in the other.

**11.19.** Suppose $Q$ is the consensus of $P_1$ and $P_2$. Prove that $P_1 + P_2 + Q = P_1 + P_2$.

Since the literals commute, we can assume without loss of generality that

$$P_1 = a_1 a_2 \cdots a_r t, \qquad P_2 = b_1 b_2 \cdots b_s t', \qquad Q = a_1 a_2 \cdots a_r b_1 b_2 \cdots b_s$$

Now, $Q = Q(t + t') = Qt + Qt'$. Because $Qt$ contains $P_1, P_1 + Qt = P_1$; and because $Qt'$ contains $P_2$, $P_2 + Qt' = P_2$. Hence

$$P_1 + P_2 + Q = P_1 + P_2 + Qt + Qt' = (P_1 + Qt) + (P_2 + Qt') = P_1 + P_2$$

**11.20.** Let $E = xy' + xyz' + x'yz'$. Find: (*a*) the prime implicants of $E$; (*b*) a minimal sum for $E$.

(*a*) Apply Algorithm 11.9A (consensus method) as follows:

$$\begin{aligned} E &= xy' + xyz' + x'yz' + xz' && \text{(Consensus of } xy' \text{ and } xyz'\text{)} \\ &= xy' + x'yz' + xz' && (xyz' \text{ includes } xz') \\ &= xy' + x'yz' + xz' + yz' && \text{(Consensus of } x'yz' \text{ and } xz'\text{)} \\ &= xy' + xz' + yz' && (x'yz' \text{ includes } yz') \end{aligned}$$

Neither step in the consensus method can now be applied. Hence $xy'$, $xz'$, and $yz'$ are the prime implicants of $E$.

(*b*) Apply Algorithm 11.9B. Write each prime implicant of $E$ in complete sum-of-products form obtaining:

$$\begin{aligned} xy' &= xy'(z + z') = xy'z + xy'z' \\ xz' &= xz'(y + y') = xyz' + xy'z' \\ yz' &= yz'(x + x') = xyz' + x'yz' \end{aligned}$$

Only the summands $xyz'$ and $xy'z'$ of $xz'$ appear among the other summands and hence $xz'$ can be eliminated as superfluous. Thus $E = xy' + yz'$ is a minimal sum for $E$.

**11.21.** Let $E = xy + y't + x'yz' + xy'zt'$. Find: (*a*) the prime implicants of $E$; (*b*) a minimal sum for $E$.

(*a*) Apply Algorithm 11.9A (consensus method) as follows:

$$\begin{aligned}
E &= xy + y't + x'yz' + xy'zt' + xzt' && \text{(Consensus of } xy \text{ and } xy'zt'\text{)}\\
&= xy + y't + x'yz' + xzt' && (xy'zt' \text{ includes } xzt')\\
&= xy + y't + x'yz' + xzt' + yz' && \text{(Consensus of } xy \text{ and } x'yz'\text{)}\\
&= xy + y't + xzt' + yz' && (x'yz' \text{ includes } yz')\\
&= xy + y't + xzt' + yz' + xt && \text{(Consensus of } xy \text{ and } y't\text{)}\\
&= xy + y't + xzt' + yz' + xt + xz && \text{(Consensus of } xzt' \text{ and } xt\text{)}\\
&= xy + y't + yz' + xt + xz && (xzt' \text{ includes } xz)\\
&= xy + y't + yz' + xt + xz + z't && \text{(Consensus of } y't \text{ and } yz'\text{)}
\end{aligned}$$

Neither step in the consensus method can now be applied. Hence the prime implicants of $E$ are $xy, y't, yz', xt, xz$, and $z't$.

(*b*) Apply Algorithm 11.9B. Write each prime implicant in complete sum-of-products form and then delete one by one those which are superfluous, i.e. those whose summands appear among the other summands. This finally yields

$$E = y't + xz + yz'$$

as a minimal sum for $E$.

## KARNAUGH MAPS

**11.22.** Find the fundamental product $P$ represented by each basic rectangle in the Karnaugh map in Fig. 11-14.

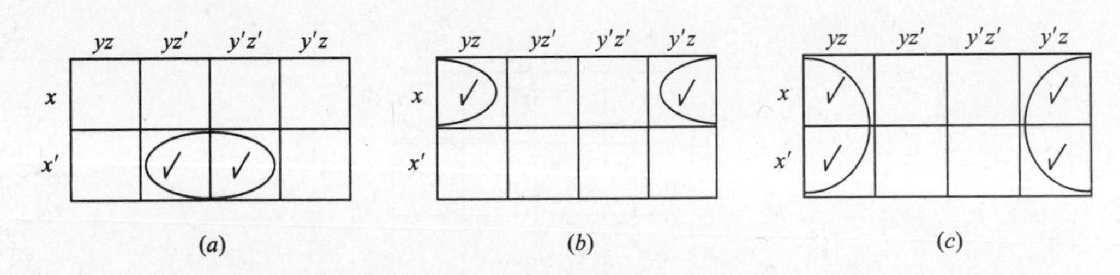


**Fig. 11-14**

In each case find those literals which appear in all the squares of the basic rectangle; then $P$ is the product of such literals.

(*a*) $x'$ and $z'$ appear in both squares; hence $P = x'z'$.

(*b*) $x$ and $z$ appear in both squares; hence $P = xz$.

(*c*) Only $z$ appears in all four squares; hence $P = z$.

**11.23.** Let $R$ be a basic rectangle in a Karnaugh map for four variables $x, y, z, t$. State the number of literals in the fundamental product $P$ corresponding to $R$ in terms of the number of squares in $R$.

$P$ will have 1, 2, 3, or 4 literals according as $R$ has 8, 4, 2, or 1 squares.

**11.24.** Find the fundamental product $P$ represented by each basic rectangle $R$ in the Karnaugh map in Fig. 11-15.

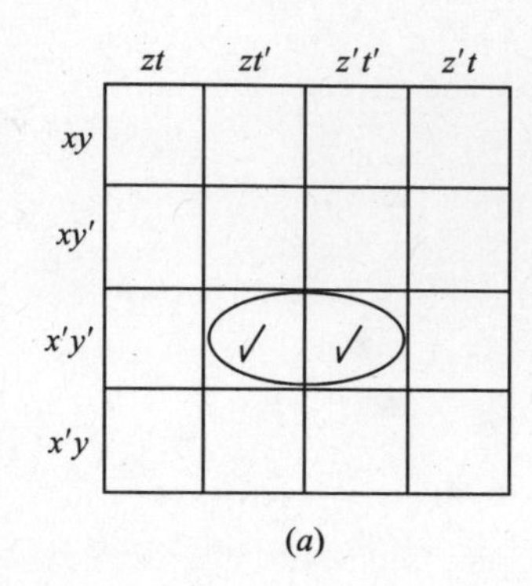


(a)

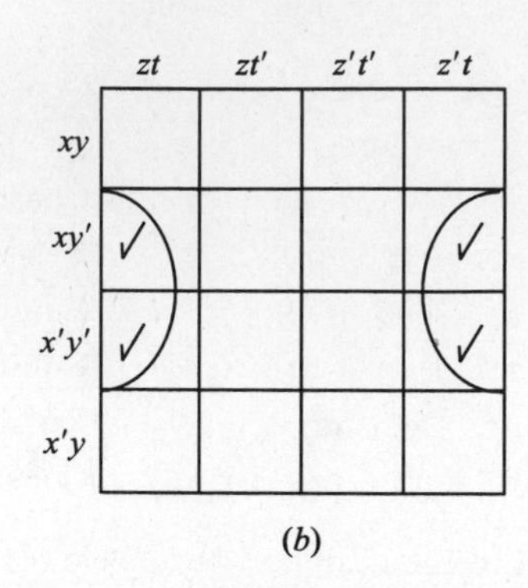


(b)

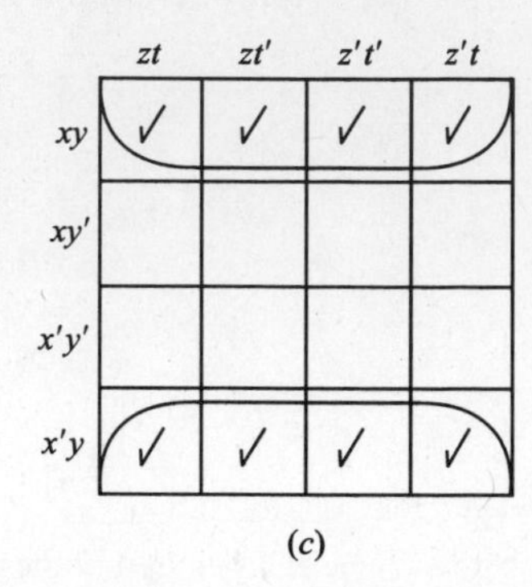


(c)

**Fig. 11-15**

In each case find those literals which appear in all the squares of the basic rectangle; then $P$ is the product of such literals. (Problem 11.23 indicates the number of such literals in $P$.)

(a) There are two squares in $R$, so $P$ has three literals. Specifically, $x', y', t'$ appear in both squares; hence $P = x'y't'$.

(b) There are four squares in $R$, so $P$ has two literals. Specifically, only $y'$ and $t$ appear in all four squares; hence $P = y't$.

(c) There are eight squares in $R$, so $P$ has only one literal. Specifically, only $y$ appears in all eight squares; hence $P = y$.

**11.25.** Let $E$ be the Boolean expression given in the Karnaugh map in Fig. 11-16.

(a) Write $E$ in its complete sum-of-products form. (b) Find a minimal form for $E$.

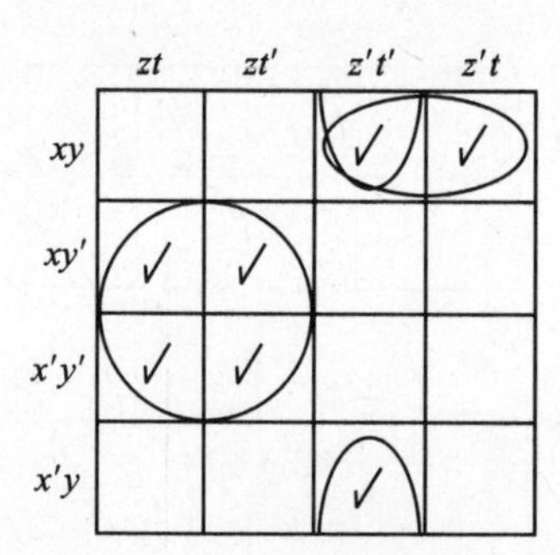


**Fig. 11-16**

(a) List the seven fundamental products checked to obtain

$$E = xyz't' + xyz't + xy'zt + xy'zt' + x'y'zt + x'y'zt' + x'yz't'$$

(b) The two-by-two maximal basic rectangle represents $y'z$ since only $y'$ and $z$ appear in all four squares. The horizontal pair of adjacent squares represents $xyz'$, and the adjacent squares overlapping the top and bottom edges represent $yz't'$. As all three rectangles are needed for a minimal cover,

$$E = y'z + xyz' + yz't'$$

is the minimal sum for $E$.

**11.26.** Consider the Boolean expressions $E_1$ and $E_2$ in variables $x, y, z, t$ which are given by the Karnaugh maps in Fig. 11-17. Find a minimal sum for (*a*) $E_1$; (*b*) $E_2$.

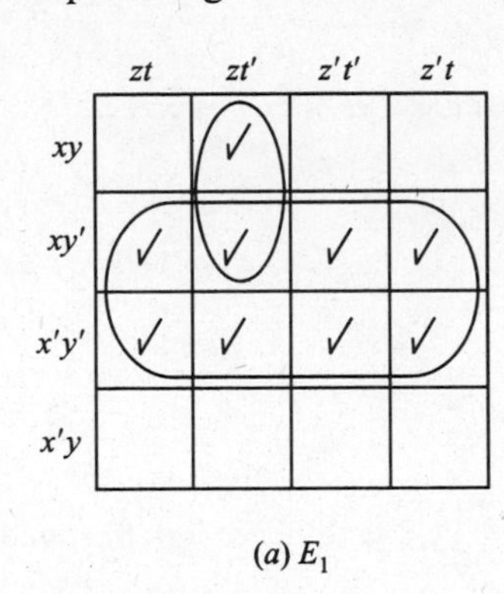


(*a*) $E_1$

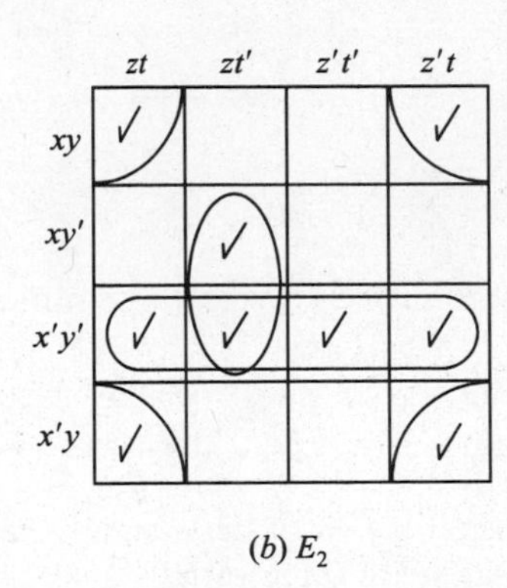


(*b*) $E_2$

**Fig. 11-17**

(*a*) Only $y'$ appears in all eight squares of the two-by-four maximal basic rectangle, and the designated pair of adjacent squares represents $xzt'$. As both rectangles are needed for a minimal cover, thus the following is the minimal sum for $E_1$:

$$E_1 = y' + xzt'$$

(*b*) The four corner squares form a two-by-two maximal basic rectangle which represents $yt$, since only $y$ and $t$ appear in all the four squares. The four-by-one maximal basic rectangle represents $x'y'$, and the two adjacent squares represent $y'zt'$. As all three rectangles are needed for a minimal cover, hence the following is the minimal sum for $E_2$:

$$E_2 = yt + x'y' + y'zt'$$

**11.27.** Consider the Boolean expressions $E_1$ and $E_2$ in variables $x, y, z, t$ which are given by the Karnaugh maps in Fig. 11-18. Find a minimal sum for: (*a*) $E_1$; (*b*) $E_2$.

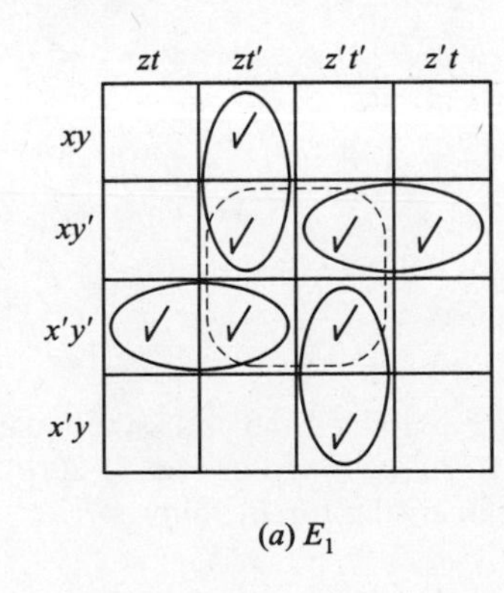


(*a*) $E_1$

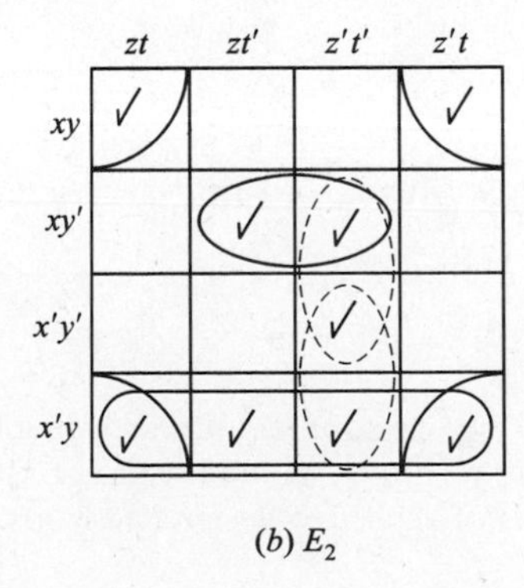


(*b*) $E_2$

**Fig. 11-18**

(*a*) There are five prime implicants, designated by the four loops and the dashed circle. However, the dashed circle is not needed to cover all the squares, whereas the four loops are required. Thus the four loops give the minimal sum for $E_1$; that is,

$$E_1 = xzt' + xy'z' + x'y'z + x'z't'$$

(*b*) There are five prime implicants, designated by the five loops of which two are dashed. Only one of the two dashed loops is needed to cover the square $x'y'z't'$. Thus there are two minimal sums for $E_2$ as follows:

$$E_2 = x'y + yt + xy't' + y'z't' = x'y + yt + xy't' + x'z't'$$

**11.28.** Use a Karnaugh map to find a minimal sum for:

(*a*) $E_1 = x'y'z' + x'yz' + xy'z + xyz'$.

(*b*) $E_2 = x'yz' + x'yz + xy'z + xyz' + xyz$.

Each term in $E_1$ and $E_2$ contains the three variables $x, y, z$, and hence it corresponds to a square in the Karnaugh map (with three variables).

(*a*) Checking the appropriate squares gives the Karnaugh map in Fig. 11-19(*a*). There are three prime implicants, as indicated by the three loops, which form a minimal cover of $E_1$. Thus a minimal form for $E_1$ follows:

$$E_1 = yz' + x'z' + xy'z$$

(*b*) The Karnaugh map appears in Fig. 11-19(*b*). There are two prime implicants, as indicated by the two loops, which form a minimal cover of $E_2$. Thus a minimal form for $E_2$ follows:

$$E_2 = xz + y$$

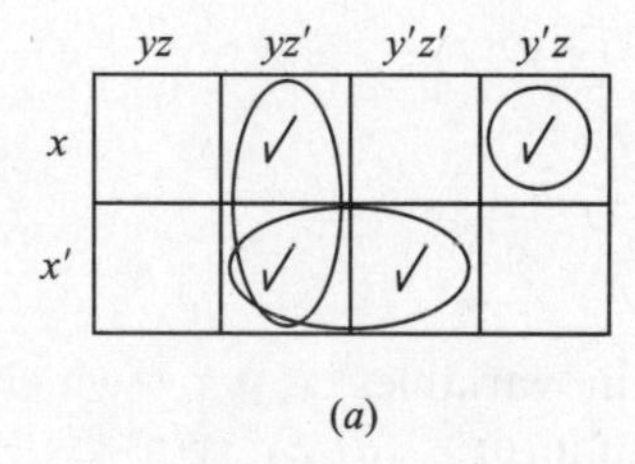

(*a*)

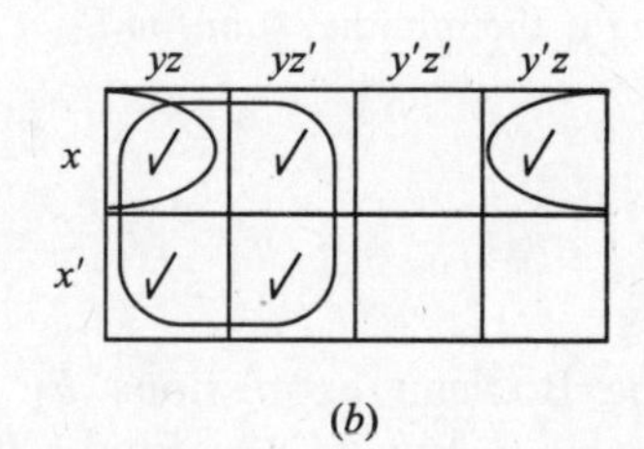

(*b*)

**Fig. 11-19**

**11.29.** Use a Karnaugh map to find a minimal sum for:

(*a*) $E_1 = x'yz + x'yz't + y'zt' + xyzt' + xy'z't'$.

(*b*) $E_2 = y't' + y'z't + x'y'zt + yzt'$.

(*a*) Check the two squares corresponding to each of $x'yz$ and $y'zt'$, and check the square corresponding to each of $x'yz't$, $xyzt'$, and $xy'z't'$. This gives the Karnaugh map in Fig. 11-20(*a*). A minimal cover consists of the three designated loops. Thus a minimal sum for $E_1$ follows:

$$E_1 = zt' + xy't' + x'yt$$

(*b*) Check the four squares corresponding to $zt'$, check the two squares corresponding to each of $y'z't$ and $yzt'$, and check the square corresponding to $x'y'zt$. This gives the Karnaugh map in Fig. 11-20(*b*). A minimal cover consists of the three designated maximal basic rectangles. Thus a minimal sum for $E_2$ follows:

$$E_2 = zt' + xy't' + x'yt$$

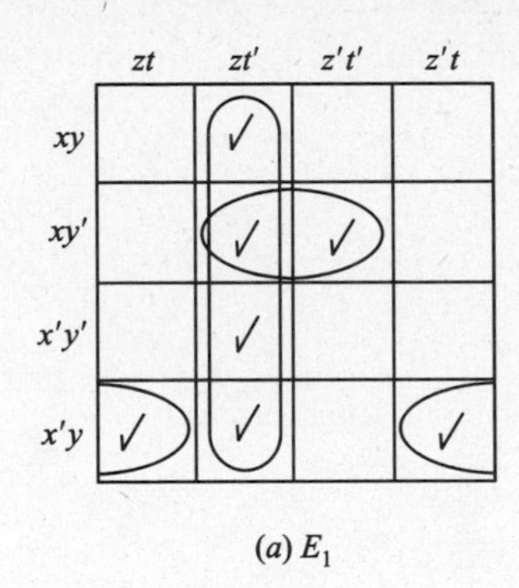


(a) $E_1$

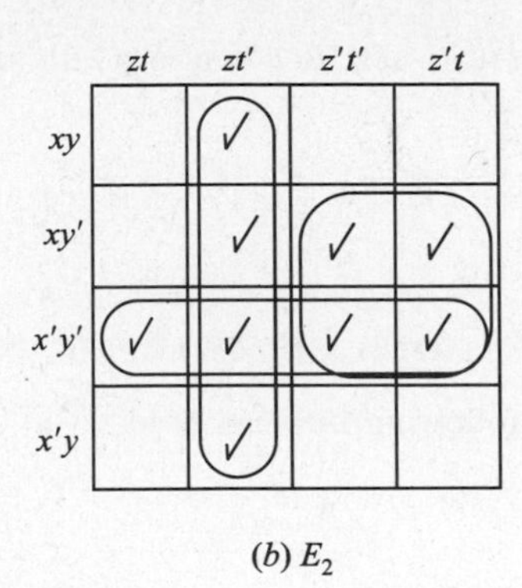


(b) $E_2$

**Fig. 11-20**

# Supplementary Problems

## BOOLEAN ALGEBRAS

**11.30.** Write the dual of each Boolean expression:

(a) $a(a' + b) = ab$; (b) $(a + 1)(a + 0) = a$; (c) $(a + b)(b + c) = ac + b$.

**11.31.** Consider the lattices $\mathbf{D}_m$ of divisors of $m$ (where $m > 1$).

(a) Show that $\mathbf{D}_m$ is a Boolean algebra if and only if $m$ is *square-free*, that is, $m$ is a product of distinct primes.

(b) If $\mathbf{D}_m$ is a Boolean algebra, show that the atoms are the distinct prime divisors of $m$.

**11.32.** Consider the following lattices: (a) $\mathbf{D}_{20}$; (b) $\mathbf{D}_{55}$; (c) $\mathbf{D}_{99}$; (d) $\mathbf{D}_{130}$. Which of them are Boolean algebras, and what are their atoms?

**11.33.** Consider the Boolean algebra $\mathbf{D}_{110}$.

(a) List its elements and draw its diagram. (b) Find all its subalgebras.

(c) Find the number of sublattices with four elements. (d) Find the set $A$ of atoms of $\mathbf{D}_{110}$.

(e) Give the isomorphic mapping $f$: $\mathbf{D}_{110} \to \mathscr{P}(A)$ as defined in Theorem 11.6.

**11.34.** Let $B$ be a Boolean algebra. Show that: (a) For any $x$ in $B$, $0 \le x \le 1$. (b) $a < b$ if and only if $b' < a'$.

**11.35.** An element $x$ in a Boolean algebra is called a *maxterm* if the identity 1 is its only successor. Find the maxterms in the Boolean algebra $\mathbf{D}_{210}$ pictured in Fig. 11-13.

**11.36.** Let $B$ be a Boolean algebra. (a) Show that complements of the atoms of $B$ are the maxterms. (b) Show that any element $x$ in $B$ can be expressed uniquely as a product of maxterms.

**11.37.** Let $B$ be a 16-element Boolean algebra and let $S$ be an 8-element subalgebra of $B$. Show that two of the atoms of $S$ must be atoms of $B$.

**11.38.** Let $B = (B, +, *, ', 0, 1)$ be a Boolean algebra. Define an operation $\triangle$ on $B$ (called the symmetric difference) by

$$x \triangle y = (x * y') + (x' * y)$$

Prove that $R = (B, \triangle, *)$ is a commutative Boolean ring.

**11.39.** Let $R = (R, +, \cdot)$ be a Boolean ring with identity $1 \neq 0$. Define

$$x' = 1 + x, \quad x + y = x + y + x \cdot y, \quad x * y = x \cdot y$$

Prove that $B = (R, +, *, ', 0, 1)$ is a Boolean algebra.

## BOOLEAN EXPRESSIONS, PRIME IMPLICANTS

**11.40.** Reduce the following Boolean products to either 0 or a fundamental product:

(*a*) $xy'zxy'$; (*b*) $xyz'sy'ts$; (*c*) $xy'xz'ty'$; (*d*) $xyz'ty't$

**11.41.** Express each Boolean expression $E(x, y, z)$ as a sum-of-products and then in its complete sum-of-products form:

(*a*) $E = x(xy' + x'y + y'z)$. (*b*) $E = (x + y'z)(y + z')$. (*c*) $E = (x' + y)' + y'z$.

**11.42.** Express each Boolean expression $E(x, y, z)$ as a sum-of-products and then in its complete sum-of-products form:

(*a*) $E = (x'y)'(x' + xyz')$. (*b*) $(x + y)'(xy')'$. (*c*) $E - y(x + yz)'$.

**11.43.** Find the consensus $Q$ of the fundamental products $P_1$ and $P_2$ where:

(*a*) $P_1 = xy'z$, $P_2 = xyt$

(*b*) $P_1 = xyz't'$, $P_2 = xzt'$

(*c*) $P_1 = xy'zt$, $P_2 = xyz'$

(*d*) $P_1 = xy't$, $P_2 = xzt$

**11.44.** For any Boolean sum-of-products expression $E$, we let $E_L$ denote the number of literals in $E$ (counting multiplicity) and $E_S$ denote the number of summands in $E$. Find $E_L$ and $E_S$ for each of the following:

(*a*) $E = xyz't + x'yt + xy'zt$. (*b*) $E = xyzt + xt' + x'y't + yt$

**11.45.** Apply the consensus method (Algorithm 11.9A) to find the prime implicants of each Boolean expression:

(*a*) $E_1 = xy'z' + x'y + x'y'z' + x'yz$.

(*b*) $E_2 = xy' + x'z't + xyzt' + x'y'zt'$.

(*c*) $E_3 = xyzt + xyz't' + xz't' + x'y'z' + x'yz't$.

**11.46.** Find a minimal sum-of-products form for each of the Boolean expressions in Problem 11.45.

## KARNAUGH MAPS

**11.47.** Find all possible minimal sums for each Boolean expression $E$ given by the Karnaugh maps in Fig. 11-21.

|  | $yz$ | $yz'$ | $y'z'$ | $y'z$ |
|---|---|---|---|---|
| $x$ | ✓ |  | ✓ | ✓ |
| $x'$ | ✓ | ✓ |  |  |

(*a*)

|  | $yz$ | $yz'$ | $y'z'$ | $y'z$ |
|---|---|---|---|---|
| $x$ | ✓ |  | ✓ | ✓ |
| $x'$ | ✓ | ✓ |  | ✓ |

(*b*)

|  | $yz$ | $yz'$ | $y'z'$ | $y'z$ |
|---|---|---|---|---|
| $x$ | ✓ |  |  | ✓ |
| $x'$ | ✓ | ✓ | ✓ | ✓ |

(*c*)

**Fig. 11-21**

**11.48.** Find all possible minimal sums for each Boolean expression $E$ given by the Karnaugh maps in Fig. 11-22.

| | $zt$ | $zt'$ | $z't'$ | $z't$ |
|---|---|---|---|---|
| $xy$ | | ✓ | | ✓ |
| $xy'$ | ✓ | ✓ | | ✓ |
| $x'y'$ | | ✓ | | |
| $x'y$ | ✓ | ✓ | ✓ | ✓ |

(*a*)

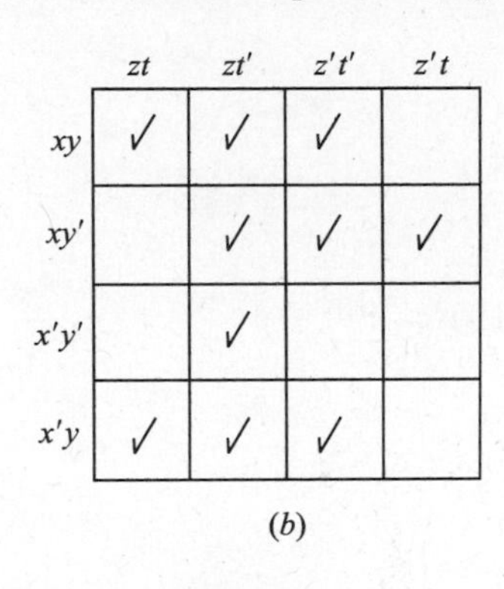

| | $zt$ | $zt'$ | $z't'$ | $z't$ |
|---|---|---|---|---|
| $xy$ | ✓ | ✓ | ✓ | |
| $xy'$ | | ✓ | ✓ | ✓ |
| $x'y'$ | | ✓ | | |
| $x'y$ | ✓ | ✓ | ✓ | |

(*b*)

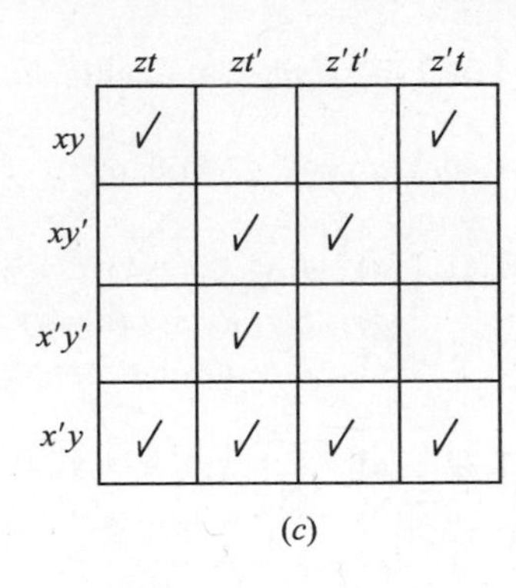

| | $zt$ | $zt'$ | $z't'$ | $z't$ |
|---|---|---|---|---|
| $xy$ | ✓ | | | ✓ |
| $xy'$ | | ✓ | ✓ | |
| $x'y'$ | | ✓ | | |
| $x'y$ | ✓ | ✓ | ✓ | ✓ |

(*c*)

**Fig. 11-22**

**11.49.** Use a Karnaugh map to find a minimal sum for each Boolean expression:

(*a*) $E = xy + x'y + x'y'$. (*b*) $E = x + x'yz + xy'z'$.

**11.50.** Use a Karnaugh map to find a minimal sum for each Boolean expression:

(*a*) $E = y'z + y'z't' + z't$. (*b*) $E = y'zt + xzt' + xy'z'$.

**11.51.** Show that the sum of two adjacent products will be a fundamental product with one fewer literal.

# Answers to Supplementary Problems

**11.30.** (*a*) $a + a'b = a + b$; (*b*) $a \cdot 0 + a \cdot 1 = a$; (*c*) $ab + bc = (a + c)b$

**11.32.** (*b*) $\mathbf{D}_{55}$; atoms 5 and 11; (*d*) $\mathbf{D}_{130}$; atoms 2, 5 and 13

**11.33.** (*a*) There are eight elements 1, 2, 5, 10, 11, 22, 55, 110. See Fig. 11-23(*a*).
(*b*) There are five subalgebras: $\{1, 110\}$, $\{1, 2, 55, 110\}$, $\{1, 5, 22, 110\}$, $\{1, 10, 11, 110\}$, $\mathbf{D}_{110}$.
(*c*) There are fifteen sublattices which include the above four three subalgebras.
(*d*) $A = \{2, 5, 11\}$
(*e*) See Fig. 11-23(*b*).

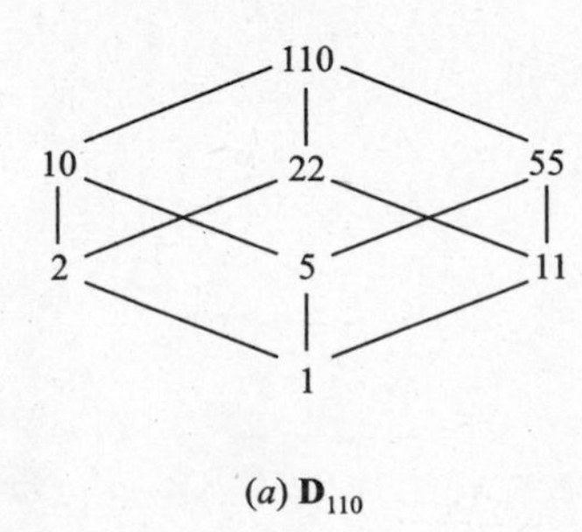


(*a*) $\mathbf{D}_{110}$

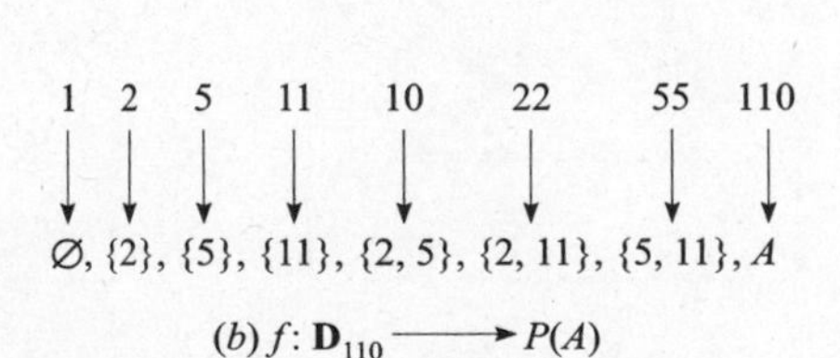


(*b*) $f: \mathbf{D}_{110} \longrightarrow P(A)$

**Fig. 11-23**

**11.35.** Maxterms: 30, 42, 70, 105

**11.36.** (*b*) *Hint*: Use duality.

**11.40.** (*a*) $xy'z$; (*b*) 0; (*c*) $xy'z't$; (*d*) 0

**11.41.** (*a*) $E = xy' + xy'z = xy'z' + xy'z$
(*b*) $E = xy + xz' = xyz + xyz' + xy'z'$
(*c*) $E = xy' + y'z = xy'z + xy'z' + x'y'z$

**11.42.** (*a*) $E = xyz' + x'y' = xyz' + x'y'z + x'y'z'$
(*b*) $E = x'y' = x'y'z + x'y'z'$
(*c*) $E = x'yz'$

**11.43.** (*a*) $Q = xzt$; (*b*) $Q = xyt'$; (*c*) and (*d*) Does not exist.

**11.44.** (*a*) $E_L = 11$, $E_S = 3$; (*b*) $E_L = 11$, $E_S = 4$

**11.45.** (*a*) $x'y$, $x'z'$, $y'z'$; (*b*) $xy'$, $xzt'$, $y'zt'$, $x'z't$, $y'z't$; (*c*) $xyzt$, $xz't'$, $y'z't'$, $x'y'z'$, $x'z't$

**11.46.** (*a*) $E = x'y + x'z'$
(*b*) $E = xy' + xzt' + x'z't + y'z't$
(*c*) $E = xyzt + xz't' + x'y'z' + x'z't$

**11.47.** (*a*) $E = xy' + x'y + yz = xy' + x'y + xz'$
(*b*) $E = xy' + x'y + z$
(*c*) $E = x' + z$

**11.48.** (*a*) $E = x'y + zt' + xz't + xy'z = x'y + zt' + xz't + xy't$
(*b*) $E = yz + yt' + zt' + xy'z'$
(*c*) $E = x'y + yt + xy't' + x'zt = x'y + yt + xy't' + y'zt$

**11.49.** (*a*) $E = x' + y$; (*b*) $E = xz' + yz$

**11.50.** (*a*) $E = y' + z't$; (*b*) $E = xy' + zt' + y'zt$

# Index